PAUL ERDŐS
(1913–1996)

A Memorial Tribute

Paul Erdős (1913–1996), giving a lecture in Madras, India in January 1984,
when he was Ramanujan Visiting Professor.

1998 KLUWER ACADEMIC PUBLISHERS
Boston/U.S.A.   Dordrecht/Holland   London/U.K.

THE RAMANUJAN JOURNAL

EDITOR-IN-CHIEF

Professor Krishnaswami Alladi
Department of Mathematics
University of Florida
Gainesville, FL 32611, USA

COORDINATING EDITORS

Professor Bruce Berndt
Department of Mathematics
University of Illinois
Urbana, IL 61801, USA

Professor Frank Garvan
Department of Mathematics
University of Florida
Gainesville, FL 32611, USA

EDITORIAL BOARD

Professor George Andrews
Department of Mathematics
The Pennsylvania State University
University Park, PA 16802, USA

Professor Richard Askey
Department of Mathematics
University of Wisconsin
Madison, WI 53706, USA

Professor Frits Beukers
Mathematics Institute
Rijksuniversiteit te Utrecht
3508 TA Utrecht
The Netherlands

Professor Jonathan Borwein
Simon Fraser Centre for Experimental
  and Constructive Mathematics
Department of Mathematics and
  Statistics
Simon Fraser University
Burnaby, B.C., V5A 156, Canada

Professor Peter Borwein
Simon Fraser Centre for Experimental
  and Constructive Mathematics
Department of Mathematics and
  Statistics
Simon Fraser University
Burnaby, B.C., V5A 156, Canada

Professor David Bressoud
Department of Mathematics and
  Computer Science
Macalester College
St. Paul, MN 55105, USA

Professor Peter Elliott
Department of Mathematics
University of Colorado
Boulder, CO 80309, USA

Professor Paul Erdös
Mathematics Institute
Hungarian Academy of Sciences
Budapest, Hungary

Professor George Gasper
Department of Mathematics
Northwestern University
Evanston, IL 60208, USA

Professor Dorian Goldfeld
Department of Mathematics
Columbia University,
New York, NY 10027, USA

Professor Basil Gordon
Department of Mathematics
University of California
Los Angeles, CA 90024, USA

Professor Andrew Granville
Department of Mathematics
University of Georgia
Athens, GA 30602, USA

Professor Adolf Hildebrand
Department of Mathematics
University of Illinois
Urbana, IL 61801, USA

Professor Mourad Ismail
Department of Mathematics
University of South Florida
Tampa, FL 33620, USA

Professor Marvin Knopp
Department of Mathematics
Temple University
Philadelphia, PA 19122, USA

Professor James Lepowsky
Department of Mathematics
Rutgers University
New Brunswick, NJ 08903, USA

Professor Lisa Lorentzen
Division of Mathematical Sciences
The Norwegian Institute of
  Technology
N-7034 Trondheim-NTH, Norway

Professor Jean-Louis Nicolas
Department of Mathematics
Université Claude Bernard
Lyon 1, 69622 Villeurbanne Cedex,
France

Professor Alfred van der Poorten
School of MPCE
Macquarie University
NSW 2109, Australia

Professor Robert Rankin
Department of Mathematics
University of Glasgow
Glasgow, G12 8QW, Scotland

Professor Gerald Tenenbaum
Institut Élie Cartan
Université Henri Poincaré Nancy 1
BP 239, F-54506 Vandoeuvre Cedex,
France

Professor Michel Waldschmidt
Université P et M Curie (Paris VI)
Mathematiques UFR 920
F-75252 Paris Cedex, France

Professor Don Zagier
Max Planck Institüt
  für Mathematik
5300 Bonn 1, Germany

Professor Doron Zeilberger
Department of Mathematics
Temple University
Philadelphia, PA 19122, USA

ISSN: 1382-4090

THE RAMANUJAN JOURNAL

Volume 2, Nos. 1/2, 1998

ISBN 978-1-4419-5058-1

Distributors for North, Central and South America:
Kluwer Academic Publishers
101 Philip Drive
Assinippi Park
Norwell, Massachusetts 02061 USA

Distributors for all other countries:
Kluwer Academic Publishers
Distribution Centre
Post Office Box 322
3300 AH Dordrecht, THE NETHERLANDS

Library of Congress Cataloging-in-Publication Data

A C.I.P. Catalogue record for this book is available
from the Library of Congress.

THE RAMANUJAN JOURNAL 2, 5–6 (1998)

Editorial

In September 1996, Professor Paul Erdős, one of the mathematical legends of this century, died while attending a conference in Warsaw, Poland. His death at the age of 83 marked the end of a great era, for Erdős was not only an outstanding mathematician but a very kind and generous human being, who encouraged hundreds of mathematicians over the decades, especially young aspirants to the subject. Many, including me, owe their careers to him. He was without doubt the most prolific mathematician of this century, having written more than 1000 papers, a significant proportion of them being joint papers. Even in a mathematical world, which is used to geniuses and their idiosyncracies, Erdős was considered an unusual phenomenon and was viewed with awe and adoration, just as Ramanujan evoked surprise and admiration. And like Ramanujan's mathematics, the contributions of Erdős will continue to inspire and influence research in the decades ahead.

Paul Erdős was unique in many ways. Born in Hungary in April 1913, he was a member of the Hungarian Academy of Sciences. But he did not have a job or any regular position. He was constantly on the move, criss-crossing the globe several times during a year, visiting one university after the other giving lectures. Somehow in his wordwide travels, like migrating birds, he managed to hover around the isotherm 70°F. So he visited Calgary in the summers, California in the winters and Florida in February/March. He seldom stayed at one place for more than two weeks except, perhaps, in his native Hungary where he returned periodically between his travels. And he did this every year for the past half a century or more! To be in constant demand at universities throughout the world, one should not only be an unending source of new ideas but should also have the ability to interact with persons of varying tastes and abilities. Erdős was superbly suited to this task. This is what kept his furious productivity going till the very end. In a long and distinguished career starting in 1931, Erdős made fundamental contributions to many branches of mathematics, most notably, Number Theory, Combinatorics, Graph Theory, Analysis, Set Theory and Geometry. He was the champion of the "elementary method", often taming difficult questions by ingenious elementary arguments.

Erdős began his illustrious career as a mathematician with a paper in 1932 on prime numbers. Interestingly, it was through this paper that he first became aware of Ramanujan's work. Ramanujan was a strong influence and inspiration for him from then on as he himself said in an article written for the Ramanujan centennial.

Two of Erdős's greatest accomplishments were the elementary proof of the prime number theorem, proved simultaneously and independently by Atle Selberg, and the Erdős-Kac theorem which gave birth to Probabilistic Number Theory. The Erdős-Kac theorem itself was an outcome of the famous Hardy-Ramanujan paper of 1917 on the number of prime factors of an integer. For these contributions, Erdős was awarded the Cole Prize of the American Mathematical Society in 1952. In 1983 he was awarded the Wolf Prize for his lifelong contributions to mathematics and he joined the ranks of other illustrious winners of this prize like Kolmogorov and Andre Weil. He was elected member of the National Academy of Sciences of USA and also elected Foreign Member of The Royal Society. He is also the recipient of numerous honorary doctorates from universities around the world.

What did Erdős do with his income and prize money? Erdős, who was a bachelor all his life, was wedded to mathematics which he pursued with a passion. Erdős had no desire for any material possessions and was saintly in his attitude towards life. He often used to say that property was a nuisance. During his visits to universities and institutes of higher learning, he was paid honoraria for his lectures. After keeping what was necessary to pay for his travel and living expenses, he would give away the remaining amount either in the form of donations to educational organizations or as prizes for solutions to mathematical problems he posed. I should emphasize that Erdős was without doubt the greatest problem proposer in history. During his lectures worldwide, he posed several problems and offered prize money ranging from $50 to $1000, depending on the difficulty of the problem. This was one way in which he spotted and encouraged budding mathematicians. It has often been mentioned about Ramanujan that his greatness was not only due to the remarkable results he proved, but also due to the many important questions that arose from his work. Similarly, Erdős will not only be remembered for the multitude of theorems he proved, but also for the numerous problems he raised.

When the idea to start the Ramanujan Journal was put forth, Erdős was very supportive. When he was invited to serve on the Editorial Board, he agreed very graciously. Had he been alive, he would have been delighted to see the first issue of the journal appear in January 1997. But before he died, he contributed a paper to the journal written jointly with Carl Pomerance and Andras Sárkőzy which appeared in volume 1, issue 3, in July 1997. Erdős will be sorely missed by the entire mathematical community, especially by those who got to know him closely. The Ramanujan Journal is proud to dedicate the first two issues of volume 2 to his memory.

Based on the success of the Ramanujan Journal, Kluwer Academic Publishers decided to launch the new book series *Developments in Mathematics* this year. This book series will publish research monographs, conference proceedings and contributed volumes in areas similar to those of The Ramanujan Journal. It was felt that it would be worthwhile to offer the Erdős special issues also in book (hard cover) form for those who may wish to purchase them separately. We are pleased that this book is the opening volume of Developments in Mathematics. Thus the new book series is off to a fine start with a volume of such high quality.

In preparing the Erdős memorial issues (volume), I had the help of Peter Elliott, Andrew Granville and Gerald Tenenbaum of The Ramanujan Journal editorial board. My thanks to them in particular, and more generally to the other members of the editorial board for their support. Finally we are grateful to the various authors for their contributions. By publishing the Erdős memorial issues in The Ramanujan Journal and in Developments in Mathematics, we are paying a fitting tribute to Erdős and Ramanujan both of whom are legends of twentieth century mathematics.

Krishnaswami Alladi
Editor-in-Chief

THE RAMANUJAN JOURNAL 2, 7–20 (1998)

Euler's Function in Residue Classes

THOMAS DENCE   tdence@ashland.edu
Department of Mathematics, Ashland University, Ashland, OH 44805

CARL POMERANCE,   carl@ada.math.uga.edu
Department of Mathematics, University of Georgia, Athens, GA 30602

Dedicated to the memory of Paul Erdős

Received June 28, 1996; Accepted October 16, 1996

Abstract. We discuss the distribution of integers n with $\varphi(n)$ in a particular residue class, showing that if a residue class contains a multiple of 4, then it must contain infinitely many numbers $\varphi(n)$. We get asymptotic formulae for the distribution of $\varphi(n)$ in the various residue classes modulo 12.

Key words:

1991 Mathematics Subject Classification:

1. Introduction

Let φ denote Euler's arithmetic function, which counts the number of positive integers up to n that are coprime to n. Given a residue class $r \bmod m$ must there be infinitely values of $\varphi(n)$ in this residue class? Let $N(x, m, r)$ denote the number of integers $n \leq x$ with $\varphi(n) \equiv r \bmod m$. If there are infinitely many Euler values in the residue class $r \bmod m$, can we find an asymptotic formula for $N(x, m, r)$ as $x \to \infty$? It is to these questions that we address this paper.

Since $\varphi(n)$ is even for each integer $n > 2$, we immediately see that if the residue class $r \bmod m$ does not contain any even numbers, then it cannot contain infinitely many values of $\varphi(n)$. Is this the only situation where we cannot find infinitely many Euler values? We conjecture that this is the case.

Conjecture. *If the residue class r mod m contains an even number then it contains infinitely many numbers $\varphi(n)$.*

This conjecture is a consequence of Dirichlet's theorem on primes in arithmetic progressions and the following elementary assertion: *If the residue class r mod m contains an even number, then there are integers a, k with $k \geq 0$ and $(a, m) = 1$ such that $a^k(a - 1) \equiv r$ mod m.* We have not been able to prove or disprove this assertion, though we conjecture it is true.

We can prove the following result.

Theorem 1.1. *If the residue class r mod m contains a multiple of 4 then it contains infinitely many numbers $\varphi(n)$.*

The proof is an elementary application of Dirichlet's theorem on primes in an arithmetic progression, and is inspired by an argument in a paper of Narkiewicz [6].

One relevant result from [6] is that if m is coprime to 6 and r is coprime to m, then there are infinitely many Euler values in the residue class r mod m. In particular, it is shown that asymptotically $1/\varphi(m)$ of the integers n with $\varphi(n)$ coprime to m have $\varphi(n) = r$ mod m. From this it is a short step to get an asymptotic formula for $N(x, m, r)$ for such pairs m, r.

In fact, for any specific pair m, r it seems possible to decide if $N(x, m, r)$ is unbounded and to obtain an asymptotic formula in case it is. We shall illustrate the kinds of methods one might use for such a project in the specific case $m = 12$.

We only have to consider the even residue classes mod 12. By Dirichlet's theorem we immediately see that the residue classes 0, 4, 6, 10 mod 12 each contain infinitely many φ-values, since there are infinitely many primes in each of the residue classes 1, 5, 7, 11 mod 12. This leaves $r = 2$ and 8. If p is an odd prime $\equiv 2$ mod 3, then $\varphi(4p) \equiv 8$ mod 12, so 8 mod 12 contains infinitely many φ-values. As noticed in [3], the residue class 2 mod 12 is tougher for φ to occupy. But if $p \equiv 11$ mod 12 and p is prime, then $\varphi(p^2) \equiv 2$ mod 12, so occupied it is.

Now we turn to estimating $N(x, 12, r)$ for r even. We begin with examining the numerical data in Table 1. Perhaps the most striking feature of Table 1 is the paucity of integers n with $\varphi(n) \equiv 2$ mod 12. This behavior was already noticed in [3], and it was shown there that the set of such integers has asymptotic density 0. Another observation that one might make is that the numbers for the 0 residue class keep growing as a percentage of the whole, from 30% at 100 to over 73% at 10^7. Though their contribution decreases as a percentage of the whole, the columns for 4 and 8 grow briskly, and seem to keep in approximately the same ratio. And the columns for 6 and 10 seem to be about equal. Can anything be proved concerning these observations? We prove the following theorem.

Theorem 1.2. *We have, as $x \to \infty$,*

$$N(x, 12, 0) \sim x, \tag{1.1}$$

Table 1. The number of $n \leq x$ with $\varphi(n)$ in a particular residue class modulo 12.

x	0	2	4	6	8	10
10^2	30	3	21	17	18	9
10^3	511	6	185	84	145	67
10^4	6114	13	1651	511	1233	476
10^5	66646	32	15125	3761	10743	3691
10^6	703339	81	140155	30190	96165	30068
10^7	7300815	208	1313834	253628	878141	253372

$$N(x, 12, 2) \sim \left(\frac{1}{2} + \frac{1}{2\sqrt{2}}\right)\frac{\sqrt{x}}{\log x}, \tag{1.2}$$

$$N(x, 12, 4) \sim c_1 \frac{x}{\sqrt{\log x}}, \tag{1.3}$$

$$N(x, 12, 6) \sim \frac{3}{8}\frac{x}{\log x}, \tag{1.4}$$

$$N(x, 12, 8) \sim c_2 \frac{x}{\sqrt{\log x}}, \tag{1.5}$$

$$N(x, 12, 10) \sim \frac{3}{8}\frac{x}{\log x}, \tag{1.6}$$

where $c_1 \doteq .6109136202$ is given by

$$c_1 = \frac{\sqrt{2\sqrt{3}}}{3\pi} c_3^{-1/2}(2c_3 + c_4), \tag{1.7}$$

with

$$c_3 = \prod_{\substack{p \text{ prime} \\ p \equiv 2(3)}} \left(1 + \frac{1}{p^2 - 1}\right), \quad c_4 = \prod_{\substack{p \text{ prime} \\ p \equiv 2(3)}} \left(1 - \frac{1}{(p+1)^2}\right), \tag{1.8}$$

and $c_2 \doteq .3284176245$ is given by the same expression as for c_1, except that $2c_3 + c_4$ is replaced by $2c_3 - c_4$.

The case of 0 mod 12 follows from a more general result of Erdős.

Theorem (Erdős). *For any positive integer m, $N(x, m, 0) \sim x$ as $x \to \infty$.*

We have not been able to find the first place this result appears but the proof follows from the fact that the sum of the reciprocals of the primes $p \equiv 1 \bmod m$ is infinite, so that almost all integers n are divisible by such a prime. But if such a prime p divides n, then $\varphi(n) \equiv 0 \bmod m$.

There is a fairly wide literature on the distribution in residue classes of values of multiplicative functions, in fact there is a monograph on the subject by Narkiewicz [7]. However, but for Narkiewicz's result above, and a result of Delange [2] (which can be used to give an asymptotic formula for the number of n up to x for which $\varphi(n)$ is not divisible by a fixed integer m), the problems considered here appear to be new.

We begin now with the proof of Theorem 1.2, giving the proof of Theorem 1.1 at the end of the paper. In the sequel, the letter p shall always denote a prime.

2. The residue classes 2, 6 and 10 mod 12

Let $S_{m,r}$ denote the set of integers n with $\varphi(n) \equiv r \bmod m$. So $N(x, m, r)$ counts how many members $S_{m,r}$ has up to x.

We begin by explicity describing $S_{12,r}$ for $r = 2$, 6 and 10. This is easy since these residue classes are contained in the class 2 mod 4, so that $S_{12,r} \subset S_{4,2}$ for $r = 2$, 6, 10. The set $S_{4,2}$ is particularly simple, consisting of numbers p^k, where p is a prime that is 3 mod 4, the doubles of these numbers, and the number 4.

We have,

$$S_{12,2} = \{3, 4, 6\} \cup \{n : n \text{ or } n/2 = p^{2k} \text{ where } p \equiv 11 \bmod 12\},$$

$$S_{12,6} = \{n : n \text{ or } n/2 = p^k \text{ where } p \equiv 7 \bmod 12, \text{ or } n \text{ or } n/2 = 3^k \text{ where } k \geq 2\},$$

$$S_{12,10} = \{n : n \text{ or } n/2 = p^{2k+1} \text{ where } p \equiv 11 \bmod 12\}.$$

We thus get (1.2), (1.4) and (1.6) of Theorem 1.2 using the prime number theorem for arithmetic progressions. For a reference on this theorem, see [1], Ch. 20.

3. Reduction to the modulus 3

Note that $n > 2$ and $\varphi(n) \equiv 1 \bmod 3$ if and only if $\varphi(n) \equiv 4$ or 10 mod 12. Further, $\varphi(n) \equiv 2 \bmod 3$ if and only if $\varphi(n) \equiv 2$ or 8 mod 12. In light of (1.2) and (1.6) of Theorem 1.2, it will suffice for (1.3) and (1.5) to show the following theorem.

Theorem 3.1. *As $x \to \infty$, we have*

$$N(x, 3, 1) \sim c_1 \frac{x}{\sqrt{\log x}}, \tag{3.1}$$

$$N(x, 3, 2) \sim c_2 \frac{x}{\sqrt{\log x}} \tag{3.2}$$

where c_1 and c_2 are given in Theorem 1.2.

Also note that $\varphi(n) \not\equiv 0 \bmod 3$ if and only if 9 does not divide n and n is not divisible by any prime $p \equiv 1 \bmod 3$. We begin our proof of Theorem 3.1 by first considering numbers n not divisible by 3. It is an easy leap from these numbers to the general case.

Let S_i be the set of integers n not divisible by 3 for which $\varphi(n) \equiv i \bmod 3$, for $i = 1$, 2. Further, let $N_i(x)$ be the number of members of S_i up to x, for $i = 1$, 2. Then the following result is immediate.

Lemma 3.2. *For $i = 1$, 2 and $x > 0$ we have*

$$N(x, 3, 1) = N_1(x) + N_2(x/3),$$

$$N(x, 3, 2) = N_2(x) + N_1(x/3).$$

Indeed, using the notation of Section 2, we have $n \in S_{3,1}$ and $n \leq x$ if and only if $n \in S_1$, $n \leq x$ or $n = 3m$ where $m \in S_2$, $m \leq x/3$. We have a similar characterization of the members of $S_{3,2}$ up to x.

Every natural number n has a unique decomposition as qf where $q = q(n)$ is the largest squarefull divisor of n and $f = f(n) = n/q$ is squarefree. (We say an integer is squarefull if it is divisible by p^2 whenever it is divisible by p.) For example, for the integer $n = 2200 = 2^3 \cdot 5^2 \cdot 11$, we have $q = q(2200) = 2^3 \cdot 5^2 = 200$ and $f = f(2200) = 11$.

Suppose n is only divisible by primes $\equiv 2$ mod 3 and write $n = qf$ as above. Then $\varphi(n) = \varphi(q)\varphi(f)$ and $\varphi(f) \equiv 1$ mod 3, so that

$$\varphi(n) \equiv \varphi(q) \bmod 3. \tag{3.3}$$

Let $\mathcal{F}$ denote the set of squarefree integers each of whose prime factors is 2 mod 3. Then $\mathcal{F} \subset \mathcal{S}_1$. Let $\mathcal{Q}$ denote the set of squarefull integers each of whose prime factors is 2 mod 3. From (3.3) we have the following lemma.

Lemma 3.3. *The set $\mathcal{S}_1$ is the disjoint union of the sets $q\mathcal{F}$ where $q \in \mathcal{S}_1 \cap \mathcal{Q}$. The set $\mathcal{S}_2$ is the disjoint union of the sets $q\mathcal{F}$ where $q \in \mathcal{S}_2 \cap \mathcal{Q}$.*

Of course, by $q\mathcal{F}$ we mean the set of integers qf where $f \in \mathcal{F}$.

4. A theorem of Landau and some consequences

In [5], Landau gives a more general theorem of which the following is a special case.

Theorem (Landau). *There is a positive constant c such that the number of integers $n \le x$ divisible only by primes $\equiv 2$ mod 3 is $\sim cx/\sqrt{\log x}$ as $x \to \infty$.*

We shall identify the constant c in Landau's theorem in Section 6.

We now deduce the following consequence of Landau's theorem. Let $\mathcal{N}$ denote the set of integers divisible only by primes $\equiv 2$ mod 3. Recall that $\mathcal{Q}$ is the set of squarefull numbers in $\mathcal{N}$.

Proposition 4.1. *For any subset $\mathcal{Q}_0$ of $\mathcal{Q}$, we have*

$$\sum_{\substack{n \le x \\ n \in \mathcal{N} \\ q(n) \in \mathcal{Q}_0}} 1 \sim cc_3^{-1} \frac{x}{\sqrt{\log x}} \sum_{q \in \mathcal{Q}_0} \frac{1}{q} \prod_{p \mid q} \frac{p}{p+1}$$

as $x \to \infty$, where c is the constant in Landau's theorem and c_3 is given in (1.8).

Note that in the special case $\mathcal{Q}_0 = \{1\}$, Proposition 4.1 asserts that the number of members of $\mathcal{F}$ up to x is $\sim cc_3^{-1}x/\sqrt{\log x}$ as $x \to \infty$.

Proof of Proposition 4.1: From Landau's theorem there is a constant c_5 such that for all $x > 1$,

$$\sum_{\substack{n \le x \\ n \in \mathcal{N}}} 1 \le c_5 \frac{x}{\sqrt{\log x}}. \tag{4.1}$$

Also, since the number of squarefull numbers $\leq x$ is $O(\sqrt{x})$, it follows that there is a constant c_6 such that for all $x > 1$,

$$\sum_{\substack{q>x \\ q\in Q}} \frac{1}{q} \leq \frac{c_6}{\sqrt{x}}. \tag{4.2}$$

From (4.1) and (4.2) we deduce the following: For each $\epsilon > 0$, there are numbers N, x_0 such that if $x \geq x_0$, then

$$\sum_{\substack{n\leq x \\ n\in \mathcal{N} \\ q(n)>N}} 1 \leq \epsilon \frac{x}{\sqrt{\log x}}. \tag{4.3}$$

Indeed, the sum in (4.3) is

$$\sum_{\substack{n\leq x, \\ n\in \mathcal{N} \\ N<q(n)\leq\sqrt{x}}} 1 + \sum_{\substack{n\leq x, \\ n\in \mathcal{N} \\ q(n)>\sqrt{x}}} 1 \leq c_5 \sum_{\substack{N<q\leq\sqrt{x} \\ q\in Q}} \frac{x/q}{\sqrt{\log(x/q)}} + \sum_{\substack{q>\sqrt{x} \\ q\in Q}} \frac{x}{q}$$

$$\leq 2c_5 \sum_{\substack{q>N \\ q\in Q}} \frac{x/q}{\sqrt{\log x}} + c_6 x^{3/4}$$

$$\leq 2c_5 c_6 \frac{x}{\sqrt{N}\sqrt{\log x}} + c_6 x^{3/4}.$$

Therefore, (4.3) follows by taking N and x sufficiently large.

Next note that for a fixed $d \in \mathcal{N}$, it follows from Landau's theorem that

$$\sum_{\substack{n\leq x \\ n\in \mathcal{N} \\ d|n}} 1 \sim c\frac{x/d}{\sqrt{\log(x/d)}} \sim \frac{c}{d}\cdot\frac{x}{\sqrt{\log x}} \tag{4.4}$$

as $x \to \infty$. Let $P(m)$ denote the largest prime factor of m when $m > 1$ and let $P(1) = 1$. Thus for any positive integer N,

$$\sum_{\substack{n\leq x, \\ n\in \mathcal{N} \\ (q(n),N!)=1}} 1 = \sum_{\substack{m\in \mathcal{N} \\ P(m)\leq N}} \mu(m) \sum_{\substack{n\leq x \\ n\in \mathcal{N} \\ m^2|n}} 1, \tag{4.5}$$

where μ is the Möbius function. Indeed, $m^2 \mid n$ if and only if $m^2 \mid q(n)$, so that $\sum \mu(m)$ for $m^2 \mid q(n)$, $P(m) \leq N$, is 0 whenever $(q(n), N!) > 1$ and 1 otherwise. Putting (4.4) and (4.5) together, we get that

$$\sum_{\substack{n\leq x, \\ n\in \mathcal{N} \\ (q(n),N!)=1}} 1 \sim \frac{cx}{\sqrt{\log x}} \sum_{\substack{m\in \mathcal{N} \\ P(m)\leq N}} \frac{\mu(m)}{m^2} = \frac{cx}{\sqrt{\log x}} \prod_{\substack{p\equiv 2(3) \\ p\leq N}} \left(1-\frac{1}{p^2}\right) \tag{4.6}$$

as $x \to \infty$.

We now use (4.3), (4.6) and the convergence of the infinite product $\prod_{p\equiv 2(3)}(1-1/p^2)$ (to the limit c_3^{-1}) to see that the proposition holds in the case $\mathcal{Q}_0 = \{1\}$. By a similar argument we can get an asymptotic formula for the number of $n \leq x$ with $n \in \mathcal{N}$, n squarefree and $(n, m) = 1$, where $m \in \mathcal{N}$ is fixed. This involves removing the factors $(1 - 1/p^2)$ from the infinite product corresponding to the primes $p \mid m$ and replacing them with $(1 - 1/p)$. That is, we should introduce the factor $p/(p + 1)$. We have for fixed $m \in \mathcal{N}$,

$$\sum_{\substack{n \leq x \\ n \in \mathcal{N} \\ q(n)=1 \\ (n,m)=1}} 1 \sim cc_3^{-1} \frac{x}{\sqrt{\log x}} \prod_{p \mid m} \frac{p}{p+1} \qquad (4.7)$$

as $x \to \infty$.

We are now ready to establish the general case of the proposition. To say that $n \in \mathcal{N}$ and $q(n) = q$ is to say that $n = n_1 q$ where $n_1 \in \mathcal{N}$, n_1 is squarefree and $(n_1, q) = 1$. Thus, from (4.7) we have for a fixed $q \in \mathcal{Q}_0$ that

$$\sum_{\substack{n \leq x \\ n \in \mathcal{N} \\ q(n)=q}} 1 \sim cc_3^{-1} \frac{x}{\sqrt{\log x}} \frac{1}{q} \prod_{p \mid q} \frac{p}{p+1}$$

as $x \to \infty$. Now using (4.3) and the convergence of the sum $\sum_{q \in \mathcal{Q}_0} 1/q$, we get that

$$\sum_{\substack{n \leq x \\ n \in \mathcal{N} \\ q(n) \in \mathcal{Q}_0}} 1 \sim cc_3^{-1} \frac{x}{\sqrt{\log x}} \sum_{q \in \mathcal{Q}_0} \frac{1}{q} \prod_{p \mid q} \frac{p}{p+1}$$

as $x \to \infty$, which is what we wished to prove. $\qquad\qquad\square$

5. The sums S_1 and S_2

We can now get asymptotic estimates for the quantities $N_i(x)$, $i = 1, 2$, that count the number of members of $\mathcal{S}_i$ up to x. Recall that $\mathcal{S}_i$ is the set of integers n not divisible by 3 for which $\varphi(n) \equiv i \bmod 3$. From Lemma 3.3 and Proposition 4.1 we immediately get that

$$N_i(x) \sim cc_3^{-1} \frac{x}{\sqrt{\log x}} \sum_{q \in \mathcal{S}_i \cap \mathcal{Q}} \frac{1}{q} \prod_{p \mid q} \frac{p}{p+1} \qquad (5.1)$$

as $x \to \infty$ for $i = 1, 2$.

Let

$$S_i = \sum_{q \in \mathcal{S}_i \cap \mathcal{Q}} \frac{1}{q} \prod_{p \mid q} \frac{p}{p+1} \qquad (5.2)$$

for $i = 1$, 2. Thus from (1.8), Lemma 3.2, (5.1) and (5.2) we get that as $x \to \infty$,

$$N(x, 3, 1) \sim cc_3^{-1}\left(S_1 + \frac{1}{3}S_2\right)\frac{x}{\sqrt{\log x}},$$

$$N(x, 3, 2) \sim cc_3^{-1}\left(S_2 + \frac{1}{3}S_1\right)\frac{x}{\sqrt{\log x}}. \tag{5.3}$$

We shall show the following result.

Proposition 5.1. *With c_3, c_4 defined in (1.8), we have*

$$S_1 + \frac{1}{3}S_2 = \frac{2}{3}c_3 + \frac{1}{3}c_4, \quad S_2 + \frac{1}{3}S_1 = \frac{2}{3}c_3 - \frac{1}{3}c_4.$$

Proposition 5.1 serves a numerical purpose, since it is easier to estimate the infinite products c_3, c_4 than the sums S_1, S_2.

Proof of Proposition 5.1: One can get a simple expression for $S_1 + S_2$. Since $q^{-1}\prod_{p|q} p/(p+1)$ is a multiplicative function of q, we have

$$S_1 + S_2 = \sum_{q \in Q} \frac{1}{q}\prod_{p|q}\frac{p}{p+1}$$

$$= \prod_{p \equiv 2(3)}\left(1 + \frac{p}{p+1}\sum_{a=2}^{\infty}\frac{1}{p^a}\right)$$

$$= \prod_{p \equiv 2(3)}\left(1 + \frac{1}{p^2 - 1}\right) = c_3. \tag{5.4}$$

Let $\omega(m)$ denote the number of distinct prime factors of m and let $\Omega(m)$ denote the number of prime factors of m counted with multiplicity. Then for $q \in Q$ we have

$$\varphi(q) = q\prod_{p|q}\left(1 - \frac{1}{p}\right) \equiv (-1)^{\Omega(q)}(-1)^{\omega(q)} \bmod 3.$$

Thus,

$$S_1 - S_2 = \sum_{q \in Q}(-1)^{\Omega(q)+\omega(q)}\frac{1}{q}\prod_{p|q}\frac{p}{p+1}$$

$$= \prod_{p \equiv 2(3)}\left(1 - \frac{p}{p+1}\sum_{a=2}^{\infty}\frac{(-1)^a}{p^a}\right)$$

$$= \prod_{p \equiv 2(3)}\left(1 - \frac{1}{(p+1)^2}\right) = c_4. \tag{5.5}$$

Proposition 5.1 follows immediately from (5.4) and (5.5). $\square$

We now say a few words on the numerical estimation of the products c_3 and c_4 in (5.4) and (5.5). Both products converge quadratically, in fact, better than quadratically, since they are over primes. However, we can hasten the convergence, making them even easier to calculate. Let

$$\bar{c}_3 = \prod_{p \neq 2(3)} \left(1 + \frac{1}{p^2 - 1} \right),$$

so that

$$c_3\bar{c}_3 = \prod_p \left(1 + \frac{1}{p^2 - 1} \right) = \prod_p \left(1 + \frac{1}{p^2} + \frac{1}{p^4} + \cdots \right) = \sum_{n=1}^{\infty} \frac{1}{n^2} = \frac{\pi^2}{6}.$$

Then

$$c_3 = \sqrt{\frac{c_3\bar{c}_3}{\bar{c}_3/c_3}} = \frac{\pi}{\sqrt{6}} \sqrt{\frac{c_3}{\bar{c}_3}}. \tag{5.6}$$

Let $\chi_1(n)$ be ± 1 when $n \equiv \pm 1 \bmod 3$, respectively, and note that

$$\frac{c_3}{\bar{c}_3} = \frac{9}{8} \prod_{p \neq 3} \left(1 + \frac{1}{p^2 - 1} \right)^{\chi_1(p)}.$$

This last product converges considerably faster than do the separate products c_3 and $\bar{c}_3$, and it is via this product and (5.6) that we get the estimate $c_3 \doteq 1.4140643909$. It is now a simple matter to estimate c_4 since we have that $c_4 = (c_4c_3)/c_3$, where

$$c_4c_3 = \prod_{p \equiv 2(3)} \left(1 + \frac{2p + 1}{(p^2 - 1)(p + 1)^2} \right)$$

converges cubically. By means of our estimation for c_3 and an estimation for c_4c_3, we get that $c_4 \doteq .8505360177$.

From (1.2), (1.6), (5.3), (5.4) and (5.5) we have

$$\frac{N(x, 12, 4)}{N(x, 12, 8)} \sim \frac{N(x, 3, 1)}{N(x, 3, 2)} \sim \frac{3S_1 + S_2}{S_1 + 3S_2} = \frac{2c_3 + c_4}{2c_3 - c_4},$$

as $x \to \infty$. The ratio of the modulo 3 counts converges to this limit more rapidly than the ratio of the modulo 12 counts, as can be seen numerically in Table 1. This is due to the modulo 12 ratio leaving out the residue class 10 mod 12, which is negligible asymptotically, but not so at small levels.

6. The calculation of Landau's constant c

In this section we shall show the following.

Proposition 6.1. *The number c in Landau's theorem is $\sqrt{2c_3\sqrt{3}}/\pi$, where c_3 is given in* (1.8).

Note that Theorem 3.1 (and so (1.3) and (1.5) of Theorem 1.2) follows immediately from (5.3), Propositions 5.1 and 6.1.

Proof of Proposition 6.1:   Using a theorem of Wirsing [8], we have that

$$c = \frac{\frac{1}{2}Ke^{-\gamma/2}}{\Gamma\left(\frac{3}{2}\right)} = \frac{Ke^{-\gamma/2}}{\sqrt{\pi}}, \tag{6.1}$$

where γ is Euler's constant and K is the number that satisfies

$$\prod_{\substack{p \le x \\ p \equiv 2(3)}} \left(1 + \frac{1}{p-1}\right) \sim K(\log x)^{1/2} \tag{6.2}$$

as $x \to \infty$. Thus, in light of (6.1), to prove Proposition 6.1 it will suffice to show that

$$K = e^{\gamma/2}\sqrt{\frac{2c_3\sqrt{3}}{\pi}}. \tag{6.3}$$

We take the logarithm of (6.2) getting that

$$\log K + \frac{1}{2}\log\log x = \sum_{\substack{p \le x \\ p \equiv 2(3)}} \log\left(1 + \frac{1}{p-1}\right) + o(1)$$

$$= \sum_{\substack{p \le x \\ p \equiv 2(3)}} \frac{1}{p} + \sum_{p \equiv 2(3)} \left(\log\left(1 + \frac{1}{p-1}\right) - \frac{1}{p}\right) + o(1), \tag{6.4}$$

as $x \to \infty$.

Let B be the number such that

$$\sum_{\substack{p \le x \\ p \equiv 2(3)}} \frac{1}{p} = \frac{1}{2}\log\log x + B + o(1) \tag{6.5}$$

as $x \to \infty$. So from (6.4) and (6.5) we get

$$\log K = B + \sum_{p \equiv 2(3)} \left(\log\left(1 + \frac{1}{p-1}\right) - \frac{1}{p}\right) \tag{6.6}$$

We now compute the number B in (6.5). Mertens showed how to do this over 100 years ago; we follow his method. Let χ_0, χ_1 be the Dirichlet characters mod 3, where

$$\chi_0(n) = \begin{cases} 1, & \text{if 3 does not divide } n \\ 0, & \text{otherwise,} \end{cases}$$

and

$$\chi_1(n) = \begin{cases} 1, & \text{if } n \equiv 1 \bmod 3 \\ -1, & \text{if } n \equiv -1 \bmod 3 \\ 0, & \text{if } n \equiv 0 \bmod 3. \end{cases}$$

Then $(\chi_0(n) - \chi_1(n))/2$ is the characteristic function of the integers $n \equiv 2 \bmod 3$. We thus have

$$\sum_{\substack{p \le x \\ p \equiv 2(3)}} \frac{1}{p} = \frac{1}{2} \sum_{p \le x} \frac{\chi_0(p) - \chi_1(p)}{p}$$

$$= -\frac{1}{6} + \frac{1}{2} \sum_{p \le x} \frac{1}{p} - \frac{1}{2} \sum_{p \le x} \frac{\chi_1(p)}{p}. \tag{6.7}$$

From Theorem 428 in Hardy and Wright [4] we have

$$\sum_{p \le x} \frac{1}{p} = \log \log x + \gamma + \sum_p \left(\log\left(1 - \frac{1}{p}\right) + \frac{1}{p} \right) + o(1) \tag{6.8}$$

as $x \to \infty$. Since the series $\sum_p \chi_1(p)/p$ converges (as we shall soon see), it follows from (6.5), (6.7) and (6.8) that

$$B = -\frac{1}{6} + \frac{1}{2}\gamma + \frac{1}{2} \sum_p \left(\log\left(1 - \frac{1}{p}\right) + \frac{1}{p} \right) - \frac{1}{2} \sum_p \frac{\chi_1(p)}{p}. \tag{6.9}$$

To evaluate the last series, consider the L-function

$$L(s, \chi_1) = \sum_{n=1}^{\infty} \frac{\chi_1(n)}{n^s}$$

for $s > 0$. (It follows from the Abel summation formula that the series converges for $s > 0$.) Since $\chi_1(n)n^{-s}$ is a multiplicative function of n, we have

$$L(s, \chi_1) = \prod_p \left(1 + \frac{\chi_1(p)}{p^s} + \frac{\chi_1(p)^2}{p^{2s}} + \cdots \right) = \prod_p \left(1 - \frac{\chi_1(p)}{p^s} \right)^{-1}.$$

Letting $s = 1$ and taking the logarithm we get

$$\log L(1, \chi_1) = -\sum_p \log\left(1 - \frac{\chi_1(p)}{p}\right)$$

$$= \sum_p \frac{\chi_1(p)}{p} - \sum_p \left(\log\left(1 - \frac{\chi_1(p)}{p}\right) + \frac{\chi_1(p)}{p}\right). \qquad (6.10)$$

It follows from Dirichlet's class number formula (see [1], Ch. 6) that $L(1, \chi_1) = \pi/3^{3/2}$, so that from (6.1) we have

$$\sum_p \frac{\chi_1(p)}{p} = \log\left(\frac{\pi}{3^{3/2}}\right) + \sum_p \left(\log\left(1 - \frac{\chi_1(p)}{p}\right) + \frac{\chi_1(p)}{p}\right).$$

Putting this identity in (6.9), we get

$$B = -\frac{1}{6} + \frac{1}{2}\gamma - \frac{1}{2}\log\left(\frac{\pi}{3^{3/2}}\right) + \frac{1}{2}\sum_p \left(\log\left(\frac{1 - 1/p}{1 - \chi_1(p)/p}\right) + \frac{1 - \chi_1(p)}{p}\right)$$

$$= -\frac{1}{6} + \frac{1}{2}\gamma - \frac{1}{2}\log\left(\frac{\pi}{3^{3/2}}\right) + \frac{1}{2}\left(\log\left(\frac{2}{3}\right) + \frac{1}{3}\right) + \frac{1}{2}\sum_{p \equiv 2(3)} \left(\log\left(\frac{1 - 1/p}{1 + 1/p}\right) + \frac{2}{p}\right).$$

$$(6.11)$$

Thus, from (6.6) and (6.11), we have

$$\log K = \frac{1}{2}\gamma - \frac{1}{2}\log\left(\frac{\pi}{2 \cdot 3^{1/2}}\right) + \frac{1}{2}\sum_{p \equiv 2(3)} \left(2\log\left(1 + \frac{1}{p-1}\right) + \log\left(\frac{1 - 1/p}{1 + 1/p}\right)\right)$$

$$= \frac{1}{2}\gamma - \frac{1}{2}\log\left(\frac{\pi}{2 \cdot 3^{1/2}}\right) + \frac{1}{2}\sum_{p \equiv 2(3)} \log\left(\frac{p^2}{p^2 - 1}\right).$$

This gives (6.3), and so we have the proposition.   □

7.   The proof of Theorem 1.1

Given a residue class $r \bmod m$ that contains a multiple of 4, we shall show that there are integers s, t with $(s + 1)(t + 1)$ coprime to m and either $st \equiv r \bmod m$ or $st(t + 1) \equiv r \bmod m$. By Dirichlet's theorem, the former condition assures that there are infinitely many pairs of different primes p, q with $p \equiv s + 1 \bmod m$ and $q \equiv t + 1 \bmod m$. If $st \equiv r \bmod m$, then $\varphi(pq) = (p - 1)(q - 1) \equiv r \bmod m$, while if $st(t + 1) \equiv r \bmod m$, then $\varphi(pq^2) = (p - 1)(q - 1)q \equiv r \bmod m$. In either case, there are infinitely many integers n with $\varphi(n) \equiv r \bmod m$.

Say the prime factorization of m is $p_1^{a_1}, p_2^{a_2} \cdots p_k^{a_k}$. We first consider the case when $r \not\equiv 2$ mod 3. Let s, t be integers such that for each odd p_i we have

$$s \equiv \begin{cases} r \bmod p_i^{a_i}, & \text{when } r \not\equiv -1 \bmod p_i \\ 2^{-1}r \bmod p_i^{a_i}, & \text{when } r \equiv -1 \bmod p_i, \end{cases}$$

$$t \equiv \begin{cases} 1 \bmod p_i^{a_i}, & \text{when } r \not\equiv -1 \bmod p_i \\ 2 \bmod p_i^{a_i}, & \text{when } r \equiv -1 \bmod p_i. \end{cases}$$

These congruences define s and t modulo the odd part of m. Suppose m is even and $2^a \| m$. If $a = 1$, then we choose s and t so that they are even, and so we have defined them modulo m. If $a \geq 2$, then by our hypothesis, $4 \mid r$. Take

$$s \equiv \frac{r}{2} \bmod 2^a, \quad t \equiv 2 \bmod 2^a.$$

In all cases we have that $st \equiv r \bmod m$ and $(s+1)(t+1)$ is coprime to m. These facts may be verified by looking at the situation modulo each $p_i^{a_i}$. For example, suppose p_i is odd and $r \equiv -1 \bmod p_i$. By our hypothesis, p_i is not 3. Then $s + 1 \equiv 2^{-1}r + 1 \equiv 2^{-1}(r+2) \equiv 2^{-1} \not\equiv 0 \bmod p_i$ and $t + 1 \equiv 3 \not\equiv 0 \bmod p_i$. The other conditions follow similarly.

Now consider the case when $r \equiv 2 \bmod 3$. Let s, t be integers such that for each odd p_i we have

$$s \equiv \begin{cases} 2^{-1}r \bmod p_i^{a_i}, & \text{when } r \not\equiv -2 \bmod p_i \\ 6^{-1}r \bmod p_i^{a_i}, & \text{when } r \equiv -2 \bmod p_i, \end{cases}$$

$$t \equiv \begin{cases} 1 \bmod p_i^{a_i}, & \text{when } r \not\equiv -2 \bmod p_i \\ 2 \bmod p_i^{a_i}, & \text{when } r \equiv -2 \bmod p_i, \end{cases}$$

(Note that if $p_i = 3$ then $r \not\equiv -2 \bmod p_i$, so we do not need the multiplicative inverse of 6.) Again suppose $2^a \| m$. If $a = 1$ then take s and t to be even. If $a \geq 2$, then by hypothesis, $4 \mid r$. Take

$$s \equiv 3^{-1}\frac{r}{2} \bmod 2^a, \quad t \equiv 2 \bmod 2^a.$$

This time note that $st(t+1) \equiv r \bmod m$ and that $(s+1)(t+1)$ is coprime to m. This completes the proof of the theorem.

Added in proof: The conjecture in the Introduction is false; for example, consider the residue classes 302 and 790 (mod 1092). Examples such as this are discussed in "Residue classes free of values of Euler's function," by K. Ford, S. Konyagin and C. Pomerance, to appear in the Proceedings of the Number Theory Conference, Zakopane, Poland, 1997. It is shown there that asymptotically almost all numbers that are 2 (mod 4) are in a residue class free of values of Euler's function.

Acknowledgments

We thank Joseph B. Dence for his assistance in calculating $L(1, \chi_1)$ by an alternate method. We also thank the referee for informing us of [7] which led us to the papers [6] and [2], and ultimately led us to the proof of Theorem 1.1. The calculation of Table 1 and of the numbers c_1, c_2, c_3, c_4 was done with the aid of Mathematica. This paper was written while the first author was visiting The University of Georgia. The second author is supported in part by a National Science Foundation grant.

References

1. H. Davenport, *Multiplicative Number Theory*, 2nd edition, Springer-Verlag, New York, 1980.
2. H. Delange, "Sur les fonctions multiplicatives à valeurs entiers," *C. R. Acad. Sci. Paris, Série A* **283** (1976), 1065–1067.
3. J.B. Dence and T. Dence, "A surprise regarding the equation $\phi(x) = 2(6n + 1)$," *The College Math. J.* **26** (1995), 297–301.
4. G.H. Hardy and E.M. Wright, *An Introduction to the Theory of Numbers*, 5th edition, Oxford University Press, p. 351, 1979.
5. E. Landau, *Handbuch der Lehre von der verteilung der Primzahlen*, 3rd edition, Chelsea Publ. Co., pp. 668–669, 1974.
6. W. Narkiewicz, "On distribution of values of multiplicative functions in residue classes," *Acta Arith.* **12** (1966/67), 269–279.
7. W. Narkiewicz, "Uniform distribution of sequences of integers in residue classes," vol. 1087 in Lecture Notes in Math., Springer-Verlag, Berlin, 1984.
8. E. Wirsing, "Über die Zahlen, deren Primteiler einer gegeben Menge angehören," *Arch. der Math.* **7** (1956), 263–272.

THE RAMANUJAN JOURNAL 2, 21–37 (1998)

Partition Identities Involving Gaps and Weights, II

KRISHNASWAMI ALLADI   alladi@math.ufl.edu
University of Florida, Gainesville, Florida 32611

Dedicated to the memory of Professor Paul Erdős

Received May 28, 1996; Accepted September 17, 1996

Abstract. In this second paper under the same title, some more weighted representations are obtained for various classical partition functions including $p(n)$, the number of unrestricted partitions of n, $Q(n)$, the number of partitions of n into distinct parts and the Rogers-Ramanujan partitions of n (of both types). The weights derived here are given either in terms of congruence conditions satisfied by the parts or in terms of chains of gaps between the parts. Some new connections between partitions of the Rogers-Ramanujan, Schur and Göllnitz–Gordon type are revealed.

Key words: partitions, weights, gaps, Durfee squares

1991 Mathematics Subject Classification: Primary—11P83, 11P81; Secondary—05A19

1. Introduction

In a recent paper [1] we derived weighted identities for various classical partition functions and discussed some applications. Given two sets of partitions S and T with $S \subseteq T$, let $P_S(n)$ (resp. $P_T(n)$) denote the number of partitions π of n with $\pi \in S$ (resp. $\pi \in T$). Clearly $P_S(n) \leq P_T(n)$. The general problem is to determine positive integral weights $w_{S,T}(\pi) = w(\pi)$ such that

$$P_T(n) = \sum_{\pi \in S, \sigma(\pi)=n} w(\pi), \tag{1.1}$$

where here and in what follows, $\sigma(\pi)$ denotes the sum of the parts of π.

For all the identities in [1], the weights are defined multiplicatively in terms of the gaps between the parts and in some cases are powers of 2. Such weighted identities have a variety of applications including new interpretations for Schur's partition theorem and Jacobi's triple product identity, combinatorial explanations for some remarkable partition congruences, and a combinatorial proof of a deep partition theorem of Göllnitz (see [1, 2]).

In this paper we establish some more weighted identities involving classical partition functions. But the weights here are of a different nature. For instance, in Theorems 1 and 2

Research supported in part by the National Science Foundation grant DMS 9400191.

of Section 3, the two 2-adic representations for $p(n)$, the number of unrestricted partitions of n, involve weights defined multiplicatively in terms of certain congruence conditions satisfied by the parts. And in Theorem 3 of Section 4, the weights are the middle binomial coefficients. Our interest in these three identities for $p(n)$ is due to their simplicity and elegance.

Theorems 4 and 5 of Section 5 yield weighted representations for the number of Rogers-Ramanujan partitions of n (of both types) in terms of partitions whose parts differ by ≥ 4. The weights here are defined multiplicatively in terms of (maximal) chains of parts satisfying certain gap conditions and turn out to be products of Fibonacci numbers. Although various proofs of the Rogers-Ramanujan identities are known (see Andrews [4]), none are simple; in particular, there is no simple combinatorial explanation of the identities. Theorems 4 and 5 are interesting because via the Fibonacci numbers, the prime number 5 enters into the problem combinatorially. It is our hope that such approaches might eventually shed some light into the combinatorial structure of Rogers-Ramanujan identities.

Theorem 6 of Section 6 deals with a two parameter extension of a weighted identity I had obtained earlier [1] connecting partitions into distinct parts and Rogers-Ramanujan partitions. The weights in Theorem 6 are also defined using chains of gaps. Special cases of Theorem 6 yield certain well known results of Göllnitz [10].

Finally, in Section 7, a different type of weighted identity is obtained, one where the smaller function $P_S(n)$ is given as a weighted sum over partitions enumerated by the larger function $P_T(n)$. The weights here are 1, 0 and -1. In particular, Theorem 7 provides an interesting link between partitions with minimum difference ≥ 2 and having no consecutive even numbers as parts (due to Göllnitz [10] and Gordon [12]) and partitions with difference ≥ 3 between parts and with no consecutive multiples of 3 (due to Schur [14]).

2. Preliminaries

For a complex number a and a positive integer n, we let

$$(a)_n = (a; q)_n = \prod_{j=0}^{n-1}(1 - aq^j),$$

and for $|q| < 1$,

$$(a)_\infty = \lim_{n \to \infty}(a)_n = \prod_{j=0}^{\infty}(1 - aq^j).$$

Given any partition π, let $\nu(\pi)$ denote the number of parts of π and $\sigma(\pi)$, the sum of the parts. When parts of a specific type are counted, this is indicated by a subscript. Each partition π may be considered as a multi-set whose elements are positive integers.

Given two partitions π_1 and π_2, let $\pi_1 \cup \pi_2$ denote the partition (multi-set) obtained by the set theoretic union of (the multi-sets) π_1 and π_2. Next, if $\pi_1 : a_1 \geq a_2 \geq \ldots$ and $\pi_2 : b_1 \geq b_2 \geq \ldots$ are partitions with parts a_i, b_j respectively, by $\pi_1 + \pi_2$ we mean the partition whose parts are $a_i + b_i$ where the integer 0 is substituted for a_i or b_i if $i > \nu(\pi_1)$ or $i > \nu(\pi_2)$.

Every partition $\pi : b_1 \geq b_2 \geq \dots$ can be represented as a Ferrers graph where the number of nodes (equally spaced) in the ith row of the graph is b_i. We make no distinction between a partition and its Ferrers graph.

The Ferrers graph of every partition π has a largest square of nodes with one vertex of the square as the upper left hand corner of the graph as shown below:

$$
\begin{array}{ccc|c}
\cdot & \cdot & \cdot & \cdot & \cdot \\
\cdot & \cdot & \cdot & \cdot & \cdot \\
\cdot & \cdot & \cdot & \cdot & \cdot \\
\hline
\cdot & \cdot & \cdot \\
\cdot \\
\end{array}
\tag{2.1}
$$

This is called the Durfee square of the partition and is denoted by $D(\pi)$. A partition with no nodes below the Durfee square is called a primary partition.

Given a Ferrers graph (partition) π, by $\rho(\pi)$ we mean the new partition whose parts are obtained by counting nodes along hooks of the graph π. For example, the graph in (2.1) is that of the partition $\pi : 8 \geq 6 \geq 6 \geq 4 \geq 4 \geq 1$. In this case, $\rho(\pi)$ is the partition $13 \geq 8 \geq 6 \geq 2$. Note that $v(\rho(\pi)) = |D(\pi)|$, the size of the Durfee square of π.

3. 2-adic representations for $p(n)$

In this section we obtain two 2-adic identities for $p(n)$. We begin with

Theorem 1. *Let $S_{2,4}$ denote the set of all partitions into parts $\not\equiv 2 \pmod 4$. For $\pi \in S_{2,4}$, let its weight $w_2(\pi) = 2^{r+s}$, where r is the number of odd parts of π that repeat and s is the number of different multiples of 4 (not counting multiplicity) in π. Then*

$$
p(n) = \sum_{\pi \in S_{2,4}, \sigma(\pi)=n} w_2(\pi).
$$

Proof: Observe that

$$
\sum_{n=0}^{\infty} p(n)q^n = \prod_{m=1}^{\infty} \frac{1}{(1-q^m)} = \frac{1}{(q)_\infty} = \frac{1}{(q^4;q^4)_\infty (q^2;q^4)_\infty (q;q^2)_\infty}.
\tag{3.1}
$$

We now make use of a fundamental identity or Euler, namely,

$$
(-q)_\infty = \prod_{m=1}^{\infty}(1+q^m) = \prod_{m=1}^{\infty} \frac{1}{(1-q^{2m-1})} = \frac{1}{(q;q^2)_\infty}
\tag{3.2}
$$

to rewrite (3.1) as

$$
\frac{1}{(q)_\infty} = \frac{(-q^2;q^2)_\infty}{(q^4;q^4)_\infty (q;q^2)_\infty} = \frac{(-q^4;q^4)_\infty}{(q^4;q^4)_\infty} \cdot \frac{(-q^2;q^4)_\infty}{(q;q^2)_\infty}.
\tag{3.3}
$$

Next observe that in the expansion

$$\frac{1+q^{kj}}{1-q^j} = \frac{1}{1-q^j} + \frac{q^{kj}}{1-q^j} = (1+q^j+q^{2j}+\cdots) + q^{kj}(1+q^j+q^{2j}+\cdots),$$

a part j that appears at least k times is counted twice. Thus

$$\frac{(-q^4; q^4)_\infty}{(q^4; q^4)_\infty}$$

is the generating function for partitions into parts $\equiv 0 \pmod 4$, where each such partition counted with weight 2^s, if s distinct multiples of 4 occur in the partition. Similarly,

$$\frac{(-q^2; q^4)_\infty}{(q; q^2)_\infty}$$

is the generating function for partitions into odd parts where each such partition is counted with weight 2^r if exactly r of the odd parts repeat. Theorem 1 is a consequence of (3.3) and the combinatorial interpretations given above.

In a similar vein, by considering the decomposition

$$\frac{1}{(q)_\infty} = \frac{(-q)_\infty}{(q^2; q^2)_\infty} = \frac{(-q^2; q^2)_\infty}{(q^2; q^2)_\infty} \cdot (-q; q^2)_\infty \tag{3.4}$$

we get

Theorem 2. *Let $\mathcal{D}_0$ denote the set of all partitions in which the odd parts do not repeat. For $\pi \in \mathcal{D}_0$, define the weight of π to be $w_0(\pi) = 2^d$, where d is the number of distinct even integers in π. Then*

$$p(n) = \sum_{\pi \in \mathcal{D}_0, \sigma(\pi)=n} w_0(\pi)$$

It is possible to give simple combinatorial proofs of these results.

Combinatorial proof of Theorem 1: Every partition π of n can be decomposed as $(\pi_0; \pi_1; \pi_2)$, where π_0 contains the parts of π which are $\equiv 0 \pmod 4$, π_2 contains the parts of π that are $\equiv 2 \pmod 4$, and π_1, contains the odd parts of π.

Next, using the combinatorial proof Euler's identity (3.2) (see Hardy and Wright [10]) we convert π_2 into a partition π' into distinct even parts. Now decompose π' as (π_0', π_2'), where π_j' consists of the parts of π' which are $\equiv j \pmod 4$. Thus the partitions π of n are in one-to-one correspondence with vector partitions $(\pi_0; \pi_1; \pi_0'; \pi_2')$. The parts of π_2' which are distinct and $\equiv 2 \pmod 4$ may be considered instead as odd parts repeated exactly twice.

Finally given a partition $(\pi_0; \pi_1) \in S_{2,4}$, if a part of π_1 repeats, we have a choice of removing two repetitions of that part and placing them in π_2' or not do this at all. Thus we have two choices here. Similarly, given a multiple of 4 belonging to π_0, we have a choice of removing that multiple and placing it in π_0' or not do this. So, here also we have two choices.

Thus each partition $(\pi_0'; \pi_2') \in S_{2,4}$ generates 2^{r+s} vector partitions $(\pi_0; \pi_1; \pi_0'; \pi_2')$ of n. Hence summing the weights 2^{r+s} over all partitions of n belonging to $S_{2,4}$ yields $p(n)$ as in Theorem 1. □

The combinatorial proof of Theorem 2 is similar and simpler.

Remarks. Weighted identities involving powers of 2 contain combinatorial information about the distribution of the underlying partition functions modulo powers of 2. In [1] we obtained a 2-adic representation for $Q(n)$, the number of partitions of n into distinct parts in terms of partitions of n with minimal difference ≥ 3 between parts. This suggested the remarkable result that for each integer $k \geq 1$,

$$Q(n) \equiv 0(\bmod 2^k) \tag{3.5}$$

for almost all n. Indeed, by means of weighted identities, the validity of (3.5) for almost all n can be established combinatorially for small k.

The distribution of $p(n)$ modulo powers of 2 is not very well understood. One of the most interesting and difficult open problems is to decide when $p(n)$ is odd and when it is even. Even Ramanujan was interested in this question. Recently, Ono [13] has made substantial progress on this problem. Both Theorems 1 and 2 imply that $p(n)$ and $Q_0(n)$, the number of partitions of n into distinct odd parts, have the same parity. For, in determining the parity of $p(n)$, it suffices to look at the cases where $r = s = d = 0$ in Theorems 1 and 2. This leads to partitions into distinct odd parts.

4. $p(n)$ and Glaisher's 3-regular partitions

A simple generalization of Euler's identity (3.2) noticed by Glaisher [7] was

$$\prod_{j=1}^{\infty} \left(1 + q^j + q^{2j} + \cdots + q^{(k-1)j}\right) = \prod_{j=1}^{\infty} \frac{(1-q^{kj})}{(1-q^j)} = \prod_{\substack{m=1 \\ m \not\equiv 0 \,(\bmod\, k)}}^{\infty} \frac{1}{(1-q^m)}. \tag{4.1}$$

The obvious combinatorial interpretation of (4.1) is that the number of partitions of n into parts $\not\equiv 0(\bmod k)$ is equal to the number of k-regular partitions of n, namely, partitions of n whose parts repeat fewer than k times. We now prove

Theorem 3. *Let G_3 denote the set of all 3-regular partitions, namely, partitions where each part repeats less than three times. For $\pi_3 \in G_3$ define its weight as*

$$w_3(\pi_3) = \binom{r}{[\frac{r}{2}]}$$

where r is the number of parts of π that do not repeat. Then

$$p(n) = \sum_{\pi \in G_3, \sigma(\pi)=n} w_3(\pi).$$

Proof: We begin with a well known identity due to Euler:

$$\sum_{n=0}^{\infty} p(n)q^n = \sum_{v=0}^{\infty} \frac{q^{v^2}}{(q)_v^2}.$$ (4.2)

By writing

$$\frac{q^{v^2}}{(q)_v^2} = \frac{q^{v(v+1)/2}}{(q)_v} \cdot \frac{q^{v(v-1)/2}}{(q)_v},$$

we see that

$$\frac{q^{v^2}}{(q)_v^2}$$

is the generating function for vector partitions $(\pi_0'; \pi_1')$, where π_0' has v distinct parts including possibly 0, and π' is a partition into v distinct parts. These are the well known Frobenius partitions. Thus the partitions of n are in one-to-one correspondence with the Frobenius partitions $(\pi_0'; \pi_1')$ of n.

Next, given a partition $\pi_3 \in G_3$, consider a part t of π_3 which repeats, in which case it repeats exactly twice. We now place t as a part of π_0' as well as of π_1'. If there are exactly r parts of π_3 which do not repeat, then choose $[\frac{r}{2}]$ of them and place them in π_0' and the rest in π_1'. If r is odd, then $\frac{r}{2} \neq [\frac{r}{2}]$, and so include 0 as a part of π_0'. Thus each partition $\pi_3 \in G_3$ spawns $\binom{r}{[\frac{r}{2}]}$ Frobenius partitions $(\pi_0'; \pi_1')$. Thus summing the weights $w_3(\pi_3)$ over partitions π_3 of n yields $p(n)$ as in Theorem 3.   □

Remark.   The weighted identity for $p(n)$ in Theorem 3 is implicit in a paper of Gordon [11] where he shows that $p(n)$ is equal to the number of δ-partitions of n. Here, by a δ-partition, one means a special type of two-rowed partition. The proof of Theorem 3 given above using Frobenius partitions is simpler compared to the approach via δ-partitions.

5. Rogers-Ramanujan partitions

The celebrated Rogers-Ramanujan identities are

$$\sum_{n=0}^{\infty} \frac{q^{n^2}}{(q)_n} = \frac{1}{(q; q^5)_\infty (q^4; q^5)_\infty}$$ (5.1)

and

$$\sum_{n=0}^{\infty} \frac{q^{n^2+n}}{(q)_n} = \frac{1}{(q^2; q^5)_\infty (q^3; q^5)_\infty}.$$ (5.2)

The products on the right hand sides of (5.1) and (5.2) are generating functions for partitions into parts $\equiv \pm i \pmod 5$ for $i = 1, 2$. The left side of (5.1) is the generating function for partitions into parts differing by ≥ 2. We call these Rogers-Ramanujan partitions of type 1 and denote the set of all such partitions by R_1. The left side of (5.2) is the generating function of partitions into parts differing by ≥ 2 and with each part ≥ 2. These are the Rogers-Ramanujan partitions of type 2 and we denote by R_2 the set of all such partitions. Note that $R_2 \subseteq R_1$. By $\rho_i(n)$ we mean the number of Rogers-Ramanujan partitions of n of type i, for $i = 1, 2$.

The hook operation

$$\pi \to \rho(\pi) \tag{5.3}$$

introduced in Section 2 provides an important link between partitions with minimal difference $\geq k - 2$ and those with minimal difference $\geq k$ as noticed in [2]:

Lemma 1. *For $k \geq 2$, the number of primary partitions of n into parts differing by $\geq k - 2$ equals the number of partitions of n into parts differing by $\geq k$.*

Utilizing Lemma 1 and an idea which we called "the sliding operation" on Ferrers graphs [1], it is possible to obtain a weighted representation for partitions of n into parts differing by $\geq k - 2$ in terms of partitions of n into parts differing by $\geq k$. Theorem 1 of [1] is the first instance of such an identity (the case $k = 2$) giving $p(n)$ as weighted sum over partitions enumerated by $\rho_1(n)$. In obtaining this, the decomposition of the νth term in Euler's identity (4.2) in the form

$$\frac{q^{\nu^2}}{(q)_\nu^2} = \frac{q^{\nu^2}}{(q)_\nu} \cdot \frac{1}{(q)_\nu} \tag{5.4}$$

was crucial, because the first factor on the right in (5.4) is the generating function for partitions in R_1 into exactly ν parts.

The next case ($k = 3$) yields a weighted identity (see Theorem 15 of [1]) for $Q(n)$, the number of partitions of n into distinct parts, in terms of partitions of n into parts differing by ≥ 3. This weighted identity is very important in view of its many applications (see [2]). In particular, it permits a three parameter refinement which leads to a new combinatorial proof of a deep partition theorem of Göllnitz [12].

The purpose of this section is to prove Theorem 4 below which deals with the next case ($k = 4$) yielding a weighted representation for $\rho_1(n)$ in terms of partitions of n into parts differing by ≥ 4.

Theorem 4. *Let $\mathcal{D}_4$ denote the set of all partitions into parts differing by ≥ 4. For $\pi_4 : b_1 > b_2 > \cdots > b_\nu$ with $\pi_4 \in \mathcal{D}_4$, define $b_{\nu+1} = -1$. Consider all maximal chains of gaps $b_i - b_{i+1} \geq 5$ in π_4. If $b_\nu = 1$, consider only chains of such gaps among $b_1, b_2, \ldots b_{\nu-1}$. Define the weight $w_1(\pi_4)$ as follows:*

(i) *If a chain in π_4 has r gaps in it, its weight is F_{r+2} where F_r is the rth—Fibonacci number defined by $F_0 = 0$, $F_1 = 1$, $F_r = F_{r-1} + F_{r-2}$, for $r \geq 2$.*

(ii) *The weight $w_1(\pi_4)$ is the product of the weights of the chains, with the usual convention that null products have value 1.*

Then

$$\rho_1(n) = \sum_{\pi_4 \in \mathcal{D}_4, \sigma(\pi_4)=n} w_1(\pi_4).$$

There is also a similar weighted identity for $\rho_2(n)$, namely

Theorem 5. *For a partition $\pi_4 \in \mathcal{D}_4$, define its weight $w_2(\pi_4)$ as follows: If $\pi_4 : b_1 > b_2 > \cdots > b_\nu$, consider only chains of gaps ≥ 5 among $b_2, b_3, \ldots, b_\nu$. Here also we adopt the convention $b_{\nu+1} = -1$ to compute chains. Also,*
(i) *If a chain has r gaps in it, its weight is F_{r+2}.*
(ii) *The weight $w_2(\pi_4)$ is the product of the weights of its chains. Then for $n \geq 2$,*

$$\rho_2(n) = \sum_{\pi_4 \in \mathcal{D}_4, \sigma(\pi_4)=n} w_2(\pi_4).$$

We now give a combinatorial proof of Theorems 4 and 5. The proof makes use of the following well-known lemma which is easily established by induction on r:

Lemma 2. *Consider r consecutive integers $\{n, n+1, n+r-1\}$. Then there are F_{r+2} subsets T of this collection with the property that T cannot contain a pair of consecutive integers.*

Proof of Theorem 4 and 5: We give only the details in the proof of Theorem 4. The proof of Theorem 5 is similar.

Given a positive integer n, consider a primary partition π of n with $\pi \in R_1$.

$$
\begin{array}{l}
\cdot \ \cdot \ \cdot \ \cdot \ \big| \ \cdot \ \cdot \ \cdot \ \cdot \ \cdot \ \cdot \\
\cdot \ \cdot \ \cdot \ \cdot \ \cdot \ \cdot \ \cdot \ \cdot \\
\cdot \ \cdot \ \cdot \big|
\end{array}
\tag{5.5}
$$

The partition $\rho(\pi) = \pi_4$ obtained by counting nodes along hooks of π, belongs to $\mathcal{D}_4$. Consider now the selection of certain columns to the right of $\mathcal{D}(\pi)$ and the placement of these columns below $\mathcal{D}(\pi)$ as rows to form a new Ferrers graph π'. We call this a *sliding operation* ψ. Thus

$$\pi' = \psi(\pi). \tag{5.6}$$

On a given Ferrers graph π, several sliding operations can be performed to yield new graphs π'. The key invariant under the sliding operation is

$$\rho(\psi(\pi)) = \rho(\pi). \tag{5.7}$$

If we require $\pi' \in R_1$, then the following conditions have to be satisfied:
Let $\pi_4 = \rho(\pi) : b_1 > b_2 > \cdots b_\nu$. Put $b_{\nu+1} = -1$. Then

(a) If $b_\nu \neq 1$, then a column of length i can be moved if and only if $b_i - b_{i+1} \geq 5$. If $b_\nu = 1$, then a column of length i can be moved if and only if $i < \nu - 1$, and $b_i - b_{i+1} \geq 5$.

(b) Given a chain gaps $b_\ell - b_{\ell+1} \geq 5$, $\mu \leq \ell \leq \mu + r - 1$, a collection of columns of length $j_1, j_2, \ldots, j_t$ in π with $\mu \leq j_i \leq \mu + r - 1$ can be moved if and only if if $j_1, j_2, \ldots, j_t$ differ by ≥ 2.

So by Lemma 2, each chain of r gaps in $\pi_4 = \rho(\pi)$ permits a total of F_{r+2} sliding operations to be performed on π. Thus each partition $\pi_4 = \rho(\pi) \in \mathcal{D}_4$ spawns $w_1(\pi_4)$ Ferrers graphs $\pi' \in R_1$ under the sliding operation. Since every $\pi' \in R_1$ can be generated in this fashion, Theorem 4 follows by this construction.

The only difference in the proof of Theorem 5 is that for $\pi' \in R_2$ we must ensure that 1 is not a part of π'. So under the sliding operation a column of length 1 cannot be moved. This means we must ignore the difference $b_1 - b_2$ and consider only maximal chains among $b_2, b_3, \ldots, b_\nu$. This proves Theorem 5. $\qquad\square$

Remarks.

(i) For large n, almost all partitions $\pi_4 : b_1 > b_2 > \ldots$ of n, with $\pi_4 \in \mathcal{D}_4$ will have the property $b_1 - b_2 \geq 5$. If $b_1 - b_2 \geq 5$, let r be the number of gaps ≥ 5 in the maximal chain of π_4 starting from $b_1 - b_2$. Note that in computing $w_2(\pi_4)$ we ignore the difference $b_1 - b_2$ while considering chains. Thus

$$\frac{w_1(\pi_4)}{w_2(\pi_4)} = \frac{F_{r+2}}{F_{r+1}} \tag{5.8}$$

is a ratio of consecutive Fibonacci numbers. Since

$$\lim_{r \to \infty} \frac{F_{r+1}}{F_r} = \frac{1 + \sqrt{5}}{2}, \tag{5.9}$$

this suggests that

$$\lim_{n \to \infty} \frac{\rho_1(n)}{\rho_2(n)} = \frac{1 + \sqrt{5}}{2}. \tag{5.10}$$

Of course there are more direct ways to prove (5.10), for instance from the relations

$$\frac{\sum_{n=0}^{\infty} \frac{q^{n^2}}{(q)_n}}{\sum_{n=0}^{\infty} \frac{q^{n^2+n}}{(q)_n}} = \frac{\sum_{n=0}^{\infty} \rho_1(n) q^n}{\sum_{n=0}^{\infty} \rho_2(n) q^n} = 1 + \cfrac{q}{1 + \cfrac{q^2}{1 + \cfrac{q^3}{\ddots}}} = R(q) \tag{5.11}$$

and

$$\lim_{q \to 1} R(q) = 1 + \cfrac{1}{1 + \cfrac{1}{1 + \cfrac{1}{\ddots}}} = \frac{1 + \sqrt{5}}{2}. \tag{5.12}$$

(ii) One of the deepest and most interesting problems is to provide a bijection converting partitions in R_i to partitions in into parts $\equiv \pm i \pmod 5$, for $i = 1, 2$. In 1980, Garcia and Milne [6] found a bijective proof of the Rogers-Ramanujan identities, but this bijection is very intricate and non–canonical. By means of weights involving Fibonacci numbers in Theorems 4 and 5, the prime number 5 is introduced combinatorially in the study of the Rogers-Ramanujan partitions. This might eventually be helpful in understanding the role of 5 in these remarkable identities.

(iii) Utilizing the sliding operation and Lemma 1, Theorem 4 could be extended by establishing a weighted representation for partitions of n into parts differing by $k - 2$ in terms of partitions of n into parts differing by k. Here the weights would be products of integers U_r (determined by certain maximal chains), where the U_r would satisfy the recurrence

$$U_r = U_{r-1} + U_{r-(k-2)}.$$

6. Partitions into distinct parts

In [1] the following weighted identity was established connecting partitions into distinct parts and the Rogers-Ramanujan partitions of type 1.

Theorem 6. Let $Q(n)$ denote the number of partitions of n into distinct parts. Given a partition $\pi \in R_1$, $\pi : b_1 + b_2 + \cdots + b_\nu$, define its weight to be $w_R(\pi) = 2^r$, where there are exactly r gaps >2 among the odd parts of π and $b_{\nu+1} = -1$. Then

$$Q(n) = \sum_{\pi \in R_1, \sigma(\pi)=n} w_R(\pi).$$

A two parameter extension of Theorem 6 was established in [1] by considering the expansion of the product

$$(-aq; q^2)_\infty (-bq^2; q^2)_\infty. \tag{6.1}$$

We now obtain another two parameter extension of Theorem 6 (Theorem 7 below) by considering the product

$$(-aq; q^4)_\infty (-bq^3; q^4)_\infty (-q^2; q^2)_\infty \tag{6.2}$$

Theorem 7 is interesting because in addition to yielding Theorem 6 as a special case when $a = b = 1$, it yields two well known theorems of Göllnitz [10] as special cases when $a = 0$ and $b = 0$ respectively. The combinatorial proof of Theorem 7 given here is a variation and extension of the method that Bressoud [5] used to prove the Göllnitz theorems.

Theorem 7. Let $\mathcal{D}$ denote the set of all partitions into distinct parts. For $\pi' \in \mathcal{D}$, let $\nu_i(\pi')$ denote the number of parts of π' which are $\equiv i \pmod 4$.

Decompose every $\pi \in R_1$ into maximal chains of parts differing by 2. The weight of each chain is defined as follows:

(i) *If the smallest part of a chain is even, its weight is 1.*

(ii) *If a chain has r parts with smallest part 1, its weight is* $a^{r-[\frac{r}{2}]}b^{[\frac{r}{2}]}$.

(iii) *If a chain has r parts and its smallest part is odd and >1, its weight is*

$$a^{r-[\frac{r}{2}]}b^{[\frac{r}{2}]} + a^{[\frac{r}{2}]}b^{r-[\frac{r}{2}]}.$$

The weight $w_R(\pi)$ of the partition $\pi \in R_1$ is the product of the weights of its chains. Then

$$\sum_{\pi'\in\mathcal{D},\sigma(\pi')=n} a^{\nu_1(\pi')}b^{\nu_3(\pi')} = \sum_{\pi\in R_1,\sigma(\pi)=n} w_R(\pi).$$

Proof: Given a partition π' of n into distinct parts, decompose it as

$$\pi' = \pi_1 \cup \pi_2,$$

where π_1 has distinct odd parts and π_2 has distinct even parts. Thus

$$a^{\nu_1(\pi')}b^{\nu_3(\pi')} = a^{\nu_1(\pi_1)}b^{\nu_3(\pi_1)}.$$

The parts of π_1 that are $\equiv 1 \pmod 4$ have weight a, while those $\equiv 3 \pmod 4$ have weight b.

Step 1: Decompose π_2 as $\pi_3 \cup \pi_4$, where all parts of π_3 are $\leq 2\nu(\pi_1)$ and π_4 has the remaining parts.

Step 2: Consider the 2-fold conjugate of π_3 which we denote by $\pi_3^*(2)$. That is the columns of $\pi_3^*(2)$ are columns of twos adding up to the parts of π_3. Equivalently we may think of $\pi_3^*(2)$ as an ordinary Ferrers graph whose first two columns are equal, whose next two columns are equal, and so on.

Step 3: Consider the partition

$$\pi_5 = \pi_1 + \pi_3^*(2)$$

obtained by adding the number of nodes in the corresponding rows of π_1 and $\pi_3^*(2)$. Since $\nu(\pi_3^*(2)) \leq \nu(\pi_1)$ by construction, we have $\nu(\pi_1) = \nu(\pi_5)$. Each row (part) of π_5 is given the same weight as the row (part) of π_1 from which it was formed. Note however that the weights of the parts of π_5 need not be determined by the residue class (mod 4) as in the case of the parts of π_1. Equivalently, we may think of π_5 as being obtained by imbedding the columns of $\pi_3^*(2)$ into π_1.

Important observations: Let the parts of π_1 be $b_1 > b_2 > \cdots > b_r$ and those of π_5 be $c_1 > c_2 > \cdots > c_r$, all odd integers. If $\pi_3^*(2)$ has a column of twos of length ℓ, (equivalently, a pair of equal columns of length ℓ), then we are guaranteed that

$$c_\ell - c_{\ell+1} > 2. \tag{6.3}$$

So, if $c_\ell - c_{\ell+1} = 2$, then $\pi_3^*(2)$ cannot contain a column of length ℓ. While it is clear that π_1 and $\pi_3^*(2)$ give rise to π_5, it is not obvious how to construct π_1 and $\pi_3^*(2)$ from π_5. The correspondence

$$(\pi_1, \pi_3^*(2)) \longleftrightarrow \pi_5 \qquad (6.4)$$

is one-to-one if weights are attached to π_5. We now describe how $\pi_3^*(2)$ can be peeled off from π_5 once the weights are prescribed. Starting from the smallest part c_r of π_6 move upward and note the subscript ℓ_1 of the first part which is either $\equiv 1 \pmod 4$ and has weight b, or is $\equiv 3 \pmod 4$ and has weight a. This means $\pi_3^*(2)$ will have a column of length ℓ_1. Moving upward beyond c_{ℓ_1}, note the subscript ℓ_2 of the first part which is either $\equiv 1 \pmod 4$ and has weight a, or is $\equiv 3 \pmod 4$ and has weight b. This means $\pi_3^*(2)$ will have a column of length ℓ_2. Now moving upwards beyond c_{ℓ_2}, we once again note the position ℓ_3 of the first part which is either $\equiv 1 \pmod 4$ having weight b, or $\equiv 3 \pmod 4$ having weight a. This will give a column of length ℓ_3 in $\pi_3^*(2)$. Proceeding in this fashion, we can decompose π_5 into $\pi_1 + \pi_3^*(2)$.

Step 4: Write the parts of π_4 in a column in descending order and below them write the parts of π_5 in descending order to form a column C.

Step 5: Subtract 0 from the bottom element of C, 2 from the one above that, 4 from the next one above and so on, to form a new column C_1.

Step 6: Rearrange the elements of C_1 in descending order to form a column C_1^R.

Step 7: Add back the integers $0, 2, 4, \ldots$, to the elements of C_1^R from the bottom upward to form a partition $\pi \in R_1$.

The weights of the partition π will be the same as the weights of the parts of π_5 to which they correspond.

All steps 1–7 are one-to-one correspondences, and so the number of partitions of n into distinct parts π' equals the number of such weighted partitions $\pi \in R_1$. However, if $\pi \in R_1$ is unweighted, it could correspond to several such partitions $\pi' \in \mathcal{D}$ because each unweighted π_5 could spawn several pairs $(\pi_1, \pi_3^*(2))$. To complete the proof of Theorem 7, we now discuss how weights could be attached in various ways to the parts of $\pi \in R_1$.

All even parts of $\pi \in R_1$ will have weight 1. So, chains of such even parts will have weight 1. If 1 is a part of π, then its weight must be a because it cannot arise of out of an imbedding of $\pi_3^*(2)$ into π_1. So for a chain starting at 1, the weight of the next larger part, namely 3, must be b, the weight of the next part in the chain, namely 5, must be a, and so on. So, if a chain in π starting at 1 has r elements, then its weight must be

$$a^{r-[\frac{r}{2}]} b^{[\frac{r}{2}]}.$$

If the smallest part of a chain of π is an odd integer >1, then its weight could be a or b. The next part in the chain would have weight b or a, the one above that would have weight a or b, and so on. So, if this chain has r elements in it, its weight could be

$$a^{r-[\frac{r}{2}]} b^{[\frac{r}{2}]} \quad \text{or} \quad a^{[\frac{r}{2}]} b^{r-[\frac{r}{2}]}.$$

Hence these weights have to be added as in (iii) of Theorem 7 to take into account all possibilities. This completes the proof of Theorem 7.   □

As noted already, Theorem 7 reduces to Theorem 6 when $a = b = 1$.
Now consider the case $a = 0$. Hence

$$a^{r-[\frac{r}{2}]}b^{[\frac{r}{2}]} = 0, \quad \text{for } r \geq 1. \tag{6.5}$$

This means that 1 cannot occur as a part of π in Theorem 7. Also

$$a^{r-[\frac{r}{2}]}b^{[\frac{r}{2}]} + a^{[\frac{r}{2}]}b^{r-[\frac{r}{2}]} = 0, \quad \text{for } r \geq 2. \tag{6.6}$$

Therefore, consecutive odd numbers cannot occur as parts of π. Also, $a = 0$ implies that integers $\equiv 1 \pmod 4$ cannot occur as parts of $\pi' \in \mathcal{D}$. Thus the case $a = 0$ yields as a special case the following result of Göllnitz [10]:

Corollary 1. *Let $A(n; k)$ denote the number of partitions of n into distinct parts $\equiv 0, 2$ or $3 \pmod 4$, having exactly k parts $\equiv 3 \pmod 4$.*
Let $B(n; k)$ denote the number of partitions of n into parts differing by ≥ 2, all parts ≥ 2, no consecutive odd numbers as parts, and the number of odd parts is exactly k. Then

$$A(n; k) = B(n; k)$$

Similarly, if $b = 0$, then (6.5) and (6.6) hold. This yields another result of Göllnitz [10] as a special case of Theorem 7:

Corollary 2. *Let $C(n; k)$ denote the number of partitions of n into distinct parts $\equiv 0, 1$ or $2 \pmod 4$, having exactly k parts $\equiv 1 \pmod 4$.*
Let $D(n; k)$ denote the number of partitions of n into parts differing by ≥ 2, no consecutive odd numbers as parts, and the number of odd parts is k. Then

$$C(n; k) = D(n; k).$$

7. Partitions of Schur and Göllnitz-Gordon

The celebrated 1926 partition theorem of Schur [14] is

Theorem 8. *Let $T(n)$ denote the number of partitions of n into parts $\equiv \pm 1 \pmod 6$. Let $S(n)$ denote the number of partitions of n into distinct parts $\equiv \pm 1 \pmod 3$. Let $S_1(n)$ denote the number of partitions of n into parts differing by ≥ 3 such that consecutive multiples of 3 cannot occur as parts. Then*

$$T(n) = S(n) = S_1(n).$$

In [3] a two parameter generalization of Theorem 8 was obtained and we describe this now. Consider three colors a, b and $c = ab$, where a and b are primary colors and $c = ab$

is a secondary color. The integer 1 occurs only in the primary colors whereas each integer $n \geq 2$ occurs in all three colors. The symbols a_n, b_n and c_n represent the integer n in colors a, b and c respectively. We assume that the colored integers satisfy the ordering

$$a_1 < b_1 < c_2 < a_2 < b_2 < c_3 < a_3 < b_3 < \ldots \qquad (7.1)$$

By a type-1 partition of an integer n we mean a partition of n into distinct integers which could occur in any of the colors such that if i and j are consecutive parts of π, then

$$\left. \begin{array}{ll} i - j \geq 2 & \text{if } i \text{ has color } a \text{ and } j \text{ has color } b, \\ \text{or} & \text{if } i \text{ has color } c = ab. \end{array} \right\} \qquad (7.2)$$

Then by considering expansions of the product

$$(-aq)_\infty (-bq)_\infty \qquad (7.3)$$

(where here a and b are free parameters) we proved in [3] the following result:

Theorem 9. *Let $V(n)$ denote the number of vector partitions $(\pi_a; \pi_b)$ of n such that π_a has distinct parts all in color a and π_b has distinct parts all in color b.*
Let $A(n)$ denote the number of type-1 partitions of n. Then

$$V(n) = A(n).$$

Schur's theorem falls out of Theorem 9 under the substitutions

$$\left. \begin{array}{l} q \mapsto q^3 \text{ (dilation)} \\ a \mapsto aq^{-2}, b \mapsto bq^{-1} \text{ (translations)} \end{array} \right\} \qquad (7.4)$$

in which case the product in (7.3) becomes

$$(-aq; q^3)_\infty (-bq^2; q^3)_\infty,$$

the two parameter refinement of the generating function of $S(n)$ in Theorem 8.
    Now consider the substitutions

$$\left. \begin{array}{l} q \mapsto q^2 \text{ (dilation)} \\ a \mapsto aq^{-1}, b \mapsto bq^{-1} \text{ (translations)} \end{array} \right\} \qquad (7.5)$$

which convert (7.3) to

$$(-aq; q^2)_\infty (-bq; q^2)_\infty. \qquad (7.6)$$

In this case the symbols a_n and b_n are equal to $2n - 1$ in colors a and b respectively, and $c_n = ab_n$ is the integer $2n - 2$ in color ab. Also type-1 partitions are those where the parts differ by ≥ 2 with the extra condition that consecutive even integers cannot occur as parts.

We refer to these as the Göllnitz-Gordon type partitions since they occur in the well known Göllnitz-Gordon identities (see [10, 12]). Also, in these type-1 partitions, if two odd parts differ by exactly 2, then the smaller part must be of color a if the larger part has color a.

Now decompose a type-1 partition into maximal chains of parts differing by 2. Note that if a chain has r elements with $r \geq 2$, then all parts of the chain must be odd. The colors of the parts in this chain from the smallest part and moving upwards could be $a, a, \ldots, a$, or $a, a, \ldots, a, b$, or $a, a, \ldots, a, b, b$ and so on, or finally, $b, b, b, \ldots, b$. Taking all these possibilities into account, the weight of this chain should be

$$a^n + a^{n-1}b + a^{n-2}b^2 + \cdots + b^n. \tag{7.7}$$

If a chain of a type-1 partition has only one element, it weight should be $a + b$ if the element is odd, and the weight should be ab if the element is even. Finally, the weight $w_g(\pi)$ of the type-1 (Göllnitz-Gordon) partition is the product of the weights of its chains.

Now consider the choices

$$a = \zeta = e^{2\pi i/6}, \quad b = \zeta^{-1} = e^{-2\pi i/6}. \tag{7.8}$$

Note that

$$a + b = \zeta + \zeta^{-1} = 1 \quad \text{and} \quad ab = \zeta\zeta^{-1} = 1. \tag{7.9}$$

Thus

$$(-aq; q^2)_\infty(-bq; q^2)_\infty = \prod_{j=1}^{\infty}(1 + \zeta q^{2j-1})(1 + \zeta^{-1}q^{2j-1}) = \prod_{j=1}^{\infty}(1 + q^{2j-1} + q^{4j-2})$$

$$= \prod_{j=1}^{\infty}\frac{(1 - q^{6j-3})}{(1 - q^{2j-1})} = \frac{1}{(q; q^6)_\infty(q^5; q^6)_\infty} = \sum_{n=0}^{\infty}T(n)q^n$$

$$= \sum_{n=0}^{\infty}S_1(n)q^n \tag{7.10}$$

by Theorem 8.

Note also with these choices that the weights of chains of length ℓ are

$$\left. \begin{array}{ll} 1, & \text{if } \ell = 0 \text{ or } 1(\text{mod } 6), \\ 0, & \text{if } \ell = 2 \text{ or } 5(\text{mod } 6), \\ -1, & \text{if } \ell = 3 \text{ or } 4(\text{mod } 6), \end{array} \right\} \tag{7.11}$$

We then get

Theorem 10. *Let $S_1(n)$ be as in Theorem 8.*

Let G denote the set of all Göllnitz-Gordon partitions, namely partitions into parts differing by ≥ 2 with no consecutive even integers as parts. For $\pi \in G$, define $w_g(\pi)$ as the product of the weights of its chains given by (7.11), Then

$$S_1(n) = \sum_{\pi \in G, \sigma(\pi)=n} w_g(\pi).$$

Remarks.

(i) Theorem 10 is a link between partitions with minimum difference 2 and no consecutive even integers as parts and partitions with minimum difference 3 having no consecutive multiples of 3 as parts.

(ii) The choices

$$a = i, \quad b = i^{-1} \tag{7.12}$$

in (7.6) is also quite interesting. For then the product in (7.6) is

$$\prod_{j=1}^{\infty}(1 + iq^{2j-1})(1 - iq^{2j-1}) = \prod_{j=1}^{\infty}(1 + q^{4j-2}), \tag{7.13}$$

the generating function for partitions into distinct parts $\equiv 2 \pmod 4$. On the other hand, the weight of a chain of ℓ odd parts in $\pi \in G$ is

$$i^\ell + i^{\ell-1}(-i) + i^{\ell-2}(-i)^2 + \cdots + (-i)^\ell = \frac{i^{\ell+1} - (-i)^{\ell+1}}{2i}$$

$$= \begin{cases} 0, & \text{if } \ell \text{ is odd}, \\ 1, & \text{if } \ell \equiv 0 \pmod 4, \\ -1, & \text{if } \ell \equiv -2 \pmod 4. \end{cases} \tag{7.14}$$

This means we cannot have chains of odd length consisting of odd parts differing by exactly 2. Since $ab = i(-i) = 1$, all even parts will have weight 1. Thus we have

Theorem 11. *Let $Q_{2,4}(n)$ denote the number of partitions of n into distinct parts $\equiv 2$ (mod 4).*

With G as in Theorem 10, let $w_i(\pi)$ be defined as the product of the weights of its chains as in (7.14). Then

$$Q_{2,4}(n) = \sum_{\pi \in G, \sigma(\pi)=n} w_i(\pi).$$

Acknowledgments

I would like to thank George Andrews and Basil Gordon for discussions on various aspects of this paper.

References

1. K. Alladi, "Partition identities involving gaps and weights," *Trans. Amer. Math. Soc.* **349** (1997), 5001–5019.
2. K. Alladi, "A combinatorial correspondence related to Göllnitz' (Big) partition theorem and applications," *Trans. Amer. Math. Soc.* **349** (1997), 2721–2735.
3. K. Alladi and B. Gordon, "Generalizations of Schur's partition theorem," *Manus. Math.* **79** (1993), 113–126.
4. G.E. Andrews, *The Theory of Partitions*, Encyclopedia of Mathematics and its Applications, Addison-Wesley, Reading, 1976, vol. 2.
5. D.M. Bressoud, "On a partition theorem of Göllnitz," *J. Reine Angew. Math.* **305** (1979), 215–217.
6. A. Garcia and S. Milne, "A Rogers-Ramanujan bijection," *J. Comb. Th. Ser. A* **31** (1981), 289–339.
7. J.W.L. Glaisher, "A theorem in partitions," *Messenger of Math.* **12** (1883), 158–170.
8. G.H. Hardy, *Ramanujan*, Cambridge Univ. Press, Cambridge, 1940.
9. G.H. Hardy and E.M. Wright, *An Introduction to the Theory of Numbers*, 4th edition, Oxford University Press, London and New York, 1960.
10. H. Göllnitz, "Partitionen mit Differenzenbedingungen," *J. Reine Angew. Math.* **225** (1967), 154–190.
11. B. Gordon, "Two new representations of the partition function," *Proc. Amer. Math. Soc.* **13** (1962), 869–873.
12. B. Gordon, "Some continued fractions of the Rogers-Ramanujan type," *Duke Math. J.* **32** (1965), 741–748.
13. K. Ono, "Parity of the partition function," *Elec. Res. Announ. Amer. Math. Soc.* **1** (1995), 35–42.
14. I. Schur, "Zur Additiven Zahlentheorie," *Gesammelte Abhandlungen*, Springer Verlag, Berlin, 1973, vol. 2, pp. 43–50.

THE RAMANUJAN JOURNAL 2, 39–45 (1998)

The Voronoi Identity via the Laplace Transform

ALEKSANDAR IVIĆ   aleks@ivic.matf.bg.ac.yu, aivic@rgf.rgf.bg.ac.yu
Katedra Matematike RGF-a, Universitet u Beogradu, Đušina 7, 11000 Beograd, Serbia, Yugoslavia

Dedicated to the memory of Paul Erdős, who proved and conjectured more than anyone else

Received January 16, 1997; Accepted June 2, 1997

Abstract. The classical Voronoi identity

$$\Delta(x) = -\frac{2}{\pi} \sum_{n=1}^{\infty} d(n) \left(\frac{x}{n}\right)^{1/2} \left(K_1(4\pi\sqrt{xn}) + \frac{\pi}{2} Y_1(4\pi\sqrt{xn}) \right)$$

is proved in a relatively simple way by the use of the Laplace transform. Here $\Delta(x)$ denotes the error term in the Dirichlet divisor problem, $d(n)$ is the number of divisors of n and K_1, Y_1 are the Bessel functions. The method of proof may be used to yield other identities similar to Voronoi's.

Key words: Voronoi identity, number of divisors, Laplace transform, Bessel functions, Fourier coefficients of cusp forms

1991 Mathematics Subject Classification: Primary 11N37; Secondary 33C10, 44A10

Let as usual the error term $\Delta(x)$ in the Dirichlet divisor problem be defined as

$$\Delta(x) := \sum_{n \le x}' d(n) - x(\log x + 2\gamma - 1) - \frac{1}{4} \quad (x > 0).$$

Here $d(n) = \sum_{\delta|n} 1$ denotes the number of divisors of n, $\sum'$ means that the last term in the sum is halved if x is an integer, and $\gamma = 0.577\ldots$ is Euler's constant. The classical Voronoi identity (see [1–3, 6–9]) states that

$$\Delta(x) = -\frac{2}{\pi} \sum_{n=1}^{\infty} d(n) \left(\frac{x}{n}\right)^{1/2} \left(K_1(4\pi\sqrt{xn}) + \frac{\pi}{2} Y_1(4\pi\sqrt{xn}) \right), \tag{1}$$

where K_1 and Y_1 are the Bessel functions in standard notation (see Watson [10] for definitions and properties). It is well-known that the series in (1) is boundedly, but not absolutely convergent. It is uniformly convergent in every interval $[x_1, x_2]$ $(0 < x_1 < x_2)$ which contains no integers. The proofs of (1) are usually long and difficult. It is the aim of this note to provide a relatively short proof of (1) by the use of Laplace transforms. The proof, which seems to be new, may be used to derive other identities similar to (1). To achieve this we shall need the integral representation

$$\left(\frac{x}{n}\right)^{1/2}\left(K_1(4\pi\sqrt{xn}) + \frac{\pi}{2}Y_1(4\pi\sqrt{xn})\right)$$

$$= \frac{1}{4\pi^2 ni}\int_{(1)}\Gamma(w)\Gamma(w-1)\cos^2\left(\frac{\pi w}{2}\right)(2\pi\sqrt{xn})^{2-2w}\,dw, \tag{2}$$

where

$$\int_{(c)} = \lim_{T\to\infty}\int_{c-iT}^{c+iT}$$

For a proof of (2) see, for example, [6, p. 87]. Note that in (2) we may shift the line of integration to $\Re e\, w = c$, $0 < c < 1$, and that the integral is absolutely convergent for $0 < c < 1/2$, since $|\cos w| \le \cosh v$, $\Gamma(w) \ll |v|^{u-1/2}e^{-\pi|v|/2}$ for $w = u + iv$. If

$$\mathcal{L}[f(x)] = \int_0^\infty f(x)e^{-sx}\,dx$$

is the (one-sided) Laplace transform of $f(x)$, then for $\Re e\, s > 0$

$$\mathcal{L}[\Delta(x)] = \int_0^\infty\left(\sum_{n\le x}{}'d(n) - x(\log x + 2\gamma - 1) - \frac{1}{4}\right)e^{-sx}\,dx$$

$$= \sum_{n=1}^\infty d(n)\int_n^\infty e^{-sx}\,dx + \frac{\log s - \gamma}{s^2} - \frac{1}{4s}$$

$$= \frac{1}{s}\sum_{n=1}^\infty d(n)e^{-sn} + \frac{\log s - \gamma}{s^2} - \frac{1}{4s}$$

$$= \frac{1}{2\pi is}\int_{(2)}\zeta^2(w)\Gamma(w)s^{-w}\,dw + \frac{\log s - \gamma}{s^2} - \frac{1}{4s}$$

$$= \frac{1}{2\pi is}\int_{(1/2)}\zeta^2(w)\Gamma(w)s^{-w}\,dw - \frac{1}{4s}. \tag{3}$$

Here we used the well-known Mellin integral

$$e^{-z} = \frac{1}{2\pi i}\int_{(c)}\Gamma(w)w^{-z}\,dw \quad (c > 0, \Re e\, z > 0),$$

and the series representation

$$\zeta^2(s) = \sum_{n=1}^\infty d(n)n^{-s} \quad (\Re e\, s > 1).$$

Change of summation and integration was justified by absolute convergence, and in the last step the residue theorem was used together with

$$\zeta(s) = \frac{1}{s-1} + \gamma + \gamma_0(s-1) + \cdots, \qquad \Gamma(s) = 1 - \gamma(s-1) + \cdots \quad (s \to 1).$$

Now we invoke the functional equation (see [2, Ch. 2] or [6, Ch. 1])

$$\zeta(s) = \chi(s)\zeta(1-s), \quad \chi(s) = 2^s \pi^{s-1} \sin\left(\frac{\pi s}{2}\right)\Gamma(1-s), \tag{4}$$

and shift the line of integration in the last integral in (3) to $\Re w = -\frac{1}{2}$. By the residue theorem ($\zeta(0) = -\frac{1}{2}$) and absolute convergence we obtain

$$\mathcal{L}[\Delta(x)] = \sum_{n=1}^{\infty} \frac{d(n)}{sn}\left[\frac{1}{2\pi i}\int_{(-1/2)} \chi^2(w)\left(\frac{n}{s}\right)^w \Gamma(w)\,dw\right]$$

$$= \sum_{n=1}^{\infty} \frac{d(n)}{\pi^2 sn}\left[\frac{1}{2\pi i}\int_{(-1/2)} (2\pi\sqrt{n})^{2w}\sin^2\left(\frac{\pi w}{2}\right)\Gamma^2(1-w)\Gamma(w)\,\frac{dw}{s^w}\right]. \tag{5}$$

To transform further (5) we need that

$$\mathcal{L}[x^w] = \frac{\Gamma(w+1)}{s^{w+1}} \quad (\Re w > -1)$$

and the functional equation for the gamma-function, namely $s\Gamma(s) = \Gamma(s+1)$. We obtain

$$\mathcal{L}[\Delta(x)] = -\sum_{n=1}^{\infty} \frac{d(n)}{\pi^2 sn}\left[\frac{1}{2\pi i}\int_{(-1/2)} (2\pi\sqrt{n})^{2w}\sin^2\left(\frac{\pi w}{2}\right)\right.$$
$$\left. \times \Gamma(1-w)\Gamma(-w)\Gamma(w+1)\frac{dw}{s^w}\right]$$

$$= -\sum_{n=1}^{\infty} \frac{d(n)}{\pi^2 sn}\left[\frac{1}{2\pi i}\int_{(3/4)} (2\pi\sqrt{n})^{2w}\sin^2\left(\frac{\pi w}{2}\right)\right.$$
$$\left. \times \Gamma(1-w)\Gamma(-w)\Gamma(w+1)\frac{dw}{s^w}\right]$$

$$= -\sum_{n=1}^{\infty} \frac{d(n)}{\pi^2 n}\mathcal{L}\left[\frac{1}{2\pi i}\int_{(3/4)} (2\pi\sqrt{n})^{2w}\sin^2\left(\frac{\pi w}{2}\right)\right.$$
$$\left. \times \Gamma(1-w)\Gamma(-w)x^w\,dw\right] \quad (1-w = z)$$

$$= -\sum_{n=1}^{\infty} \frac{d(n)}{\pi^2 n}\mathcal{L}\left[\frac{1}{2\pi i}\int_{(1)} (2\pi\sqrt{xn})^{2-2z}\cos^2\left(\frac{\pi z}{2}\right)\Gamma(z)\Gamma(z-1)\,dz\right]$$

$$= -\frac{2}{\pi}\sum_{n=1}^{\infty} d(n)\mathcal{L}\left[\left(\frac{x}{n}\right)^{1/2}\left(K_1(4\pi\sqrt{xn}) + \frac{\pi}{2}Y_1(4\pi\sqrt{xn})\right)\right]$$

$$= \mathcal{L}\left[-\frac{2}{\pi}\sum_{n=1}^{\infty} d(n)\left(\frac{x}{n}\right)^{1/2}\left(K_1(4\pi\sqrt{xn}) + \frac{\pi}{2}Y_1(4\pi\sqrt{xn})\right)\right]$$

by using Cauchy's theorem and (2), provided that we can justify the fact that

$$\int_0^\infty e^{-sx} f(x)\, dx = -\frac{2}{\pi} \sum_{n=1}^\infty d(n) \mathcal{L}\left[\left(\frac{x}{n}\right)^{1/2} \left(K_1(4\pi\sqrt{xn}) + \frac{\pi}{2} Y_1(4\pi\sqrt{xn})\right)\right], \quad (6)$$

where

$$f(x) := -\frac{2}{\pi} \sum_{n=1}^\infty d(n) \left(\frac{x}{n}\right)^{1/2} \left(K_1(4\pi\sqrt{xn}) + \frac{\pi}{2} Y_1(4\pi\sqrt{xn})\right). \quad (7)$$

We also used

$$\frac{1}{2\pi i s} \int_{(3/4)} (2\pi\sqrt{n})^{2w} \sin^2\left(\frac{\pi w}{2}\right) \Gamma(1-w)\Gamma(-w)\Gamma(w+1) \frac{dw}{s^w}$$

$$= \mathcal{L}\left[\frac{1}{2\pi i} \int_{(3/4)} (2\pi\sqrt{n})^{2w} \sin^2\left(\frac{\pi w}{2}\right) \Gamma(1-w)\Gamma(-w) x^w\, dw\right], \quad (8)$$

which follows from the absolute convergence of the integrals. Thus, assuming that (6) is true, we have shown that

$$\mathcal{L}[\Delta(x)] = \mathcal{L}\left[-\frac{2}{\pi} \sum_{n=1}^\infty d(n) \left(\frac{x}{n}\right)^{1/2} \left(K_1(4\pi\sqrt{xn}) + \frac{\pi}{2} Y_1(4\pi\sqrt{xn})\right)\right] = \mathcal{L}[f(x)].$$

Suppose that $x_0 \notin \mathbb{N}$. Then both $\Delta(x_0)$ and $f(x_0)$ are continuous at $x = x_0$. Hence by the uniqueness theorem for Laplace transforms (Doetsch [4, Ch. 2]) it follows that (1) holds for $x = x_0$. But if $x \in \mathbb{N}$, then the validity of (1) follows from the validity of (1) when $x \notin \mathbb{N}$, as shown, for example, by Jutila [7].

To establish (6) we shall use the crude bound $f(x) \ll x$. This easily follows if we write the series in (7) as a Stieltjes integral involving $\Delta(x)$, use integration by parts, the elementary bound $\Delta(x) \ll x^{1/2}$, and the asymptotic formulas (see [10])

$$K_\nu(x) = \left(\frac{\pi}{2x}\right)^{1/2} e^{-x} \left(1 + O\left(\frac{1}{x}\right)\right) \quad (x \to \infty) \quad (9)$$

$$Y_\nu(x) = \left(\frac{2}{\pi x}\right)^{1/2} \left[\sin\left(x - \frac{\pi\nu}{2} - \frac{\pi}{4}\right) + \frac{4\nu^2 - 1}{8x} \cos\left(x - \frac{\pi\nu}{2} - \frac{\pi}{4}\right)\right]$$

$$+ O(x^{-5/2}) \quad (x \to \infty). \quad (10)$$

Then we note that, for $N \geq 1, \sigma = \Re s$,

$$\int_0^\infty e^{-sx} f(x)\, dx = \int_0^N e^{-sx} f(x)\, dx + O\left(\left(\frac{N}{\sigma} + \frac{1}{\sigma^2}\right) e^{-\sigma N}\right),$$

and the O-term tends to zero as $N \to \infty$ since $\Re e\, s > 0$. Since the series defining $f(x)$ is boundedly convergent, it may be integrated termwise over any finite interval. Hence

$$\int_0^N e^{-sx} f(x)\, dx = -\frac{2}{\pi} \sum_{n=1}^\infty d(n) \int_0^N e^{-sx} \left(\frac{x}{n}\right)^{1/2} \left(K_1(4\pi\sqrt{xn}) + \frac{\pi}{2} Y_1(4\pi\sqrt{xn})\right) dx$$

$$= -\frac{2}{\pi} \sum_{n=1}^\infty d(n) \int_0^\infty e^{-sx} \left(\frac{x}{n}\right)^{1/2} \left(K_1(4\pi\sqrt{xn}) + \frac{\pi}{2} Y_1(4\pi\sqrt{xn})\right) dx$$

$$+ S(N, s),$$

where

$$S(N, s) := \frac{2}{\pi} \sum_{n=1}^\infty d(n) \int_N^\infty e^{-sx} \left(\frac{x}{n}\right)^{1/2} \left(K_1(4\pi\sqrt{xn}) + \frac{\pi}{2} Y_1(4\pi\sqrt{xn})\right) dx. \quad (11)$$

Thus, to prove (6) it is sufficient to show that

$$\lim_{N \to \infty} S(N, s) = 0. \quad (12)$$

But since

$$\frac{d}{dz}[z^\nu K_\nu(z)] = -z^\nu K_{\nu-1}(z), \qquad \frac{d}{dz}[z^\nu Y_\nu(z)] = z^\nu Y_{\nu-1}(z),$$

it follows that

$$\left(\frac{x}{n}\right)^{1/2} \left(K_1(4\pi\sqrt{xn}) + \frac{\pi}{2} Y_1(4\pi\sqrt{xn})\right) = \frac{d}{dx}\left[\frac{x}{2\pi n}\left(\frac{\pi}{2} Y_2(4\pi\sqrt{xn}) - K_2(4\pi\sqrt{xn})\right)\right].$$

Thus using integration by parts, (9) and (10) we obtain

$$S(N, s) = \frac{2}{\pi} \sum_{n=1}^\infty d(n)\left(e^{-sN} \frac{N}{2\pi n}\left(K_2(4\pi\sqrt{Nn}) - \frac{\pi}{2}(Y_2(4\pi\sqrt{Nn})\right)\right.$$

$$\left. - s \int_N^\infty e^{-sx} \frac{x}{2\pi n}\left(K_2(4\pi\sqrt{xn}) - \frac{\pi}{2}(Y_2(4\pi\sqrt{xn})\right) dx\right)$$

$$\ll e^{-\sigma N}\left(1 + \frac{|s|}{\sigma}\right) N^{3/4} \sum_{n=1}^\infty d(n) n^{-5/4} = \zeta^2\left(\frac{5}{4}\right)\left(1 + \frac{|s|}{\sigma}\right) e^{-\sigma N} N^{3/4}$$

and (12) follows since $\sigma > 0$. This completes the proof of (1).

It should be remarked that the foregoing method may be used to furnish other classical identities that are analogous to (1) (see Berndt [1] for general identities of this type). In particular, this is true of Hardy's identity (see [5])

$$P(x) = \sum_{n=1}^\infty r(n)\left(\frac{x}{n}\right)^{1/2} J_1(2\pi\sqrt{xn}), \quad P(x) = \sum_{n \le x}{}' r(n) - \pi x + 1, \quad r(n) = \sum_{n=a^2+b^2} 1,$$

$$(13)$$

and of

$$A(x) = (-1)^{k/2} x^{k/2} \sum_{n=1}^{\infty} a(n) n^{-k/2} J_k(2\pi\sqrt{xn}), \quad A(x) = \sum_{n\leq x}{}' a(n), \qquad (14)$$

where $a(n)$ is the nth Fourier coefficient of a cusp form of weight k ($k \geq 12$ is an even integer) for the full modular group. In both (13) and (14), $x > 0$ and the series are boundedly convergent and uniformly convergent in any closed interval free of integers, similarly to the series in (1). If one uses the above method of Laplace transforms, then the basis of the analysis is the formula

$$\mathcal{L}[x^{\nu/2} J_\nu(2\sqrt{ax})] = a^{\nu/2} s^{-\nu-1} e^{-a/s} \qquad (\Re e\, \nu > -1, \Re e\, s > 0). \qquad (15)$$

In the case of (14) one uses the properties of modular forms, and in the case of (13) the functional equation

$$R\left(\frac{1}{z}\right) = z\pi R(\pi^2 z) + z\pi - 1, \qquad (16)$$

where for $\Re e\, z > 0$

$$R(z) := \sum_{n=1}^{\infty} r(n) e^{-nz} = \left(\sum_{m=-\infty}^{\infty} e^{-m^2 z}\right)^2 - 1.$$

Then (16) is an easy consequence of the classical theta-formula (see, e.g., Chandrasekharan [2] for proof)

$$\sum_{m=-\infty}^{\infty} e^{-m^2 z} = z^{-1/2} \sum_{m=-\infty}^{\infty} e^{-\pi m^2/z} \qquad (\Re e\, z > 0).$$

Alternatively, one may use the functional equations for the Dirichlet series generated by $r(n)$ and $a(n)$, respectively, which are analogous to the functional equation for $\zeta^2(s)$. Using (16) we obtain, for $\Re e\, s > 0$,

$$\mathcal{L}[P(x)] = \int_0^{\infty} \left(\sum_{n\leq x}{}' r(n) - \pi x + 1\right) e^{-sx}\, dx$$

$$= \sum_{n=1}^{\infty} r(n) \int_n^{\infty} e^{-sx}\, dx - \frac{\pi}{s^2} + \frac{1}{s} = \frac{1}{s} R(s) - \frac{\pi}{s^2} + \frac{1}{s}$$

$$= \frac{1}{s}\left\{\frac{\pi}{s}\left[R\left(\frac{\pi^2}{s}\right) + 1\right] - 1\right\} - \frac{\pi}{s^2} + \frac{1}{s} = \frac{\pi}{s^2} \sum_{n=1}^{\infty} r(n)\, e^{-\pi^2 n/s}.$$

As in the analysis that established (8) we have, using (15) with $\nu = 1$ and $a = \pi^2 n$,

$$\mathcal{L}[P(x)] = \sum_{n=1}^{\infty} r(n) \mathcal{L}\left[\left(\frac{x}{n}\right)^{1/2} J_1(2\pi\sqrt{xn})\right] = \mathcal{L}\left[\sum_{n=1}^{\infty} r(n)\left(\frac{x}{n}\right)^{1/2} J_1(2\pi\sqrt{xn})\right]. \qquad (17)$$

From (17) one deduces (13) much in the same way as one obtained (1) from (8). The derivation of (14) is in similar lines.

References

1. B.C. Berndt, "Identities involving the coefficients of a class of Dirichlet series I," *Trans. Amer. Math. Soc.* **137** (1969), 345–359.
2. K. Chandrasekharan, *Arithmetical Functions*, Springer-Verlag, Berlin, 1970.
3. A.L. Dixon and W.L. Ferrar, "Lattice point summation formulae," *Quart. J. Math.* (Oxford) **2** (1931), 31–54.
4. G. Doetsch, *Handbuch der Laplace-Transformation*, Band I, Birkhäuser Verlag, Basel und Stuttgart, 1950.
5. G.H. Hardy, "The average order of the arithmetical functions $P(x)$ and $\Delta(x)$," *Proc. London Math. Soc.* **15**(2) (1916), 192–213.
6. A. Ivić, *The Riemann Zeta-Function*, John Wiley & Sons, New York, 1985.
7. M. Jutila, "A method in the theory of exponential sums," Lecture Notes, Tata Institute of Fundamental Research, vol. 80, Bombay, 1987 (distr. by Springer-Verlag, Berlin).
8. T. Meurman, "A simple proof of Voronoi's identity," *Astérisque* **209** (1992), 265–274.
9. G.F. Voronoi, "Sur une fonction transcendante et ses applications à la sommation de quelques séries," *Ann. École Normale* **21**(3) (1904), 207–267, 459–534.
10. G.N. Watson, *A Treatise on the Theory of Bessel Functions*, 2nd edition, Cambridge University Press, Cambridge, 1944.

THE RAMANUJAN JOURNAL 2, 47–54 (1998)

The Residue of $p(N)$ Modulo Small Primes

KEN ONO   ono@math.psu.edu
*School of Mathematics, Institute for Advanced Study, Princeton, New Jersey 08540; and Department
of Mathematics, Penn State University, University Park, Pennsylvania 16802*

Dedicated to the memory of Paul Erdős

Received January 23, 1997; Accepted December 12, 1997

Abstract. For primes ℓ we obtain a simple formula for $p(N)$ (mod ℓ) as a weighted sum over ℓ-square affine partitions of N. When $\ell \in \{3, 5, 7, 11\}$, the weights are explicit divisor functions. The Ramanujan congruences modulo 5, 7, 11, 25, 49, and 121 follow immediately from these formulae.

Key words: partitions and q-series

1991 Mathematics Subject Classification: 11P83

On several occasions Professor Erdős asked me whether or not anyone has proved a good theorem regarding the parity of $p(N)$, the unrestricted partition function. Although there are numerous papers on the subject (see [5–7, 10, 15, 16, 20]), including two of my own, I must confess that little is really known. He was interested in the conjecture [18] that the number of $N \leq X$ for which $p(N)$ is even is $\sim \frac{1}{2} X$, and more generally he was interested in the distribution of $p(N)$ (mod ℓ) for primes ℓ. The difficulty of such problems appears to be that there is no known good method of computing $p(N)$ (mod ℓ) apart from mild variations of Euler's recurrence. Here we give an alternate method for computing $p(N)$ (mod ℓ) which does not depend on recurrences. Perhaps these formulae shed light on these difficult questions.

A partition of N is called a t-*core* if none of the hook numbers of the associated Ferrers-Young diagram are multiples of t, and their number is denoted $C(t, N)$. These partitions are important in the representation theory of permutation groups and finite general linear groups (see [2, 4, 8, 9, 11–13, 17]). Its generating function is

$$f(t, q) := \sum_{N=0}^{\infty} C(t, N)q^N = \prod_{n=1}^{\infty} \frac{(1 - q^{tn})^t}{(1 - q^n)}. \tag{1}$$

If ℓ is prime, then a partition $\Lambda = (\lambda_1, \lambda_2, \ldots)$ of N is called ℓ-*affine* (also ℓ-ary) if each λ_i is a power of ℓ. Such partitions are important in representation theory, and are used to

The author is supported by National Science Foundation grants DMS-9304580 and DMS-9508976, and NSA
grant MSPR-Y012.

compute McKay numbers of certain classical groups (see [11–13]). Here we will need a subclass of these partitions, the ℓ-*square affine* partitions. A partition Λ is ℓ-square affine if each λ_i is an even power of ℓ.

Throughout this note a_i and n_i will denote nonnegative integers, d a positive integer, p a prime, and $(\frac{\bullet}{p})$ the Legendre symbol modulo p, where $(\frac{n}{p}) = 0$ if $n \equiv 0 \pmod{p}$. Furthermore, we recall that $\eta(z) := q^{1/24} \prod_{n=1}^{\infty}(1 - q^n)$ with $q := e^{2\pi i z}$ is Dedekind's weight $1/2$ modular cusp form.

Proposition 1. *If ℓ is prime and $N < \ell^{2s+2}$, then*

$$p(N) \equiv \sum_{a_0+a_1\ell^2+\cdots+a_s\ell^{2s}=N} C(\ell, a_0)C(\ell, a_1) \cdots C(\ell, a_s) \pmod{\ell}.$$

Proof: If k is a nonnegative integer, then

$$f(\ell^{k+1}, q) = \prod_{n=1}^{\infty} \frac{(1 - q^{\ell^k n})^{\ell^k}}{(1 - q^n)} \cdot \frac{(1 - q^{\ell^{k+1}n})^{\ell^{k+1}}}{(1 - q^{\ell^k n})^{\ell^k}} \equiv f(\ell^k, q) \cdot \prod_{n=1}^{\infty} \frac{(1 - q^{\ell^{2k+1}n})^{\ell}}{(1 - q^{\ell^{2k}n})} \pmod{\ell}$$

$$= f(\ell^k, q) \cdot f(\ell, q^{\ell^{2k}}).$$

Therefore, $f(\ell^{k+1}, q) \equiv f(\ell, q) \cdot f(\ell, q^{\ell^2}) \cdots f(\ell, q^{\ell^{2k}}) \pmod{\ell}$, and so by (1) we obtain

$$\sum_{N=0}^{\infty} C(\ell^{k+1}, N)q^N \equiv \prod_{n=1}^{\infty} \frac{(1 - q^{\ell^{2k+2}n})}{(1 - q^n)}$$

$$\equiv \left(\sum_{N=0}^{\infty} C(\ell, N)q^N\right) \cdot \left(\sum_{N=0}^{\infty} C(\ell, N)q^{\ell^2 N}\right)$$

$$\cdots \left(\sum_{N=0}^{\infty} C(\ell, N)q^{\ell^{2k} N}\right) \pmod{\ell}. \tag{2}$$

Therefore, if $N < \ell^{2k+2}$, then

$$p(N) \equiv \sum_{a_0+a_1\ell^2+\cdots+a_k\ell^{2k}=N} C(\ell, a_0)C(\ell, a_1) \cdots C(\ell, a_k) \pmod{\ell}.$$

It is easy to see that the indices consist precisely of the ℓ-square affine partitions of N.
$\square$

The following result was obtained earlier by Hirschhorn in [5].

Theorem 1. *If $N < 4^{s+1}$, then*

$$p(N) \equiv \#\left\{(n_0, n_1, \ldots, n_s) \,\middle|\, \frac{1}{2}\sum_{i=0}^{s} 4^i \left(n_i^2 + n_i\right) = N\right\} \pmod{2}.$$

Proof: The result follows from Proposition 1 and the following well-known q-series identity:

$$\sum_{N=0}^{\infty} C(2, N)q^N = \prod_{n=1}^{\infty} \frac{(1 - q^{2n})^2}{(1 - q^n)} = \sum_{n=0}^{\infty} q^{\frac{n^2+n}{2}}.$$

□

Theorem 2. *If $N < 9^{s+1}$, then*

$$p(N) \equiv \sum_{a_0+9a_1+\cdots+9^s a_s=N} \sigma_3(a_0)\sigma_3(a_1)\cdots\sigma_3(a_s) \pmod{3},$$

where $\sigma_3(n) := \sum_{d|3n+1} d$.

Proof: The result follows from Proposition 1 and the following Eisenstein series identity [4]:

$$\frac{\eta^3(9z)}{\eta(3z)} = \sum_{N=0}^{\infty} C(3, N)q^{3N+1} = \sum_{n=0}^{\infty} \sum_{d|3n+1} \left(\frac{d}{3}\right) q^{3n+1}.$$

□

Theorem 3. *If $N < 25^{s+1}$, then*

$$p(N) \equiv \sum_{a_0+25a_1+\cdots+25^s a_s=N} \sigma_5(a_0)\sigma_5(a_1)\cdots\sigma_5(a_s) \pmod{5},$$

where $\sigma_5(n) := (n + 1)\sum_{d|n+1} d$.

Proof: The result follows from Proposition 1 and the identity (see [3, 4])

$$\frac{\eta^5(5z)}{\eta(z)} = \sum_{N=0}^{\infty} C(5, N)q^{N+1} = \sum_{n=1}^{\infty} \sum_{d|n} \left(\frac{d}{5}\right) \cdot \frac{n}{d} \cdot q^n.$$

□

Theorem 4. *If $N < 49^{s+1}$, then*

$$p(N) \equiv \sum_{a_0+49a_1+\cdots+49^s a_s=N} \sigma_7(a_0)\sigma_7(a_1)\cdots\sigma_7(a_s) \pmod{7},$$

where $\sigma_7(n) := (n + 2)\sum_{d|n+2}(2d + nd + 6d^3)$.

Proof: It is well known that [3]

$$\frac{\eta^7(7z)}{\eta(z)} = \sum_{N=0}^{\infty} C(7, N)q^{N+2} = \frac{1}{8}\sum_{n=1}^{\infty} \sum_{d|n} \left(\frac{d}{7}\right) \cdot \frac{n^2}{d^2}q^n - \frac{1}{8}\eta^3(z)\eta^3(7z).$$

Since $\eta^3(z)\eta^3(7z) \equiv \sum_{n=1}^{\infty} \tau(n)q^n$ (mod 7) where $\tau(n)$ is Ramanujan's tau-function, the result now follows by Proposition 1 and the Lehmer congruence [21]

$$\tau(n) \equiv n \sum_{d|n} d^3 \quad \text{(mod 7)}.$$

□

Theorem 5. *If* $N < 121^{s+1}$, *then*

$$p(N) \equiv \sum_{a_0+121a_1+\cdots+121^s a_s=N} \sigma_{11}(a_0)\sigma_{11}(a_1)\cdots\sigma_{11}(a_s) \quad \text{(mod 11)},$$

where

$$\sigma_{11}(n) := A(n+5) + 3(n+5)\sum_{d|n+5}(2d^7 + (n+5)^5 d^7 + 7(n+5)^3 d),$$

and

$$A(m) := \begin{cases} 0 & \text{if } \mathrm{ord}_{11}(m) \geq 1, \\ 0 & \text{if } \mathrm{ord}_p(m) \equiv 1 \pmod 2 \text{ for some } \left(\frac{p}{11}\right) = -1, \\ 3m^2\prod_j(\delta_j+1) & \text{if } m = \prod_{(\frac{p_i}{11})=-1} p^{2\delta_i} \prod_{(\frac{p_j}{11})=1} p_j^{\delta_j}. \end{cases}$$

Proof: Here $\eta^{11}(11z)/\eta(z)$ is a weight 5 holomorphic modular form with respect to $\Gamma_0(11)$ with Nebentypus character $(\frac{-11}{\bullet})$. Define the cusp forms $C_1(z)$, $C_2(z)$ and $C_3(z)$ by

$$C_1(z) := \sum_{n=1}^{\infty}\sum_{d|n}\left(\frac{d}{11}\right)\cdot\frac{n^4}{d^4}\cdot q^n - 1275\frac{\eta^{11}(11z)}{\eta(z)},$$

$C_2(z) := C_1(z) \mid T_3$, and $C_3(z) := C_1(z) \mid T_2$. Here T_p is the usual Hecke operator with Nebentypus $(\frac{-11}{\bullet})$. The three newforms in S_5 $(11, (\frac{-11}{\bullet}))$ are

$$N_1(z) := \sum_{n=1}^{\infty} a(n)q^n = \frac{3}{85}C_1(z) + \frac{1}{85}C_2(z) = q + 7q^3 + 16q^4 - 49q^5 - \cdots,$$

$$N_2(z) := \frac{15-\sqrt{-30}}{1275}\cdot\left(-7C_1(z) + C_2(z) + \frac{\sqrt{-30}}{3}C_3(z)\right)$$

$$= q + \sqrt{-30}q^2 - 3q^3 - \cdots,$$

$$N_3(z) := \frac{15+\sqrt{-30}}{1275}\cdot\left(-7C_1(z) + C_2(z) - \frac{\sqrt{-30}}{3}C_3(z)\right)$$

$$= q - \sqrt{-30}q^2 - 3q^3 - \cdots.$$

Furthermore, it turns out that

$$\frac{\eta^{11}(11z)}{\eta(z)} = \sum_{N=0}^{\infty} C(11, N) q^{N+5}$$

$$= \frac{1}{1275} \sum_{n=1}^{\infty} \sum_{d|n} \left(\frac{d}{11}\right) \cdot \frac{n^4}{d^4} \cdot q^n - \frac{1}{150} N_1(z)$$

$$+ \frac{15 + \sqrt{-30}}{5100} N_2(z) + \frac{15 - \sqrt{-30}}{5100} N_3(z). \qquad (3)$$

The forms $N_2(z)$ and $N_3(z)$ are complex conjugates and if $B(z) = \sum_{n=1}^{\infty} b(n) q^n$ is

$$B(z) := \frac{15 + \sqrt{-30}}{30} N_2(z) + \frac{15 - \sqrt{-30}}{30} N_3(z) = q - 2q^2 - 3q^3 - 14q^4 + \cdots,$$

then using the methods of Sturm and Swinnerton-Dyer [19, 21] we obtain

$$b(n) \equiv \left(8n + 4n\left(\frac{n}{11}\right)\right) \sum_{d|n} d^7 \pmod{11}.$$

Therefore, by (3) we obtain

$$C(11, N) \equiv 3a(N + 5) + 3(N + 5) \sum_{d|N+5} \left(2d^7 + \left(\frac{N+5}{11}\right) d^7\right.$$

$$\left. + 7(N + 5)^3 \left(\frac{d}{11}\right) d^6\right) \pmod{11}.$$

Completing the proof simply requires formulae for $a(n) \pmod{11}$. Since $N_1(z)$ is a newform it turns out that $a(1) = 1$ and

$$a(n)a(m) = a(nm) \qquad\qquad\qquad \text{if } \gcd(n, m) = 1, \qquad (4)$$

$$a(p^{k+1}) = a(p)a(p^k) - \left(\frac{-11}{p}\right) p^4 a(p^{k-1}) \quad \text{if } k \geq 1. \qquad (5)$$

The form $N_1(z)$ has complex multiplication by $\mathbb{Q}(\sqrt{-11})$, and we find that for primes p

$$a(p) = \begin{cases} 121 & \text{if } p = 11, \\ 0 & \text{if } p \equiv 2, 6, 7, 8, 10 \pmod{11}, \\ \dfrac{2x^4 - 132x^2 y^2 + 242 y^4}{16} & \text{if } p \equiv 1, 3, 4, 5, 9 \pmod{11} \text{ and } 4p = x^2 + 11 y^2. \end{cases}$$

Therefore, if $p \equiv 0, 2, 6, 7, 8, 10 \pmod{11}$, then $a(p) \equiv 0 \pmod{11}$. If $p \equiv 1, 3, 4, 5, 9 \pmod{11}$ and $4p = x^2 + 11y^2$, then $a(p) \equiv 7x^4 \pmod{11}$. Since $x^2 \equiv 4p \pmod{11}$, we find that $a(p) \equiv 2p^2 \pmod{11}$. Using (4) and (5) it is now an easy exercise to verify that $A(n) \equiv 3a(n) \pmod{11}$ for every $n > 1$. The result follows from Proposition 1. $\qquad\square$

Corollary 1. *For every nonnegative integer n*

$$p(5n + 4) \equiv 0 \pmod{5},$$
$$p(7n + 5) \equiv 0 \pmod{7},$$
$$p(11n + 6) \equiv 0 \pmod{11}.$$

Proof: The congruences modulo 5 and 7 follow from the observation that $(n + 1) \mid \sigma_5(n)$ and $(n + 2) \mid \sigma_7(n)$. The congruence modulo 11 follows from the fact that $\sigma_{11}(n) \equiv A(n + 5) \pmod{n + 5}$ and $A(n) \equiv 0 \pmod{11}$ if $\text{ord}_{11}(n) \geq 1$.   □

It also turns out that the Ramanujan congruences modulo 25, 49, and 121 follow easily from the proofs of Theorems 3, 4 and 5.

Theorem 6. *For every nonnegative integer n*

$$p(25n + 24) \equiv 0 \pmod{25},$$
$$p(49n + 47) \equiv 0 \pmod{49},$$
$$p(121n + 116) \equiv 0 \pmod{121}.$$

Proof: If $\ell = 5, 7$, or 11, and $\delta(\ell) := \frac{\ell^2 - 1}{24}$, then it is easy to verify using the information from the proofs of Theorems 3, 4, and 5 that

$$C(\ell, \ell^2 N - \delta(\ell)) \equiv 0 \pmod{\ell^2} \tag{6}$$

for every positive integer N. Moreover, it is easy to see that the above Ramanujan congruences are equivalent to the assertion that

$$C(\ell^2, \ell^2 N - \delta(\ell)) \equiv 0 \pmod{\ell^2} \tag{7}$$

for every positive integer N. Define integers $B(\ell, N)$ by

$$\sum_{N=0}^{\infty} B(\ell, N) q^{\ell N} := \left(\sum_{N=0}^{\infty} C(\ell, N) q^{\ell N} \right)^{\ell}.$$

Since

$$\sum_{N=0}^{\infty} C(\ell^2, N) q^N = \left(\sum_{N=0}^{\infty} C(\ell, N) q^N \right) \cdot \left(\sum_{N=0}^{\infty} B(\ell, N) q^{\ell N} \right),$$

we find that

$$C(\ell^2, \ell^2 N - \delta(\ell)) = \sum_{k \geq 0} C(\ell, \ell^2 N - \delta(\ell) - \ell k) B(\ell, k). \tag{8}$$

Since $C(\ell, \ell^2 N - \delta(\ell) - \ell k) \equiv 0 \pmod{\ell}$, and $B(\ell, k) \equiv 0 \pmod{\ell}$ if $k \not\equiv 0 \pmod{\ell}$, we find that

$$C(\ell^2, \ell^2 N - \delta(\ell)) \equiv \sum_{k \geq 0} C(\ell, \ell^2 N - \delta(\ell) - \ell^2 k) B(\ell, \ell k) \pmod{\ell^2}.$$

However, by (6) we obtain the Ramanujan congruences mod 25, 49, and 121. $\qquad\square$

Concluding remarks

There are analogs of these results where ℓ-affine partitions replace ℓ-square affine partitions. I chose to use the ℓ-square affine partitions because the weighted sums involve fewer terms, and the formulae for $3 \leq \ell \leq 11$ only involve divisor sums rather than values of Hecke Grössencharacters. Nevertheless, there is some interest in working out formulae for $p(N)$ (mod ℓ) via ℓ-affine partitions.

Recently, there has been a lot of interest in the *method of weighted words* as developed by Alladi and Gordon. These works, some joint with Andrews, lead to combinatorial explanations of identities where two seemingly unrelated partition functions are shown to be equal (see [1]). Here we exhibited $p(N)$ (mod ℓ) as a *weighted* sum over ℓ-square affine partitions of N where the weights are products of values of the ℓ-core partition function. Perhaps this resonates with the Alladi-Gordon method and can be viewed as an example of a mod ℓ theory.

References

1. K. Alladi, "The method of weighted words and applications to partitions," *Number Theory*, Paris, 1992–1993, London Math. Soc. Lecture Notes, vol. 215, 1995, pp. 1–36.
2. P. Fong and B. Srinivasan, "The blocks of finite general linear groups and unitary groups," *Invent. Math.* **69** (1982), 109–153.
3. F. Garvan, D. Kim, and D. Stanton, "Cranks and t-cores," *Invent. Math.* **101** (1990), 1–17.
4. A. Granville and K. Ono, "Defect zero p-blocks for finite simple groups," *Trans. Amer. Math. Soc.* **348**(1) (1996), 331–347.
5. M. Hirschhorn, "On the residue mod 2 and mod 4 of $p(n)$," *Acta Arith.* **38** (1980), 105–109.
6. M. Hirschhorn, "On the parity of $p(n)$ II," *J. Combin. Theory (A)* **62** (1993), 128–138.
7. M. Hirschhorn and M. Subbarao, "On the parity of $p(n)$," *Acta Arith.* **50**(4) (1988), 355–356.
8. G. James and A. Kerber, *The Representation Theory of the Symmetric Group*, Addison-Wesley, Reading, 1979.
9. A. Klyachko, "Modular forms and representations of symmetric groups, integral lattices and finite linear groups," *Zap. Nauchn. Sem. Leningrad Otdel. Mat. Inst. Steklov* **116** (1982).
10. O. Kolberg, "Note on the parity of the partition function," *Math. Scand.* **7** (1959), 377–378.
11. H. Nakamura, "On some generating functions for McKay numbers—Prime power divisibilities of the hook products of Young diagrams," *J. Math. Sci., U. Tokyo* **1** (1994), 321–337.
12. J. Olsson, "Remarks on symbols, hooks and degrees of unipotent characters," *J. Comb. Th. (A)* **42** (1986), 223–238.
13. J. Olsson, "Combinatorics and representations of finite groups," University Essen Lecture Notes, vol. 20, 1993.
14. K. Ono, "On the positivity of the number of partitions that are t-cores," *Acta Arith.* **66**(3) (1994), 221–228.
15. K. Ono, "Parity of the partition function in arithmetic progressions," *J. Reine Angew. Math.* **472** (1996), 1–15.
16. K. Ono, "Odd values of the partition function," *Disc. Math.* **169** (1997), 263–268.
17. K. Ono and L. Sze, "4-core partitions and class numbers," *Acta Arith.* **65** (1997), 249–272.

18. T.R. Parkin and D. Shanks, "On the distribution of parity in the partition function," *Math. Comp.* **21** (1967), 466–480.
19. J. Sturm, "On the congruence of modular forms," Springer Lecture Notes, vol. 1240, Springer-Verlag, 1984.
20. M. Subbarao, "Some remarks on the partition function," *Amer. Math. Monthly* **73** (1966), 851–854.
21. H.P.F. Swinnerton-Dyer, "On ℓ-adic Galois representations and congruences for coefficients of modular forms," Springer Lecture Notes, vol. 350, 1973.

THE RAMANUJAN JOURNAL 2, 55–58 (1998)

A Small Maximal Sidon Set

IMRE Z. RUZSA   ruzsa@math-inst.hu
Mathematical Institute of the Hungarian Academy of Sciences, Budapest, Pf. 127, H-1364 Hungary

To the memory of Paul Erdős, my master and mentor

Received February 5, 1997; Accepted March 19, 1997

Abstract. We construct a Sidon set $A \subset [1, N]$ which has $\ll (N \log N)^{1/3}$ elements and which is maximal in the sense that the inclusion of any other integer from $[1, N]$ destroys the Sidon property.

Key words: Sidon set

1991 Mathematics Subject Classification: Primary—11B75, Secondary—05B10, 11B13, 11B34

A set A of integers is a *Sidon set*, if all the sums $a + a'$, $a, a' \in A$ are distinct. We say that a finite Sidon set $A \subset [1, N]$ is *maximal* (for this N) if there is no Sidon set A' such that

$$A \subset A' \subset [1, N], \quad A' \neq A.$$

An easy counting argument shows that for a maximal Sidon set we have always $|A| \gg N^{1/3}$. Erdős, Sárközy, and Sós [2] asked whether this can be improved. We show that this is not far from optimal.

Theorem. *There is a maximal Sidon set in* $[1, N]$ *such that*

$$|A| \ll (N \log N)^{1/3}.$$

Proof: Select a prime p (it will be around $(N \log N)^{1/3}$) and put $q = 1 + p + p^2$. There is a Sidon set of size $p + 1$ modulo q, that is, there is a set

$$B = \{b_0, b_1, \ldots, b_p\}, \quad B \subset [1, q]$$

such that the sums $b_i + b_j$ all have distinct residues modulo q. (See, for instance, Halberstam-Roth [3].)

If we take such a set B, then for arbitrary integers $d_0, \ldots, d_p$ the numbers in

$$A_0 = \{a_i\}, \quad a_i = b_i + d_i q$$

Supported by Hungarian National Foundation for Scientific Research, Grant No. T 017433.

form a Sidon set. This set satisfies $A_0 \subset [1, N]$ if

$$0 \leq d_i \leq M - 1, \quad M = \left[\frac{N}{q}\right].$$

Any such set can be extended to a maximal Sidon set A. We can include a number m to A_0 and still have a Sidon set if and only if none of the equations

$$
\begin{aligned}
m &= a_u + a_v - a_w, \\
2m &= a_u + a_v
\end{aligned}
\tag{1}
$$

has a solution. We will achieve that (1) is solvable for most integers $1 \leq m \leq N$ and this will keep the size of A down.

Equation (1) can certainly hold only if

$$m \equiv b_u + b_v - b_w \pmod{q}. \tag{2}$$

Lemma. *Suppose that $m \not\equiv b_i \pmod{q}$ for all i. Then there is a sequence (u_i, v_i, w_i) of triplets of integers, $1 \leq i \leq I$, such that each $u = u_i$, $v = v_i$, $w = w_i$ is a solution of congruence (2), for $i \neq j$ the sets $\{u_i, v_i, w_i\}$ and $\{u_j, v_j, w_j\}$ are disjoint and $I \geq p/8$.*

Proof: Since the numbers $b_i - b_j$, $i \neq j$ are all incongruent modulo q and their number is $q - 1$, they represent each nonzero residue exactly once. Thus for each u there is exactly one pair (v, w) such that (2) holds. This shows that there are altogether p solutions. Similarly, for a fixed v there is exactly one (u, w), and for a given w there are at most two pairs (u, v).

Now let (u_i, v_i, w_i), $1 \leq i \leq I$ be a maximal selection of pairwise disjoint triplets of solutions. Each triplet (u, v, w) excludes the following triplets: itself; (v, u, w); at most two with u in the third position; at most two with v in the third position; one with w in the first position and one with w in the second. These are 8 excluded triplets altogether (with the selected triplets included), thus $8I \geq p$. $\square$

Now we construct A_0. For each i we select d_i randomly from the numbers $0, \ldots, M - 1$, so that all M^{p+1} possibilities are equally probable.

We estimate the probability that (1) holds for a number m such that $m \not\equiv b_i \pmod{q}$. Equation (1) is equivalent to (2) and

$$d_u + d_v - d_w = \frac{m - (b_u + b_v - b_w)}{q} = m'. \tag{3}$$

Since $2 - 2q \leq m - (b_u + b_v - b_w) \leq N + q - 2$, the integer m' satisfies $-1 \leq m' \leq M + 1$.

In (2) and (3), w must be different from u and v, but u and v may coincide. If they are all distinct, then the probability that (3) holds is

$$\mathbf{P}(d_u + d_v - d_w = m') = S/M^3,$$

where S denotes the number of triplets (x, y, z) such that $x + y - z = m', 0 \leq x, y, z \leq M - 1$. If $u = v$, then this probability is S'/M^2, where S' is the number of pairs (x, z) such that $2x - z = m', 0 \leq x, z \leq M - 1$.

Routine calculations show that (for $M > M_0$) uniformly in the range $-1 \leq m' \leq M + 1$ we have $S \geq cM^2$ and $S' \geq cM$ with a positive absolute constant c, thus

$$\mathbf{P}(d_u + d_v - d_w = m') \geq c/M.$$

If we apply this for the selected disjoint triplets (u_i, v_i, w_i), then we have a collection of independent events and we obtain

$$\mathbf{P}\big(d_{u_i} + d_{v_i} - d_{w_i} \neq m' \text{ for all } 1 \leq i \leq I\big) \leq \left(1 - \frac{c}{M}\right)^I \leq e^{-cI/M}.$$

We have

$$e^{-cI/M} \leq \exp - \frac{cp}{8M} \leq \exp - \frac{cpq}{8N} \leq \exp - \frac{cp^3}{8N}.$$

Thus, if we take p so that $p > (CN \log N)^{1/3}$ with $C = 8/c$, then this probability will be $< 1/N$. By Chebyshev's theorem, we can achieve this with $p \ll (N \log N)^{1/3}$. Consequently, for any m such that $m \not\equiv b_u \pmod{q}$ we have

$$\mathbf{P}((1) \text{ is not solvable}) < 1/N$$

and hence

$$\mathbf{P}((1) \text{ is solvable for all } m \not\equiv b_u \pmod{q}) > 0.$$

The set A_0 corresponding to such a choice will be a Sidon set; it may not be maximal, but it has the following property. Any Sidon set

$$A \supset A_0, \quad A \subset [1, N]$$

is of the form $A = A_0 \cup A_1$, where each element $a \in A_1$ satisfies $a \equiv b_u \pmod{q}$ for some u.

Now extend this A_0 to a maximal Sidon set A. We estimate the size of A_1. For each $a \in A_1$, consider the unique u such that $a \equiv b_u \pmod{q}$. For this u, the number $a - a_u = a - b_u - q d_u$ is a multiple of q. These numbers are all distinct by the Sidon property of A, and they are in the interval $(1 - N, N - 1)$. Their number is $|A_2|$, thus we have

$$|A_2| \leq 1 + 2N/q \ll N^{1/3},$$

thus

$$|A| = |A_1| + |A_2| \ll (N \log N)^{1/3}$$

as claimed. $\square$

Remark. Let $g(N)$ denote the minimal cardinality of a maximal Sidon set in $[1, N]$. Thus we know that

$$N^{1/3} \ll g(N) \ll (N \log N)^{1/3}. \tag{4}$$

If the right side gave the correct order of magnitude, this would immediately yield the celebrated theorem of Ajtai, Komlós, and Szemerédi [1] on the existence of an infinite Sidon set which has always $\gg (N \log N)^{1/3}$ elements up to N. At present I do not even have a heuristic argument that would indicate which side of (4) gives the truth.

References

1. M. Ajtai, J. Komlós, and E. Szemerédi, "A dense infinite Sidon sequence," *European J. Comb.* **2** (1981), 1–11.
2. P. Erdős, A. Sárközy, and Vera T. Sós, "On sum sets of Sidon sets I.," *J. Number Theory* **47** (1994), 329–347.
3. H. Halberstam and K.F. Roth, *Sequences*, Clarendon, London, 1996 (2nd edition, Springer-Verlag, New York–Berlin, 1983).

THE RAMANUJAN JOURNAL 2, 59–66 (1998)

Sums and Products from a Finite Set of Real Numbers

KEVIN FORD   ford@math.utexas.edu
Department of Mathematics, University of Texas, Austin, Texas 78712

Dedicated to the memory of Paul Erdős

Received February 21, 1997; Accepted June 9, 1997

Abstract. If A is a finite set of positive integers, let $E_h(A)$ denote the set of h-fold sums and h-fold products of elements of A. This paper is concerned with the behavior of the function $f_h(k)$, the minimum of $|E_h(A)|$ taken over all A with $|A| = k$. Upper and lower bounds for $f_h(k)$ are proved, improving bounds given by Erdős, Szemerédi, and Nathanson. Moreover, the lower bound holds when we allow A to be a finite set of arbitrary positive real numbers.

Key words: sequences, sums, products

1991 Mathematics Subject Classification: Primary—11B75

For finite sets of real numbers A and B, define

$$A + B = \{a + b : a \in A, b \in B\}, \quad AB = \{ab : a \in A, b \in B\}.$$

More generally, if $h \geq 2$ define

$$hA = \{a_1 + \cdots + a_h : a_i \in A\}, \quad A^h = \{a_1 \cdots a_h : a_i \in A\}.$$

Erdős [2] conjectured that for any finite set A of positive integers,

$$|E_h(A)| \gg_\varepsilon |A|^{h-\varepsilon}, \tag{1}$$

where

$$E_h(A) = hA \cup A^h.$$

In other words, no set A can have simultaneously few sums and few products. Notice that trivially

$$\frac{1}{2}(|hA| + |A^h|) \leq |E_h(A)| \leq |hA| + |A^h|. \tag{2}$$

Our chief interest here is the behavior of the function

$$f_h(k) = \min\{|E_h(A)| : |A| = k, A \subset \mathbb{N}\}.$$

Erdős and Szemerédi [3] proved the nontrivial bounds

$$k^{1+\delta} \ll f_2(k) \ll k^{2-c/\log_2 k}, \tag{3}$$

where c and δ are positive constants an $\log_k x$ denotes the kth iterate of the logarithm. Nathanson [7] showed that $\delta = 1/31$ is admissible, and we note that the argument works for any finite set of positive *real* numbers. No bounds for $|E_h(A)|$ for $h \geq 3$ have been published. However, for any $a \in A$, A^h contains $a^{h-2}p$ for each $p \in A^2$ and hA contains $(h-2)a + s$ for each $s \in 2A$. Thus, by (2),

$$|E_h(A)| \geq \frac{1}{2}(|hA| + |A^h|) \geq \frac{1}{2}(|2A| + |A^2|) \geq \frac{1}{2}|E_2(A)|. \tag{4}$$

We also have

$$|E_h(A)| \leq |hA| + |A^h| \leq |A|^{h-2}(|2A| + |A^2|) \leq 2|A|^{h-2}|E_2(A)|.$$

In particular, if (1) fails for a particular h, it fails for all larger h.

When $h = 2$, (1) has been established for certain very special sets of positive integers A. Nathanson and Tenenbaum [8] proved (1) under the assumption that $|2A| \leq 3|A| - 4$ using Freiman's structure theory of set addition (see [4]). As noted by Nathanson and Jia [6], (1) can also be proved in the case where A is contained in a "short" interval of length $|A|^{o(\log_2 |A|)}$ using the fact that $\log d(n) = O(\log n/\log_2 n)$, where $d(n)$ is the number of divisors of n.

In this note, we improve the lower bound for $|E_2(A)|$ using a refinement of Nathanson's argument [7].

Theorem 1. *If A is a finite set of positive real numbers, then*

$$|E_2(A)| \geq \frac{1}{6}|A|^{1+(1/15)}.$$

A slight modification of one part of the argument produces lower bounds for $|E_h(A)|$ for $h \geq 3$ which are superior to the bound obtained by combining (4) with Theorem 1. However, the exponent only tends to 8/7 as h tends to infinity.

Theorem 2. *If A is a finite set of positive real numbers, then*

$$|E_h(A)| \gg |A|^{1+\frac{h-1}{7h+1}}.$$

Finally, we investigate how small the sets $E_h(A)$ can be. Erdős and Szemerédi proved the lower bound in (3) by taking A to be a set of sufficiently "smooth" numbers (numbers without large prime factors). Using modern results concerning the distribution of smooth numbers, we prove an analogous result for $f_h(k)$, where the "constant" c grows rapidly with h.

Theorem 3. *For each fixed h, we have*

$$f_h(k) \leq k^{h - c_h / \log_2 k + O((\log_3 k)/(\log_2 k)^2)},$$

where $c_h = h(h-1) \log h$.

The starting point for the proof of Theorems 1 and 2 is a lower bound on the number of sums and products when B is contained in a dyadic interval. In this case, Nathanson [7] showed that $|E_2(B)| \gg |B|^{16/15}$.

Lemma 1. *Suppose B is a finite set of real numbers contained in $[x, 2x]$ for some positive x. Then*

$$|2B| + |B^2| \geq \frac{7}{20} |B|^{8/7}.$$

Proof: Let $k = |B|$ and suppose $k \geq 10^7$, for otherwise the right side in the lemma is less than $4k - 2$ and the lemma is trivial. Suppose $1 \leq l < k$ and group the numbers in B as follows. Let B_1 be the set of l smallest numbers in B, let B_2 denote the set of l smallest numbers in $B \setminus B_1$, etc. This partitions B into $B_1, B_2, \ldots, B_{[k/l]}$ with $< l$ numbers left over. Let the diameter of a set be the difference between the largest and the smallest numbers in the set. Let B^* be the set B_i with smallest diameter and let d be the diameter of B^*.

Now suppose $1 \leq i < j \leq [k/l]$ with $j - i \geq 3$ and

$$b_1^*, b_2^* \in B^*, \quad b_i \in B_i, \quad b_j \in B_j.$$

Then

$$b_1^* + b_i < (b_2^* + d) + (b_j - 2d) < b_2^* + b_j \tag{5}$$

and

$$\begin{aligned}
b_j b_2^* &> (b_i + 2d) b_2^* \\
&\geq b_i (b_1^* - d) + 2d b_2^* \\
&= b_i b_1^* + d(2b_2^* - b_i) \geq b_i b_1^*.
\end{aligned} \tag{6}$$

From now on consider only the sets $B_1, B_4, B_7, \ldots$. By (5) and (6), the sets $B^* + B_i$ are distinct, as are the sets $B^* B_i$. Let

$$P_i = |B^* \cdot B_i|, \quad S_i = |B^* + B_i|. \tag{7}$$

Then

$$|2B| + |B^2| \geq \sum_{i \equiv 1 (\mathrm{mod}\ 3)} P_i + S_i. \tag{8}$$

Fix i and define

$$r(m) = |\{(b^*, b_i) : b^* b_i = m, b^* \in B^*, b_i \in B_i\}|.$$

When $r(m) > 0$, denote by (b_j^*, b_j') $(1 \le j \le r(m))$ the distinct pairs of numbers $b_j^* \in B^*, b_j' \in B_i$ with product m. Notice that $b_{j_1}^* + b_{j_2}' \in B^* + B_i$ for each of the $r(m)^2$ pairs (j_1, j_2). For each $n \in B^* + B_i$, define

$$s_m(n) = |\{(j_1, j_2) : b_{j_1}^* + b_{j_2}' = n\}|.$$

With m, n fixed there are $\frac{1}{2}(s_m(n)^2 - s_m(n)) \ge s_m(n) - 1$ quadruples (j_1, j_2, j_3, j_4) with $b_{j_1}^* < b_{j_3}^*$ and

$$
\begin{aligned}
b_{j_1}^* + b_{j_2}' &= b_{j_3}^* + b_{j_4}' = n, \\
b_{j_2}^* b_{j_2}' &= b_{j_4}^* b_{j_4}' = m.
\end{aligned}
\tag{9}
$$

On the other hand, given any four numbers $(b_{j_1}^*, b_{j_2}^*, b_{j_3}^*, b_{j_4}^*)$ in B^* with $b_{j_1}^* < b_{j_3}^*$, Eq. (9) have at most one solution b_{j_2}', b_{j_4}' and thus i, m and n are uniquely determined. If we let N_i be the number of quadruples corresponding to each i, then by (7) and the Cauchy-Schwarz inequality,

$$
\begin{aligned}
N_i &\ge \sum_m \sum_n s_m(n) - 1 \\
&\ge \sum_m (r(m)^2 - S_i) \\
&\ge l^4 / P_i - P_i S_i.
\end{aligned}
$$

Also, $N_i \ge 0$ for each i. If $b_{j_1}^* < b_{j_3}^*$, then (9) implies $b_{j_2}^* < b_{j_4}^*$ and hence

$$\sum_i N_i \le \frac{1}{4} l^4. \tag{10}$$

Define

$$I_1 = \left\{ i \equiv 1 \pmod 3 : S_i P_i^2 \ge \frac{1}{2} l^4 \right\},$$

$$I_2 = \left\{ i \equiv 1 \pmod 3 : S_i P_i^2 < \frac{1}{2} l^4 \right\}.$$

A straightforward calculation shows that

$$S_i + P_i \ge \frac{3}{2} l^{4/3} \quad (i \in I_1). \tag{11}$$

We also have $N_i \geq l^4/2P_i$ for $i \in I_2$, hence by (10),

$$\sum_{i \in I_2} \frac{1}{P_i} \leq \frac{1}{2}. \tag{12}$$

Let $M_1 = |I_1|$, $M_2 = |I_2|$ and $H = M_1 + M_2$. By (8), (11), (12) and the Cauchy-Schwarz inequality,

$$|2B| + |B^2| \geq \frac{3}{2} l^{4/3} M_1 + \sum_{i \in I_2} P_i$$

$$\geq \frac{3}{2} M_1 l^{4/3} + 2M_2^2$$

$$= \frac{3}{2} l^{4/3} (H - M_2) + 2M_2^2.$$

The right side is minimized at $M_2 = \frac{3}{8} l^{4/3}$. Since $H \geq \frac{1}{3}[k/l] \geq \frac{k}{3l} - \frac{1}{3}$, we obtain

$$|2B| + |B^2| \geq \frac{3}{2} H l^{4/3} - \frac{9}{32} l^{8/3}$$

$$\geq \frac{1}{2} k l^{1/3} - \frac{9}{32} l^{8/3} - \frac{1}{2} l^{4/3}. \tag{13}$$

Ignoring the last term, the optimal value of l is

$$l = \left[\left(\frac{2}{9} k \right)^{3/7} \right].$$

The lemma now follows from (13), since $k \geq 10^7$ and $l \geq (\frac{2}{9}k)^{3/7} - 1$. $\square$

Lemma 2. *Suppose $h \geq 2$ and that for every finite set of positive real numbers B contained in some interval $[x, 2x]$, we have $|hB| + |B^h| \geq c|B|^{1+1/u}$. Then for any finite set A of positive real numbers, we have*

$$|E_h(A)| \geq \frac{c}{2} (c h^h h!/2)^{-\frac{1}{hu+1}} |A|^{1 + \frac{h-1}{hu+1}}.$$

Proof: Let $k = |A|$ and break A into blocks

$$A_j = A \cap [2^{j-1}, 2^j) \quad (j \in \mathbb{Z}).$$

Let

$$J = \{ j : |A_j| > 0 \},$$

$$m = \sum_{j \in J} |A_j|^{1+(1/u)}.$$

For each h-tuple of numbers $a_1, a_2, \ldots, a_h \in A_j$, we have $\sum a_i \in [h2^{j-1}, h2^j)$ and $\prod a_i \in [2^{h(j-1)}, 2^{hj})$. Therefore, the sets hA_j are disjoint, as are the sets A_j^h. Hölder's inequality gives

$$k = \sum_{j \in J} |A_j| \le |J|^{\frac{1}{u+1}} \left(\sum_{j \in J} |A_j|^{1+(1/u)} \right)^{\frac{u}{u+1}} = m^{\frac{u}{u+1}} |J|^{\frac{1}{u+1}},$$

which implies $|J| \ge k^{u+1} m^{-u}$. Choose one number a_j from each nonempty set A_j and set $n = 2 + [\frac{\log(h-1)}{\log 2}]$. For $0 \le r \le n-1$, let J_r be the subset of J with $j \equiv r \pmod n$. For some r, $|J_r| \ge \frac{|J|}{n}$. Form the set $C = \{a_j : j \equiv r \pmod n\}$. Since $a_{i+n} \ge 2^{n-1} a_i \ge h a_i$ for each i, the sums of distinct h-tuples of numbers in C are distinct. It follows from (2) and the hypothesis that

$$|E_h(A)| \ge \max \left(\frac{1}{2} \sum_{j \in J} |hA_j| + |A_j^h|, \frac{|C|^h}{h!} \right)$$

$$\ge \max \left(\frac{cm}{2}, \frac{k^{hu+h} m^{-uh}}{h^h h!} \right).$$

The right side is minimized when $m^{hu+1} = 2k^{hu+h}/(ch^h h!)$, and this completes the proof.
□

Combining Lemma 1 with Lemma 2 (taking $h = 2$, $c = \frac{7}{20}$, $u = 7$) gives Theorem 1. Theorem 2 follows from (4) and Lemmas 1 and 2. Proving $f_h(k) \gg k^{\beta(h)}$ with $\beta(h)$ tending to ∞ with h will require a nontrivial extension of Lemma 1 to the case $h \ge 3$, and it is not clear how this can be accomplished.

It is curious that nowhere in the argument was it necessary to assume the set A was a set of integers. Based on this observation, we make the following conjecture.

Conjecture. *If A is a finite set of positive real numbers, then*

$$|E_h(A)| \gg_\varepsilon |A|^{h-\varepsilon}.$$

Before proving Theorem 3, we need a few definitions. A natural number n is said to be y-smooth if n is divisible by no prime factor $>y$. Denote by $\Psi(x, y)$ the number of y-smooth numbers $\le x$. Important in the study of $\Psi(x, y)$ is the Dickman function $\rho(u)$, defined for $u \ge 0$ by

$$\rho(u) = \begin{cases} 1 & (0 \le u \le 1), \\ 1 - \int_1^u \frac{\rho(v-1)}{v} \, dv & (u > 1). \end{cases}$$

We quote the following well-known results (Theorem 1.2 and Corollary 2.3 of [5]). Here we take $u = \frac{\log x}{\log y}$.

Lemma 3. *For any fixed $\varepsilon > 0$ we have*

$$\Psi(x, y) = x\rho(u)^{1+O(E(u))}$$

uniformly in the range

$$y \geq 2, 1 \leq u \leq y^{1-\varepsilon},$$

where

$$E(u) = \exp\{-(\log u)^{3/5-\varepsilon}\}.$$

Lemma 4. *Uniformly in $u \geq 3$, we have*

$$\rho(u) = \exp\left\{-u\left(\log u + \log_2 u - 1 + O\left(\frac{\log_2 u}{\log u}\right)\right)\right\}.$$

From now on assume h is fixed. In particular, constants implied by the symbol O may depend on h. Suppose x is large and set

$$\delta = \frac{2h \log h}{\log_2 x}, \qquad \alpha = \frac{h + \delta}{h - 1}.$$

Let A be the set of $(\log x)^\alpha$-smooth numbers $\leq x$. Set $k = |A| = \Psi(x, (\log x)^\alpha)$ and $u = \frac{\log x}{\alpha \log_2 x}$. By Lemmas 3 (with $\varepsilon = \min(1/2, 1 - 1/\alpha)$) and 4, we have

$$k = x\rho(u)^{1+O(E(u))}$$

$$= x \exp\left\{-\frac{\log x}{\alpha \log_2 x}(\log_2 x - \log \alpha - 1) + O(L(x))\right\}$$

$$= x^{1-1/\alpha} \exp\left\{O\left(\frac{\log x}{\log_2 x}\right)\right\},$$

where

$$L(x) = \frac{\log x \log_3 x}{(\log_2 x)^2}.$$

Consequently,

$$u = \frac{\log k}{(\alpha - 1)\log_2 k}\left(1 + O\left(\frac{1}{\log_2 k}\right)\right). \tag{14}$$

Thus,

$$|hA| \leq hx \leq k\rho(u)^{-1-O(E(u))}. \tag{15}$$

Lemma 3 also gives

$$|A^h| \le \Psi(x^h, (\log x)^\alpha) = x^h \rho(hu)^{1+O(E(hu))} = k^h \left(\frac{\rho(hu)}{\rho(u)^h}\right)^{1+O(E(u))}. \quad (16)$$

By Lemma 4 and (14), we deduce

$$\rho(u) = \exp\left\{-\frac{\log k}{\alpha-1} + \frac{1+\log(\alpha-1)}{\alpha-1}\frac{\log k}{\log_2 k} + O(L(k))\right\}$$

$$\ge \exp\left\{-(h-1)(1-\delta+O(\delta^2))\log k - (h-1)\log h\frac{\log k}{\log_2 k} + O(L(k))\right\}$$

$$\ge k^{-(h-1)} \exp\left\{h(h-1)\log h\frac{\log k}{\log_2 k} + O(L(k))\right\}.$$

Similarly, we obtain

$$\frac{\rho(hu)}{\rho(u)^h} = \exp\left\{-h(h-1)\log h\frac{\log k}{\log_2 k} + O(L(k))\right\}.$$

Combining these estimates with (15) and (16) gives

$$|hA| + |A^h| \le k^h \exp\left\{-h(h-1)\log h\frac{\log k}{\log_2 k} + O(L(k))\right\},$$

which completes the proof of Theorem 3.

Remark. Following acceptance of this paper, the author learned that G. Elekes [1] has proved $|E_2(A)| \gg |A|^{5/4}$. The method also yields $|E_h(A)| \gg |A|^{3/2-2^{-h}}$.

References

1. G. Elekes, "On the number of sums and products," *Acta Arith.* **81** (1997), 365–367.
2. P. Erdős, "Problems and results on combinatorial number theory III," *Number Theory Day*, New York 1976; Lecture Notes in Mathematics, vol. 626, Springer-Verlag, Berlin, 1977, pp. 43–72.
3. P. Erdős and Szemerédi, "On sums and products of integers," *Studies in Pure Mathematics*, To the Memory of Paul Turán (P. Erdős, L. Alpár, G. Halász, and A. Sárközy, eds.), Birkhäuser Verlag, Basel, 1983, pp. 213–218.
4. G. Freiman, "Foundations of a structural theory of set addition," *Translations of Mathematical Monographs*, Amer. Math. Soc., RI, vol. 37, 1973.
5. A. Hildebrand and G. Tenenbaum, "Integers without large prime factors," *J. Théor. Nombres Bordeaux* **5** (1993), 411–484.
6. X. Jia and M. Nathanson, "Finite graphs and the number of sums and products," preprint.
7. M. Nathanson, "On sums and products of integers," preprint.
8. M. Nathanson and G. Tenenbaum, "Inverse theorems and the number of sums and products," preprint.

THE RAMANUJAN JOURNAL 2, 67–151 (1998)

The Distribution of Totients

KEVIN FORD   ford@math.utexas.edu
Department of Mathematics, University of Texas at Austin, Austin, TX 78712

Dedicated to the memory of Paul Erdős

Received March 5, 1997; Accepted October 28, 1997

Abstract. This paper is a comprehensive study of the set of totients, i.e., the set of values taken by Euler's ϕ-function. The main functions studied are $V(x)$, the number of totients $\leq x$, $A(m)$, the number of solutions of $\phi(x) = m$ (the "multiplicity" of m), and $V_k(x)$, the number of $m \leq x$ with $A(m) = k$. The first of the main results of the paper is a determination of the true order of $V(x)$. It is also shown that for each $k \geq 1$, if there is a totient with multiplicity k then $V_k(x) \gg V(x)$. Sierpiński conjectured that every multiplicity $k \geq 2$ is possible, and we deduce this from the Prime k-tuples Conjecture. An older conjecture of Carmichael states that no totient has multiplicity 1. This remains an open problem, but some progress can be reported. In particular, the results stated above imply that if there is one counterexample, then a positive proportion of all totients are counterexamples. The lower bound for a possible counterexample is extended to $10^{10^{10}}$ and the bound $\liminf_{x \to \infty} V_1(x)/V(x) \leq 10^{-5,000,000,000}$ is shown. Determining the order of $V(x)$ and $V_k(x)$ also provides a description of the "normal" multiplicative structure of totients. This takes the form of bounds on the sizes of the prime factors of a pre-image of a typical totient. One corollary is that the normal number of prime factors of a totient $\leq x$ is $c \log \log x$, where $c \approx 2.186$. Lastly, similar results are proved for the set of values taken by a general multiplicative arithmetic function, such as the sum of divisors function, whose behavior is similar to that of Euler's function.

Key words: Euler's function, totients, distributions, Carmichael's Conjecture, Sierpiński's Conjecture

1991 Mathematics Subject Classification: Primary—11A25, 11N64

1. Introduction

Let $\mathscr{V}$ denote the set of values taken by Euler's ϕ-function (totients), i.e.,

$$\mathscr{V} = \{1, 2, 4, 6, 8, 10, 12, 16, 18, 20, 22, 24, 28, 30, \ldots\}.$$

Let

$$
\begin{aligned}
\mathscr{V}(x) &= \mathscr{V} \cap [1, x], \\
V(x) &= |\mathscr{V}(x)|, \\
\phi^{-1}(m) &= \{n : \phi(n) = m\}, \\
A(m) &= |\phi^{-1}(m)|, \\
V_k(x) &= |\{m \leq x : A(m) = k\}|.
\end{aligned}
\tag{1.1}
$$

We will refer to $A(m)$ as the multiplicity of m. This paper is concerned with the following problems:

1. What is the order of $V(x)$?
2. What is the order of $V_k(x)$ when the multiplicity k is possible?
3. What multiplicities are possible?
4. What is the normal multiplicative structure of totients?

1. The fact that $\phi(p) = p - 1$ for primes p implies $V(x) \gg x/\log x$ by the Prime Number Theorem. Pillai [25] gave the first non-trivial bound on $V(x)$, namely

$$V(x) \ll \frac{x}{(\log x)^{(\log 2)/e}}.$$

Using sieve methods, Erdős [7] improved this to

$$V(x) \ll_\varepsilon \frac{x}{(\log x)^{1-\varepsilon}}$$

and later in [8] showed

$$V(x) \gg \frac{x \log \log x}{\log x}.$$

Erdős and Hall [10, 11] sharpened the bounds further, showing

$$\frac{x}{\log x} \exp\{c_1(\log_3 x)^2\} \ll V(x) \ll \frac{x}{\log x} \exp\{c_2(\log_2 x)^{1/2}\}$$

for certain positive constants c_1 and c_2. Here and throughout this paper $\log_k x$ denotes the kth iterate of the logarithm. The upper bound was improved again by Pomerance [26], who showed that for some positive constant c_3,

$$V(x) \ll \frac{x}{\log x} \exp\{c_3(\log_3 x)^2\}.$$

The gap between c_1 and c_3 was removed by Maier and Pomerance [23], who showed that

$$V(x) = \frac{x}{\log x} \exp\{(C + o(1))(\log_3 x)^2\}, \qquad (1.2)$$

where C is a constant defined as follows. Let

$$F(x) = \sum_{n=1}^{\infty} a_n x^n, \qquad a_n = (n + 1) \log(n + 1) - n \log n - 1. \qquad (1.3)$$

Since $a_n \sim \log n$ and $a_n > 0$, it follows that $F(x)$ is defined and strictly increasing on $[0, 1)$, $F(0) = 0$ and $F(x) \to \infty$ as $x \to 1^-$. Thus, there is a unique number ϱ such that

$$F(\varrho) = 1 \quad (\varrho = 0.542598586098471021959\ldots). \qquad (1.4)$$

In addition, $F'(x)$ is strictly increasing, and

$$F'(\varrho) = 5.69775893423019267575\ldots$$

Let

$$C = \frac{1}{2|\log \varrho|} = 0.81781464640083632231\ldots \tag{1.5}$$

and

$$D = 2C(1 + \log F'(\varrho) - \log(2C)) - 3/2$$
$$= 2.17696874355941032173\ldots \tag{1.6}$$

Our main result is a determination of the true order of $V(x)$.

Theorem 1. *We have*

$$V(x) = \frac{x}{\log x} \exp\{C(\log_3 x - \log_4 x)^2 + D\log_3 x - (D + 1/2 - 2C)\log_4 x + O(1)\}.$$

2. Erdős [9] showed by sieve methods that if $A(m) = k$, then for most primes p, $A(m(p-1)) = k$. If the multiplicity k is possible, it follows immediately that $V_k(x) \gg x/\log x$. Applying the machinery used to prove Theorem 1, we can show that for each k, either $V_k(x) = 0$ for all x, or $V_k(x)$ is the same order as $V(x)$.

Theorem 2. *If there is a number d with $A(d) = k$, then*

$$V_k(x) \gg_\varepsilon d^{-1-\varepsilon} V(x) \quad (x \geq x_0(k)).$$

In other words, a positive fraction of totients have multiplicity k if the multiplicity k is possible. This suggests that the multiplicity of "most" totients is bounded. Specifically, we prove

Theorem 3. *We have*

$$\frac{|\{m \in \mathcal{V}(x) : A(m) \geq N\}|}{V(x)} = \sum_{k \geq N} \frac{V_k(x)}{V(x)} \ll N^{-1} \exp\{O(\sqrt{\log N})\}.$$

A simple modification of the proof of Theorems 1 and 2 also gives the following result concerning totients in short intervals. First, define $\pi(x)$ to be the number of primes $\leq x$. A real number θ is said to be admissible if $\pi(x + x^\theta) - \pi(x) \gg x^\theta/\log x$ with x sufficiently large. The current record is due to Baker and Harman [1], who showed that $\theta = 0.535$ is admissible.

Theorem 4. *If θ is admissible, $y \geq x^\theta$ and the multiplicity k is possible, then*

$$V_k(x + y) - V_k(x) \asymp \frac{y}{x + y} V(x + y).$$

Corollary. *For every fixed $c > 1$, $V(cx) - V(x) \asymp V(x)$.*

Erdős has asked if $V(cx) \sim cV(x)$ for each fixed $c > 1$, which would follow from an asymptotic formula for $V(x)$. The method of proof of Theorem 1, however, falls short of answering Erdős' question.

It is natural to ask what the maximum totient gaps are, in other words what is the behavior of the function $M(x) = \max_{v_i \leq x}(v_i - v_{i-1})$ if $v_1, v_2, \ldots$ denotes the sequence of totients? Can it be shown, for example, that for x sufficiently large, that there is a totient between x and $x + x^{1/2}$?

3. In 1907, Carmichael [3] announced that for every m, the equation $\phi(x) = m$ has either no solutions x or at least two solutions. In other words, no totient can have multiplicity 1. His proof of this assertion was flawed, however, and the existence of such numbers remains an open problem. In [4], Carmichael did show that no number $m < 10^{37}$ has multiplicity 1, and conjectured that no such m exists (this is now known as Carmichael's Conjecture). Klee [21] improved the lower bound for a counterexample to 10^{400}, Masai and Valette [24] improved it to $10^{10,000}$ and recently Schlafly and Wagon [31] showed that a counterexample must exceed $10^{10,000,000}$. An immediate corollary of Theorems 1 and 2 is

Theorem 5. *We have*

$$\limsup_{x \to \infty} \frac{V_1(x)}{V(x)} < 1.$$

Furthermore, Carmichael's Conjecture is equivalent to the bound

$$\liminf_{x \to \infty} \frac{V_1(x)}{V(x)} = 0.$$

Although this is a long way from proving Carmichael's Conjecture, Theorem 5 shows that the set of counterexamples cannot be a "thin" subset of $\mathscr{V}$. Either there are no counterexamples or a positive fraction of totients are counterexamples.

The basis for the computations of lower bounds for a possible counterexample is a lemma of Carmichael, generalized by Klee, which allows one to show that if $A(m) = 1$ then x must be divisible by the squares of many primes. Using the method outlined in [31] and modern computer hardware, we push the lower bound for a counterexample to Carmichael's Conjecture to the aesthetically pleasing bound $10^{10^{10}}$.

Theorem 6. *If $A(m) = 1$, then m exceeds $10^{10^{10}}$.*

As a corollary, a variation of an argument of Pomerance [27] gives the following.

Theorem 7. *We have*

$$\liminf_{x \to \infty} \frac{V_1(x)}{V(x)} \leq 10^{-5,000,000,000}.$$

The proof of these two theorems motivates another classification of totients. Let $V(x; k)$ be the number of totients up to x, all of whose pre-images are divisible by k. A trivial corollary to the proof of Theorem 2 is

Theorem 8. *If d is a totient, all of whose pre-images are divisible by k, then*

$$V(x; k) \gg_\varepsilon d^{-1-\varepsilon} V(x).$$

Thus, for each k, either $V(x; k) = 0$ for all x or $V(x; k) \gg_k V(x)$.

In the 1950s, Sierpiński conjectured that all multiplicities $k \geq 2$ are possible (see [9, 28]), and in 1961, Schinzel [29] deduced this conjecture from his well-known Hypothesis H. Schinzel's Hypothesis H [30], a generalization of Dickson's Prime k-tuples Conjecture [6], states that any set of polynomials $F_1(n), \ldots, F_k(n)$, subject to certain restrictions, are simultaneously prime for infinitely many n. Using a much simpler, iterative argument, we show that Sierpiński's Conjecture follows from the Prime k-tuples Conjecture.

Theorem 9. *The Prime k-tuples Conjecture implies that for each $k \geq 2$, there is a number d with $A(d) = k$.*

While certainly true, a proof of Hypothesis H remains elusive even in the simple case where $k = 2$ and F_1, F_2 are linear polynomials (generalized twin primes). However, combining the iterative process used to prove Theorem 9 with the theory of "almost primes", it is possible to prove Sierpińsk's Conjecture unconditionally. Details will appear in a forthcoming paper [14].

4. Establishing Theorems 1 and 2 requires a determination of what a "normal" totient looks like. This will initially take the form of a series of linear inequalities in the prime factors of a pre-image of a totient. An analysis of these inequalities reveals the normal sizes of the prime factors of a pre-image of a typical totient. To state our results, we first define

$$L_0 = L_0(x) = [2C(\log_3 x - \log_4 x)]. \tag{1.7}$$

In a simplified form, we show that for all but $o(V(x))$ totients $m \leq x$, every pre-image n satisfies

$$\log_2 q_i(n) \sim \varrho^i (1 - i/L_0) \log_2 x \quad (0 \leq i \leq L_0), \tag{1.8}$$

where $q_i(n)$ denotes the $(i + 1)$st largest prime factor of n. Recall (1.3) and let

$$g_0 = 1, \quad g_i = \sum_{j=1}^{i} a_j g_{i-j} \quad (i \geq 1). \tag{1.9}$$

We will prove in Section 3 that $\frac{1}{5} < \varrho^i g_i < \frac{1}{3}$ for all i.

Theorem 10. *Suppose x is large, and $L_0 = L_0(x)$. For $i \leq L_0 - 5$, define $\beta_i = \varrho^i(1 - i/L_0)$ and let $V(x; \mathscr{C})$ denote the number of totients $m \leq x$ with a pre-image n*

satisfying condition $\mathscr{C}$. If $\varepsilon \le i\varrho^i g_i / L_0$, then

$$V\left(x; \left|\frac{q_i(n)}{\beta_i \log_2 x} - 1\right| \ge \varepsilon\right)$$

$$\ll V(x) \exp\left\{-\frac{L_0(L_0 - i)}{2i}\varepsilon^2\left(1 + O\left(\frac{\varepsilon L_0}{i} + \frac{1}{\varepsilon}\sqrt{\frac{i}{L_0(L_0 - i)}}\log(L_0 - i)\right)\right)\right\}.$$

If $\frac{i}{5L_0} \le \varepsilon \le \frac{1}{2}$ and $i \le L_0/2$ then

$$V\left(x; \left|\frac{q_i(n)}{\beta_i \log_2 x} - 1\right| \ge \varepsilon\right)$$

$$\ll V(x) \exp\left\{-\frac{\varrho^i g_i}{1 - \varrho^i g_i}L_0\varepsilon + O(L_0\varepsilon^2 + \sqrt{L_0\varepsilon}\log(L_0\varepsilon))\right\}.$$

There are many ways of using Theorem 10 to construct a result concerning the simultaneous approximation of many of the prime factors of a normal totient. We prove a result, where the goal is to obtain near best approximations of the maximum number of prime factors. Here $\omega(m)$ denotes the number of distinct prime factors of m and $\Omega(m)$ denotes the number of prime factors of m counted with multiplicity.

Theorem 11. *Suppose $g(x)$ is an increasing function of x satisfying $g(x) = o(\log_3 x)$. For a given x, set $L_0 = L_0(x)$ and $\beta_i = \varrho^i(1 - i/L_0)$ for $0 \le i \le L_0$. Then the number of totients $m \le x$ with a pre-image n not satisfying*

$$\left|\frac{\log_2 q_i(n)}{\beta_i \log_2 x} - 1\right| \le \sqrt{\frac{i}{L_0(L_0 - i)}}\log(L_0 - i)\log g(x) \quad (1 \le i \le L_0 - g(x)) \quad (1.10)$$

and

$$L_0(x) - g(x) \le \omega(n) \le \Omega(n) \le L_0(x) + g(x)$$

is

$$\ll V(x)e^{-\frac{1}{4}\log^2 g(x)}.$$

In essence, Theorem 11 says that the set of $n \le x$ having about $L_0(x)$ prime factors distributed according to (1.10) generates almost all totients. It also says that for most totients, all of its pre-images are "virtually" square-free. The function $g(x)$ need not tend to infinity. Notice that the intervals in (1.10) are not only disjoint, but the gaps between them are rather large. In particular, this "discreteness phenomenon" means that for most totients $m \le x$, no pre-image n has any prime factors p in the intervals

$$0.999 \ge \frac{\log_2 p}{\log_2 x} \ge 0.543, \quad 0.542 \ge \frac{\log_2 p}{\log_2 x} \ge 0.295, \text{ etc.}$$

This should be compared to the distribution of the prime factors of a normal integer $n \leq x$ (e.g., Theorem 12 of [17]).

We also deduce the normal order of $\Omega(m)$ and $\omega(m)$ for totients m. If each prime $q_i(n)$ of a pre-image n is "normal" and (1.8) holds, then $\Omega(m)$ should be about

$$(1 + \varrho + \varrho^2 + \cdots) \log_2 x = \frac{\log_2 x}{1 - \varrho}.$$

Theorem 12. *Suppose* $\varepsilon = \varepsilon(x)$ *satisfies* $0 \leq \varepsilon \leq 0.8$. *Then*

$$\#\left\{ m \in \mathscr{V}(x) : \left| \frac{\Omega(m)}{\log_2 x} - \frac{1}{1 - \varrho} \right| \geq \varepsilon \right\} \ll V(x) \exp\{-K\varepsilon \log_3 x + O(\sqrt{\varepsilon \log_3 x})\},$$

where

$$K = \frac{2Ca_1(1 - \varrho)}{1 - (1 + a_1)\varrho} = 1.166277\ldots$$

Consequently, if $g(x) \to \infty$ *arbitrarily slowly, then almost all totients* $m \leq x$ *satisfy*

$$\left| \frac{\Omega(m)}{\log_2 x} - \frac{1}{1 - \varrho} \right| \leq \frac{g(x)}{\log_3 x}.$$

Moreover, the theorem holds with $\Omega(m)$ *replaced by* $\omega(m)$.

Corollary 13. *If either* $g(m) = \omega(m)$ *or* $g(m) = \Omega(m)$, *then*

$$\sum_{m \in \mathscr{V}(x)} g(m) = \frac{V(x) \log_2 x}{1 - \varrho} \left(1 + O\left(\frac{1}{\log_3 x} \right) \right).$$

By contrast, Erdős and Pomerance [12] showed that the average of $\Omega(\phi(n))$, taken over all $n \leq x$, is $\frac{1}{2}(\log_2 x)^2 + O((\log_2 x)^{3/2})$.

As the details of the proofs of these results are extremely complex and require very delicate estimating, we summarize the central ideas here. First, for most integers m, the prime divisors of m are "nicely distributed", meaning the number of prime factors of m lying between a and b is about $\log_2 b - \log_2 a$. This is a more precise version of the classical result of Hardy and Ramanujan [19] that most numbers m have about $\log_2 m$ prime factors. Take an integer n with prime factorization $p_0 p_1 \cdots$, where for simplicity we assume n is square-free, and $p_0 > p_1 > \cdots$. By sieve methods it can be shown that for most primes p, the prime divisors of $p - 1$ have the same "nice" distribution. If $p_0, p_1, \ldots$ are such "normal" primes, it follows that $\phi(n) = (p_0 - 1)(p_1 - 1) \cdots$ has about $\log_2 n - \log_2 p_1$ prime factors in $[p_1, n]$, about $2(\log_2 p_1 - \log_2 p_2)$ prime factors in $[p_2, p_1]$, and in general, $\phi(n)$ will have $k(\log_2 p_{k-1} - \log_2 p_k)$ prime factors in $[p_k, p_{k-1}]$. That is, n has k times as many prime factors in the interval $[p_k, p_{k-1}]$ as does a "normal" integer of its size. If n has many "large" prime divisors, then the prime factors of $m = \phi(n)$ will be much denser than normal, and the number, N_1, of such integers m will be "small". On the other hand, the number, N_2 of integers n with relatively few "large" prime factors is also "small". Our objective then is to precisely define these concepts of "large" and "small" so as to minimize $N_1 + N_2$.

The argument in [23] is based on the heuristic that a normal totient is generated from a number n satisfying

$$\log_2 q_i(n) \approx \varrho^i \log_2 x \tag{1.11}$$

for each i (compare with (1.8)). As an alternative to this heuristic, assuming all prime factors of a pre-image n of a totient are normal leads to consideration of a series of inequalities among the prime factors of n. We show that such n generate "most" totients. By mapping the L largest prime factors of n (excluding the largest) to a point in $\mathbb{R}^L$, the problem of determining the number of such n up to x reduces to the problem of finding the volume of a certain region of $\mathbb{R}^L$, which we call the fundamental simplex. Our result is roughly

$$V(x) \approx \frac{x}{\log x} \max_L T_L (\log_2 x)^L,$$

where T_L denotes the volume of the simplex. It turns out that the maximum occurs at $L = L_0(x) + O(1)$. Careful analysis of these inequalities reveals that "most" of the integers n for which they are satisfied satisfy (1.8). Thus, the heuristic (1.11) gives numbers n for which the smaller prime factors are too large. The crucial observation that the Lth largest prime factor ($L = L_0 - 1$) satisfies $\log_2 p_L \approx \frac{1}{L} \varrho^L \log_2 x$ is a key to determining the true order of $V(x)$.

In Section 2 we define "normal" primes and show that most primes are "normal". The set of linear inequalities used in the aforementioned heuristic are defined and analyzed in Section 3. The principal result is a determination of the volume of the simplex defined by the inequalities, which requires excursions into linear algebra and complex analysis. Section 4 is devoted to proving the upper bound for $V(x)$, and in Section 5, the lower bound for $V_k(x)$ is deduced. Together these bounds establish Theorems 1 and 2, as well as Theorems 4, 5 and 8 as corollaries. The distribution of the prime factors of a pre-image of a typical totient are detailed in Section 6, culminating in the proof of Theorems 10, 11, 12 and Corollary 13.

In Section 7, we summarize the computations giving Theorem 6 and present very elementary proofs of Theorems 7 and 9. Section 8 is concerned with the behavior of the ratios $V_k(x)/V(x)$. We prove Theorem 3 and discuss some consequences. We are unable to determine the behavior for any specific k (other than what Theorem 2 gives), but can say a few things about the behavior as $k \to \infty$. Lastly, Section 9 outlines an extension of all of these results to more general multiplicative arithmetic functions such as $\sigma(n)$, the sum of divisors function. Specifically, we prove

Theorem 14. *Suppose $f : \mathbb{N} \to \mathbb{N}$ is a multiplicative function satisfying*

$$\{f(p) - p : p \text{ prime}\} \text{ is a finite set not containing } 0, \tag{1.12}$$

$$\sum_{\substack{h \geq 16 \\ h \text{ square-full}}} \frac{\varepsilon(h)}{f(h)} \ll 1, \quad \varepsilon(h) = \exp\{\log_2 h (\log_3 h)^{20}\}. \tag{1.13}$$

Then the analogs of Theorems 1–4, 8, 10–13 and 15 hold with $f(n)$ replacing $\phi(n)$, with the exception of the dependence on d in Theorems 2 and 8, which may be different.

Theorem 15 depends on the definition of the fundamental simplex, and is not stated until Section 6.

2. Preliminary lemmata

Let $P^+(n)$ denote the largest prime factor of n and let $\Omega(n, U, T)$ denote the total number of prime factors p of n such that $U \leq p \leq T$, counted according to multiplicity. Constants implied by the Landau $O-$ and Vinogradov $\ll -$ and $\gg -$ symbols are absolute unless otherwise specified, and $c_1, c_2, \ldots$ will denote absolute constants, not depending on any parameter. Symbols in boldface type indicate vector quantities.

A small set of additional symbols will have constant meaning throughout this paper. These include the constants $\mathscr{V}, \varrho, C, D, a_i, g_i$, defined respectively in (1.1), (1.4), (1.5), (1.6), (1.3) and (1.9) as well as the constants $\mathscr{S}_L, T_L, \gamma_i, g_i^*$ and h_i^*, defined later in Section 3. Also included are the following functions: the functions defined in (1.1), $L_0(x)$ (1.7), $F(x)$ (1.3); the functions $Q(\alpha), \delta(U, T)$ and $W(x)$ defined respectively in Lemma 2.1, (2.4) and (2.6) below; and $\mathscr{S}_L(\xi), T_L(\xi), \mathscr{R}_L(\xi; x), R_L(\xi; x)$ and $x_i(n; x)$ defined in Section 3. Other variables are considered "local" and may change meaning from section to section, or from lemma to lemma.

A crucial tool in the proofs of Theorems 1 and 2 is a more precise version of the result from [23] that for most primes p, the larger prime factors of $p - 1$ are nicely distributed (see Lemma 2.8 below). We begin with three basic lemmas.

Lemma 2.1. *If $z > 0$ and $0 < \alpha < 1 < \beta$ then*

$$\sum_{k \leq \alpha z} \frac{z^k}{k!} < e^{(1-Q(\alpha))z},$$

$$\sum_{k \geq \beta z} \frac{z^k}{k!} < e^{(1-Q(\beta))z},$$

where $Q(\lambda) = \int_1^\lambda \log t \, dt = \lambda \log(\lambda) - \lambda + 1$.

Proof: We have

$$\sum_{k \leq \alpha z} \frac{z^k}{k!} = \sum_{k \leq \alpha z} \frac{(\alpha z)^k}{k!} \left(\frac{1}{\alpha}\right)^k \leq \left(\frac{1}{\alpha}\right)^{\alpha z} \sum_{k \leq \alpha z} \frac{(\alpha z)^k}{k!} < \left(\frac{e}{\alpha}\right)^{\alpha z} = e^{(1-Q(\alpha))z}.$$

The second inequality follows in the same way.   □

Lemma 2.2. *For $\alpha > 1$, the number of integers $n \leq x$ for which $\Omega(n) \geq \alpha \log_2 x$ is*

$$\ll_\alpha x(\log x)^{-Q^*(\alpha)},$$

where

$$Q^*(\alpha) = \begin{cases} Q(\alpha) & (\alpha < 2), \\ \alpha \log 2 - 1 & (\alpha \geq 2). \end{cases}$$

Proof: This can be deduced from the Theorems in Chapter 0 of [17]. □

Lemma 2.3. *The number of $n \leq x$ divisible by a number $m \geq \exp\{(\log_2 x)^2\}$ with $P^+(m) \leq m^{1/\log_2 x}$ is $\ll x/\log^2 x$.*

Proof: Let $\Psi(x, y)$ denote the number of integers $\leq x$ which have no prime factors $> y$. For x large, standard estimates of $\Psi(x, y)$ ([20], Theorem 1.1 and Corollary 2.3) give

$$\Psi\left(z, z^{1/\log_2 x}\right) \ll z \exp\{-(\log_2 x \log_3 x)/2\}$$

uniformly for $z \geq \exp\{(\log_2 x)^2\}$. The lemma follows by partial summation. □

For the statement of the basic sieve results, we adopt the notation of [16].

$$\mathscr{A} = \text{a finite set of integers,}$$
$$\mathscr{A}_d = \{a \in \mathscr{A} : d \mid a\},$$
$$\mathscr{P} = \text{a finite set of primes,}$$
$$P(z) = \prod_{\substack{p \in \mathscr{P} \\ p \leq z}} p,$$
$$v(d) = \text{a multiplicative function defined on } \{d : d \mid P(z)\},$$
$$X = \text{an approximation to } |\mathscr{A}|,$$
$$r_d = |\mathscr{A}_d| - \frac{v(d)}{d} X,$$
$$W(z) = \prod_{p \mid P(z)} \left(1 - \frac{v(p)}{p}\right),$$
$$S(\mathscr{A}, \mathscr{P}, z) = |\{a \in \mathscr{A} : (a, P(z)) = 1\}|.$$

We also need to impose certain condtions on the function $v(d)$.

$$\text{(R)} \qquad\qquad |r_d| \leq v(d),$$
$$\text{(Ω_1)} \qquad\qquad 0 \leq \frac{v(p)}{p} \leq 1 - 1/A_1,$$
$$\text{($\Omega_2(\kappa)$)} \qquad \sum_{w < p \leq z} \frac{v(p) \log p}{p} \leq \kappa \log \frac{z}{w} + A_2, \quad 2 \leq w \leq z.$$

Lemma 2.4. (R), (Ω_1), ($\Omega_2(\kappa)$) : *Let $z \leq X$ and put $u = \log X/\log z$. Then*

$$S(\mathscr{A}, \mathscr{P}, z) = XW(z)\{1 + O(e^{-\sqrt{\log X}} + e^{-u(\log u - \log\log 3u - \log \kappa - 2)})\}.$$

The implied constant depends on A_1 and A_2.

Proof: This is Theorem 2.5 of [16], commonly referred to as the Fundamental Lemma. □

Lemma 2.5. (Ω_1), $(\Omega_2(\kappa))$: We have

$$S(\mathscr{A}, \mathscr{P}, z) \ll_\kappa X W(z) + \sum_{\substack{d < z^2 \\ d \mid P(z)}} 3^{\omega(d)} |r_d|.$$

The implied constant depends on A_1 and A_2.

Proof: This is Theorem 4.1 of [16] and is a consequence of Selberg's sieve. □

Lemma 2.6. Suppose $a_1, \ldots, a_h$ are positive integers and $b_1, \ldots, b_h$ are integers such that

$$E = \prod_{i=1}^h a_i \prod_{1 \le i < j \le h} (a_i b_j - a_j b_i) \ne 0.$$

Suppose further that $\log E \le C (\log x)^3$ and $z \le x$. Then

$$\#\left\{ n \le x : P^+\left(\prod (a_i n + b_i) \right) \ge z \right\} \ll x (\log \log x)^h (\log z)^{-h},$$

where the implied constant may depend on h, C.

Proof: This follows from Lemma 2.5 (see also [16, Theorem 4.2]). □

Next, we examine the normal multiplicative structure of the shifted primes $p - 1$. Let $M(U, T, \delta, x)$ be the number of primes $p \le x$ with

$$|\Omega(p - 1, U, T) - (\log_2 T - \log_2 U)| \ge \delta \log_2 T.$$

Lemma 2.7. Uniformly in $\delta \le \frac{3}{4}$, $\log^2 x \le U < T \le x$, we have

$$M(U, T, \delta, x) \ll \frac{x (\log_2 x)^2}{\log x} (\log U \log T)^{-\delta^2/3}.$$

Proof: Assume x is sufficiently large and assume $\delta \ge \sqrt{3/\log_2 x}$, for otherwise the lemma is trivial. The number of primes $p \le x$ with $p - 1$ divisible by the square of a prime $q \ge \log^2 x$ is

$$\le \sum_{q \ge \log^2 x} \frac{x}{q^2} \ll \frac{x}{\log^2 x},$$

and, by Lemma 2.3, the number with $P^+(p-1) \le x^{1/\log_2 x}$ is $O(x/\log^2 x)$. Let N_k denote the number of remaining primes $p \le x$ satisfying $\Omega(p-1, U, T) = k$. Write $p-1 = qa$, where $q = P^+(p-1) \ge x^{1/\log_2 x}$. By Lemma 2.5,

$$
N_k \ll \sum_a \frac{x}{\phi(a)\log^2(x/a)}
$$

$$
\ll \frac{x(\log_2 x)^2}{\log^2 x} \sum_a \frac{1}{\phi(a)},
$$

where the sum is on $a \le x$ with either k or $k-1$ prime factors in $[U, T]$. Now let $u = \log_2 U$ and $t = \log_2 T$. Since x is sufficiently large,

$$
\sum_{U \le p \le T} \frac{1}{p-1} = t - u + O(1/\log U) \le t - u + 1 =: z.
$$

Set $j = k$ or $j = k - 1$. Since the primes $q \ge U$ dividing $p - 1$ occur to the first power, we have

$$
\sum_{\substack{a \le x \\ \Omega(a,U,T)=j}} \frac{1}{\phi(a)} \le \prod_{\substack{p \le x \\ p \notin [U,T]}} \left(1 + \frac{p}{(p-1)^2}\right) \frac{1}{j!} \left(\sum_{U < p \le T} \frac{1}{p-1}\right)^j
$$

$$
\ll \log x \left(\frac{\log U}{\log T}\right) \frac{z^j}{j!}.
$$

Summing on k gives

$$
M(U, T, \delta, x) \le \sum_{k \le t-u-\delta t} N_k + \sum_{k \ge t-u+\delta t} N_k + O(x/\log^2 x)
$$

$$
\ll \frac{x(\log_2 x)^2}{\log x} \frac{\log U}{\log T} \left(\sum_{k \le \alpha z} \frac{z^k}{k!} + \sum_{k \ge \beta z} \frac{z^k}{k!}\right),
$$

where

$$
\alpha = 1 - \frac{\delta t + 1}{z}, \qquad \beta = 1 + \frac{\delta t - 2}{z}.
$$

With x large, $\delta t \ge \log^2 x / \sqrt{\log_2 x} \ge 4$, so $\beta > 1$. Since $Q(1 - \lambda) > Q(1 + \lambda)$ when $\lambda > 0$, it follows that $Q(\beta) < Q(\alpha)$ provided $\alpha \ge 0$. In any case, Lemma 2.1 implies

$$
M(U, T, \delta, x) \ll \frac{x(\log_2 x)^2}{\log x} e^{-Q(\beta)z}. \tag{2.1}
$$

To estimate $-Q(\beta)z$, set

$$
\lambda = \frac{\delta t - 2}{t + 1},
$$

so that $\beta > 1 + \lambda$ and $\lambda = \delta + O(1/t)$. By the integral representation of $Q(\beta)$,

$$Q(\beta) \geq Q(1 + \lambda) + (\beta - (1 + \lambda)) \log(1 + \lambda) = -\lambda + \beta \log(1 + \lambda).$$

Using the fact that $0 \leq \delta \leq \frac{3}{4}$, we obtain

$$
\begin{aligned}
-Q(\beta)z &\leq \lambda z - \beta z \log(1 + \lambda) \\
&\leq \lambda(t - u) - (t - u + \delta t) \log(1 + \delta) + O(1) \\
&= t(\delta - \log(1 + \delta) - \delta \log(1 + \delta)) - u(\delta - \log(1 + \delta)) + O(1) \\
&\leq -\delta^2(u/3 + t/3) + O(1).
\end{aligned}
$$

The lemma now follows from (2.1). □

When $S \geq 2$, a prime p is said to be S-normal if

$$\Omega(p - 1, 1, S) \leq 2 \log_2 S \tag{2.2}$$

and for every pair of numbers (U, T) with $S \leq U < T \leq p - 1$ we have

$$|\Omega(p - 1, U, T) - (\log_2 T - \log_2 U)| < \delta(U, T) \log_2 T, \tag{2.3}$$

where

$$\delta(U, T) = \sqrt{\frac{\log_2 S}{2(\log_2 U + \log_2 T)}}. \tag{2.4}$$

Note that if $p < S$, only (2.2) is applicable.

Lemma 2.8. *Uniformly in $S \geq 2$, $x \geq e^e$, the number of primes $p \leq x$ which are not S-normal is*

$$\ll \frac{x(\log_2 x)^4}{\log x} (\log S)^{-1/6}.$$

Proof: Assume x is sufficiently large and $S \geq \log^2 x$, otherwise the lemma is trivial. By Lemmas 2.2 and 2.3, the number of primes $p \leq x$ with either $P^+(p - 1) \leq x^{1/\log_2 x}$, $\Omega(p - 1) \geq 10 \log_2 x$ or $p - 1$ divisible by the square of a prime $q \geq S$ is $O(x/\log^2 x)$. We first deal with condition (2.2). Let N_k be the number of remaining primes $p \leq x$ with $\Omega(p - 1, 1, S) = k$. Applying Lemma 2.5 as in the proof of Lemma 2.7 gives

$$
\begin{aligned}
\sum_{k \geq 2 \log_2 S} N_k &\ll \frac{x(\log_2 x)^2}{\log^2 x} \sum_{\substack{a \leq x \\ \Omega(a,1,S) \geq 2 \log_2 S - 1}} \frac{1}{\phi(a)} \\
&\ll \frac{x(\log_2 x)^2}{\log^2 x} (3/2)^{-(2 \log_2 S - 1)} \sum_{a \leq x} \frac{(3/2)^{\Omega(a,1,S)}}{\phi(a)}.
\end{aligned}
$$

Now

$$\sum_{a \leq x} \frac{(3/2)^{\Omega(a,1,S)}}{\phi(a)} \leq \prod_{S < p \leq x} \left(1 + \frac{p}{(p-1)^2}\right) \prod_{p < S} \left(1 + \frac{3p}{(p-1)(2p-3)}\right)$$
$$\ll \log x (\log S)^{1/2}.$$

Therefore the number of primes $p \leq x$ not satisfying (2.2) is

$$O\left(\frac{x(\log_2 x)^2}{\log x}(\log S)^{-0.31}\right).$$

When $S \leq x$ we next bound the number of remaining primes not satisfying (2.3) for a discrete set of pairs (U, T). Set $\xi = 1/10$ and

$$\delta'(U, T) = \delta(U, T)\left(1 - \frac{1}{\log_2 S}\right).$$

For integral k, define $t_k = \log_2 x - \xi k$ and $T_k = \exp \exp t_k$. When

$$0 \leq k < j \leq [(\log_2 x - \log_2 S)/\xi] + 1, \tag{2.5}$$

Lemma 2.7 gives

$$M(T_j, T_k, \delta'(T_j, T_k), x) \ll \frac{x(\log_2 x)^2}{\log x}(\log T_j \log T_k)^{-\delta'(T_j, T_k)^2/3}$$
$$\ll \frac{x(\log_2 x)^2}{\log x}(\log S)^{-1/6}.$$

Thus

$$\sum_{(2.5)} M(T_j, T_k, \delta'(T_j, T_k), x) \ll \frac{x(\log_2 x)^4}{\log x}(\log S)^{-1/6}.$$

Finally, we show by an interpolation argument that if $p - 1$ is not counted in any of the $M(T_j, T_k, \delta'(T_j, T_k), x)$, then (2.3) is satisfied for all permissible U, T. Suppose $s \leq U < T \leq x$ and set $t = \log_2 T$, $u = \log_2 U$, $k = [(\log_2 x - t)/\xi]$ and $j = [(\log_2 x - u)/\xi]$. Then

$$T_{j+1} < U \leq T_j, \quad T_{k+1} < T \leq T_k.$$

We also have

$$\delta'(T_{j+1}, T_k) \leq \left(1 - \frac{1}{\log_2 S}\right)\sqrt{1 + \frac{\xi}{2\log_2 S}}\delta(U, T) \leq \left(1 - \frac{1}{2\log_2 S}\right)\delta(U, T)$$

and the same upper bound for $\delta'(T_j, T_{k+1})$. From the inequality

$$\frac{t\delta(U, T)}{2\log_2 S} \geq \frac{1}{4}\sqrt{\frac{t}{\log_2 S}} \geq \frac{1}{4},$$

it follows that

$$\Omega(p-1, U, T) \leq \Omega(p-1, T_{j+1}, T_k) \leq t_k - t_{j+1} + t_k \delta'(T_{j+1}, T_k)$$
$$\leq t - u + 2\xi + (t + \xi)(1 - 1/(2\log_2 S))\delta(U, T)$$
$$\leq t - u + t\delta(U, T).$$

For the lower bound, suppose $k \geq j + 1/\xi$, for otherwise $t - u < 1$ and the lower bound in (2.3) is trivial. We then have

$$\Omega(p-1, U, T) \geq \Omega(p-1, T_j, T_{k+1}) \geq t_{k+1} - t_j - t_{k+1}\delta'(T_j, T_{k+1})$$
$$\geq t - u - 2\xi - t(1 - 1/(2\log_2 S))\delta(U, T)$$
$$\geq t - u - t\delta(U, T). \qquad \square$$

Lemma 2.9. *The number of $m \in \mathcal{V}(x)$ for which either $d^2 \mid m$ or $d^2 \mid n$ for some $n \in \phi^{-1}(m)$ and $d > Y$ is $O(x \log_2 x/Y)$.*

Proof: If $\phi(n) = m \leq x$, then from the standard result $n/\phi(n) \ll \log_2 n$, we have $n \ll x \log_2 x$. Summing over all possible d gives the result. $\qquad \square$

Now define

$$W(x) = \max_{2 \leq y \leq x} \frac{V(y) \log y}{y}. \qquad (2.6)$$

Lemma 2.10. *The number of $m \in \mathcal{V}(x)$ for which some $n \in \phi^{-1}(m)$ is divisible by a prime which is not S-normal is*

$$O\left(\frac{xW(x)(\log_2 x)^5}{\log x}(\log S)^{-1/6}\right).$$

Proof: Since $W(x) \gg 1$, we may suppose $S \geq \log^3 x$, for otherwise the lemma is trivial. Suppose p is a prime divisor of n for some $n \in \phi^{-1}(m)$. If $n = n'p$ then either $\phi(n) = (p-1)\phi(n')$ or $\phi(n) = p\phi(n')$, so in either case $\phi(n') \leq x/(p-1)$. Let $G(t)$ denote the number of primes $p \leq t$ which are not S-normal. By Lemma 2.8, the number of m in question is at most

$$\sum_p V\left(\frac{x}{p-1}\right) \ll \sum_p \frac{xW(x/(p-1))}{(p-1)\log(x/p)}$$
$$\ll xW(x) \int_2^{x/2} \frac{G(t)\,dt}{t^2 \log(x/t)}$$
$$\ll xW(x)(\log_2 x)^4(\log S)^{-1/6} \int_2^{x/2} \frac{dt}{t \log t \log(x/t)}$$
$$\ll \frac{xW(x)(\log_2 x)^5}{\log x}(\log S)^{-1/6}. \qquad \square$$

3.   The fundamental simplex

Suppose $\boldsymbol{\xi} = (\xi_0, \ldots, \xi_{L-2})$ is a vector in $\mathbb{R}^{L-1}$ satisfying either

$$1 \le \xi_i \le 1.1 \quad (0 \le i \le L-2) \tag{3.1}$$

or

$$0.9 \le \xi_i \le 1 \quad (0 \le i \le L-2). \tag{3.2}$$

Recall (1.3) and let $\mathscr{S}_L(\boldsymbol{\xi})$ to be the set of points $x_1, \ldots, x_L$ in $\mathbb{R}^L$ satisfying

$$0 \le x_L \le \cdots \le x_1 \le 1 \tag{3.3}$$

and

$$
\begin{aligned}
(I_0) \qquad & a_1 x_1 + a_2 x_2 + \cdots + a_L x_L \le \xi_0, \\
(I_1) \qquad & a_1 x_2 + a_2 x_3 + \cdots + a_{L-1} x_L \le \xi_1 x_1, \\
& \qquad\qquad \vdots \\
(I_{L-2}) \qquad & \qquad\qquad\qquad a_1 x_{L-1} + a_2 x_L \le \xi_{L-2} x_{L-2}.
\end{aligned}
$$

Define

$$T_L(\boldsymbol{\xi}) = \mathrm{Vol}(\mathscr{S}_L(\boldsymbol{\xi})).$$

For convenience, let $\mathscr{S}_L = \mathscr{S}_L((1, 1, \ldots, 1))$ (the "fundamental simplex") and $T_L = \mathrm{Vol}(\mathscr{S}_L)$. For a natural number n, let $q_0(n) \ge q_1(n) \ge \ldots$ denote the prime factors of n. When $i \ge \Omega(n)$, set $q_i(n) = 1$. For a fixed x, let

$$
x_i(n; x) =
\begin{cases}
\dfrac{\log_2 q_i(n)}{\log_2 x} & (i < \Omega(n),\, q_i(n) > 2) \\[2mm]
0 & (i \ge \Omega(n) \text{ or } q_i(n) = 2).
\end{cases}
\tag{3.4}
$$

Let $\mathscr{R}_L(\boldsymbol{\xi}; x)$ denote the set of integers n with $\Omega(n) \le L$ and

$$(x_0(n; x), x_1(n; x), \ldots, x_{L-1}(n; x)) \in \mathscr{S}_L(\boldsymbol{\xi}).$$

For large x set

$$R_L(\boldsymbol{\xi}; x) = \sum_{n \in \mathscr{R}_L(\boldsymbol{\xi}; x)} \frac{1}{\phi(n)}. \tag{3.5}$$

Heuristically, the sum $R_L(\boldsymbol{\xi}; x)$ can be related to $T_L(\boldsymbol{\xi})$ as follows. Let

$$f(z) = \sum_{\log_2 p \le z \log_2 x} \frac{1}{p} = z \log_2 x + O(1).$$

Then

$$R_L(\boldsymbol{\xi}; x) \approx \int_{\mathscr{S}_L(\boldsymbol{\xi})} df(x_1) \cdots df(x_L) \approx (\log_2 x)^L T_L(\boldsymbol{\xi}).$$

Establishing $R_L(\boldsymbol{\xi}; x) \asymp (\log_2 x)^L T_L(\boldsymbol{\xi})$ (Lemma 3.12) is not a simple task, however.

To simplify our later work, we first relate $\mathscr{S}_L(\xi)$ to $\mathscr{S}_L$. The next lemma follows immediately from the definition of $\mathscr{S}_L(\xi)$.

Lemma 3.1. *Let* $\xi_{L-1} = 1$. *If* (3.1) *holds and* $\mathbf{x} \in \mathscr{S}_L(\xi)$, *then* $\mathbf{y} \in \mathscr{S}_L$, *where* $y_i = (\xi_0 \cdots \xi_{i-1})^{-1} x_i$. *If* (3.2) *holds and* $\mathbf{y} \in \mathscr{S}_L$, *then* $\mathbf{x} \in \mathscr{S}_L(\xi)$, *where* $x_i = (\xi_0 \cdots \xi_{i-1}) y_i$.

Define

$$H(\xi) = \xi_0^L \xi_1^{L-1} \cdots \xi_{L-2}^2.$$

Corollary 3.2. *We have*

$$T_L \le T_L(\xi) \le H(\xi) T_L$$

when (3.1) *holds and*

$$H(\xi) T_L \le T_L(\xi) \le T_L$$

when (3.2) *holds.*

In applications, $H(\xi)$ will be close to 1, so we concentrate on bounding T_L.

Lemma 3.3. *We have*

$$T_L \asymp \frac{\varrho^{L(L+3)/2}}{L!} (F'(\varrho))^L.$$

Before proceeding to the proof, we note the following immediate corollary.

Corollary 3.4. *If* $H(\xi) \asymp 1$, *then*

$$T_L(\xi) \asymp \frac{\varrho^{L(L+3)/2}}{L!} (F'(\varrho))^L.$$

Furthermore, if $L = 2C(\log_3 x - \log_4 x) - \Psi$, *where* $0 \le \Psi \ll \sqrt{\log_3 x}$, *then*

$$(\log_2 x)^L T_L(\xi) = \exp\{C(\log_3 x - \log_4 x)^2 + D\log_3 x - (D + 1/2 - 2C)\log_4 x$$
$$- \Psi^2/4C - (D/2C - 1)\Psi + O(1)\}.$$

If $L = [2C(\log_3 x - \log_4 x)] - \Psi > 0$, *then*

$$(\log_2 x)^L T_L(\xi) \ll \exp\{C(\log_3 x - \log_4 x)^2 + D\log_3 x - (D + 1/2 - 2C)\log_4 x$$
$$- \Psi^2/4C - (D/2C - 1)\Psi\}.$$

Proof: The second and third parts follows from (1.5), (1.6) and Stirling's formula. □

Below is the primary tool needed for Lemma 3.3.

Lemma 3.5. *Suppose* $\mathbf{v}_0, \mathbf{v}_1, \ldots, \mathbf{v}_L$ *are vectors in* $\mathbb{R}^L$, *any* L *of which are linearly independent. Suppose further that*

$$\mathbf{v}_0 + \sum_{i=1}^{L} b_i \mathbf{v}_i = 0, \tag{3.6}$$

where $b_i > 0$ *for every* i. *Also suppose* $\alpha > 0$. *The volume,* V, *of the region defined by*

$$\{\mathbf{v}_i \cdot \mathbf{x} \le 0 \, (1 \le i \le L), \mathbf{v}_0 \cdot \mathbf{x} \le \alpha\}$$

is

$$V = \frac{\alpha^L}{L!(b_1 b_2 \cdots b_L)|\det(\mathbf{v}_1, \ldots, \mathbf{v}_L)|}.$$

Proof: We may assume without loss of generality that $\alpha = b_1 = b_2 = \cdots = b_L = 1$, for the general case follows by suitably scaling the vectors $\mathbf{v}_i$. Aside from the point $\mathbf{0}$, the other vertices of the simplex are $\mathbf{p}_1, \ldots, \mathbf{p}_L$, where $\mathbf{p}_i$ is the unique vector satisfying

$$\mathbf{p}_i \cdot \mathbf{v}_j = 0 \quad (1 \le j \le L, j \ne i);$$
$$\mathbf{p}_i \cdot \mathbf{v}_0 = 1. \tag{3.7}$$

Taking the dot product of $\mathbf{p}_i$ with each side of (3.6) yields

$$\mathbf{v}_i \cdot \mathbf{p}_i = -1, \tag{3.8}$$

so $\mathbf{p}_i$ lies in the region $\{\mathbf{v}_i \cdot \mathbf{x} \le 0\}$. The given region is thus an L-dimensional "hypertetrahedron" with volume $|\det(\mathbf{p}_1, \ldots, \mathbf{p}_L)|/L!$. From (3.8) we have

$$(\mathbf{p}_1, \ldots, \mathbf{p}_L)(\mathbf{v}_1, \ldots, \mathbf{v}_L)^T = -I,$$

where I is the identity matrix. Taking determinants gives the lemma.

Having $2L-2$ inequalities defining $\mathscr{S}_L$ creates complications estimating T_L, so we devise a scheme where only $L+1$ inequalities are considered at a time, thus allowing the use of Lemma 3.5. The numbers b_i occurring in that lemma will come from the sequence $\{g_i\}$ (see (1.9)), about which we need precise growth information.

Lemma 3.6. *We have* $g_i = \frac{\varrho^{-i}}{\varrho F'(\varrho)} - \gamma_i$, *where* $\gamma_1 > \gamma_2 > \cdots > 0$ *and* $\sum_{i=1}^{\infty} \gamma_i = O(1)$.

The proof of Lemma 3.6 is rather technical, based on contour integration and properties of $F(x)$ and its analytic continuation. The details are deferred to the end of this section.

Let $\mathbf{e}_1, \ldots, \mathbf{e}_L$ denote the standard basis for $\mathbb{R}^L$, so that $\mathbf{e}_i \cdot \mathbf{x} = x_i$. For $1 \le i \le L-1$, set $\mathbf{u}_i = \mathbf{e}_i - \mathbf{e}_{i+1}$. For $1 \le i \le L-2$, set

$$\mathbf{v}_i = -\mathbf{e}_i + \sum_{j=1}^{L-i} a_j \mathbf{e}_{i+j} \tag{3.9}$$

and also set

$$\mathbf{v}_0 = \sum_{j=1}^{L} a_j \mathbf{e}_j, \quad \mathbf{v}_{L-1} = -\mathbf{e}_{L-1} + \mathbf{e}_L, \quad \mathbf{v}_L = -\mathbf{e}_L.$$

For convenience, define

$$h_0 = -1, \quad h_i = g_{i-1} - g_i \quad (i \geq 1) \tag{3.10}$$

and

$$g_0^* = 1, \quad g_i^* = g_i + (1 - a_1)g_{i-1}, \quad h_0^* = -1, \quad h_i^* = h_i + (1 - a_1)h_{i-1}. \tag{3.11}$$

Thus, for $1 \leq j \leq L - 2$, inequality (I_j) may be abbreviated as $\mathbf{v}_j \cdot \mathbf{x} \leq 0$. Also, inequality (I_0) is equivalent to $\mathbf{v}_0 \cdot \mathbf{x} \leq 1$ and the inequality $x_{L-1} \geq x_L \geq 0$ is represented by $\mathbf{v}_{L-1} \cdot \mathbf{x} \leq 0$ and $\mathbf{v}_L \cdot \mathbf{x} \leq 0$. The sequence $\{g_i\}$ can be thought of as the inverse of the sequence $\{a_i\}$. A straightforward calculation gives

$$\mathbf{e}_i = -\sum_{j=i}^{L-1} g_{j-i} \mathbf{v}_j - g_{L-i}^* \mathbf{v}_L. \tag{3.12}$$

It follows that

$$\mathbf{v}_0 + \sum_{j=1}^{L-1} g_j \mathbf{v}_j + g_L^* \mathbf{v}_L = \mathbf{0} \tag{3.13}$$

and

$$\mathbf{u}_i = \sum_{j=i}^{L-1} h_{j-i} \mathbf{v}_j + h_{L-i}^* \mathbf{v}_L \quad (1 \leq i \leq L - 2). \tag{3.14}$$

We now have the ingredients for the proof of Lemma 3.3. The basic idea is that inequalities (I_0)–(I_{L-2}) by themselves determine a region which is only slightly larger than $\mathscr{S}_L$. In other words, the inequalities $1 \geq x_1 \geq \cdots \geq x_{L-1} \geq x_L$ are relatively insignificant. Set

$$\mathscr{S}_L^* = \{\mathbf{v}_0 \cdot \mathbf{x} \leq 1; \mathbf{v}_i \cdot \mathbf{x} \leq 0 \quad (1 \leq i \leq L)\},$$
$$u_0 = \mathscr{S}_L^* \cap \{x_1 \geq 1\}, \tag{3.15}$$
$$u_i = \mathscr{S}_L^* \cap \{x_i \leq x_{i+1}\} \quad (1 \leq i \leq L - 2),$$

and

$$T_L^* = \mathrm{Vol}(\mathscr{S}_L^*), \quad V_i = \mathrm{Vol}(u_i) \quad (0 \leq i \leq L - 2). \tag{3.16}$$

Evidently,

$$\mathscr{S}_L = \mathscr{S}_L^* \setminus \bigcup_{i=0}^{L-2} u_i,$$

so that

$$T_L^* - \sum_{i=0}^{L-2} V_i \leq T_L \leq T_L^*.$$ (3.17)

Since $|\det(v_1, \ldots, v_L)| = 1$, Lemma 3.5 and (3.13) give

$$T_L^* = \frac{1}{L!(g_1 \cdots g_{L-1})g_L^*}.$$ (3.18)

For the remaining argument, assume L is sufficiently large. We shall show that

$$\sum_{i=0}^{L-2} V_i < 0.6T_L^*,$$ (3.19)

which, combined with (3.17), (3.18) and Lemma 3.6, proves Lemma 3.3.

Lemma 3.7. *We have*

$$V_0 \ll (4/5)^L T_L^*.$$

Proof: The condition $x_1 \geq 1$ combined with $v_0 \cdot x \leq 1$ implies $u \cdot x \leq 0$, where

$$u = v_0 - e_1.$$ (3.20)

In preparation for the application of Lemma 3.5, we first express v_0 as a linear combination of $u, v_2, \ldots, v_L$. By (3.12) and (3.13),

$$u = \sum_{j=1}^{L-1}(g_{j-1} - g_j)v_j + (g_{L-1}^* - g_L^*)v_L.$$

The representation (3.13) is unique, thus

$$v_0 + \frac{a_1}{1 - a_1}u + \sum_{j=2}^{L} b_j v_j = 0,$$

where

$$b_j = g_j + \frac{a_1}{1 - a_1}(g_j - g_{j-1}) \quad (2 \leq j \leq L - 1),$$
$$b_L = g_L^* + \frac{a_1}{1 - a_1}(g_L^* - g_{L-1}^*).$$

In addition, $|\det(u, v_2, \ldots, v_L)| = (1 - a_1)$. Therefore, by Lemma 3.5,

$$V_0 \ll \frac{1}{L!(b_2 b_3 \cdots b_L)}.$$

Lemma 3.6 implies $b_j > (5/4)g_j$ for large j and the lemma now follows from (3.18). $\square$

Lemma 3.8. *For $i \geq 1$, we have*

$$V_i = \frac{1}{(1-a_1)L!(g_1 \cdots g_{i-1})A_i B_i} \prod_{j=i+2}^{L-1} \left(\frac{1}{g_j - B_i h_{j-i}} \right) \frac{1}{g_L^* - B_i h_{L-i}^*},$$

where

$$A_i = g_i + \frac{g_{i+1}}{1-a_1}, \quad B_i = \frac{g_{i+1}}{1-a_1}.$$

Proof: In u_i we have

$$a_1 x_{i+1} + a_2 x_{i+2} + \cdots + a_{L-i} x_L \leq x_i \leq x_{i+1},$$

which implies

$$x_{i+1} \geq \frac{1}{1-a_1}(a_2 x_{i+2} + \cdots + a_{L-i} x_L)$$
$$\geq x_{i+2} + a_2 x_{i+3} + \cdots + a_{L-i-1} x_L.$$

The condition $v_{i+1} \cdot x \leq 0$ is therefore implied by the other inequalities defining u_i, which means

$$V_i = \mathrm{Vol}\{v_0 \cdot x \leq 1; v_j \cdot x \leq 0 \ (1 \leq j \leq L, j \neq i+1); u_i \cdot x \leq 0\}.$$

We note $|\det(v_1, \ldots, v_i, u_i, v_{i+2}, \ldots, v_L)| = (1 - a_1)$. It is also easy to show from (3.14) that

$$0 = v_0 + \sum_{j=1}^{i-1} g_j v_j + A_i v_i + B_i u_i + \sum_{j=i+2}^{L} b_j v_j,$$

where

$$b_j = g_j - B_i h_{j-i} \quad (i+2 \leq j \leq L-1), \quad b_L = g_L^* - B_i h_{L-i}^*.$$

An application of Lemma 3.5 now completes the determination of V_i. □

We now deduce numerical estimates for V_i / T_L^*. Adopting the notation of Lemma 3.8 and using Lemma 3.6 gives

$$g_i + g_{i+1}/(1-a_1) > 4.003 g_i \quad (i \geq 1),$$
$$g_j - B_i h_{j-i} > 1.443 g_j \quad (i \text{ large, say } i \geq L - 100),$$
$$g_j - B_i h_{j-i} > 1.161 g_j \quad (i \geq 1),$$
$$g_L^* - B_i h_{L-i}^* > 1.443 g_L^* \quad (i < L - 2),$$
$$g_L^* - B_i h_2^* > 1.196 g_L^* \quad (i = L - 2).$$

From these bounds it follows that

$$V_{L-2}/T_L^* < (4.003 \times 1.196)^{-1},$$
$$V_i/T_L^* < (4.003 \times 1.443^{L-i-1})^{-1} \quad (L - 99 \leq i \leq L - 3),$$
$$V_i/T_L^* < (4.003 \times 1.443^{99} \times 1.161^{L-i-100})^{-1} \quad (1 \leq i \leq L - 100).$$

Therefore,

$$\sum_{i=0}^{L-2} V_i/T_L^* < O((4/5)^L) + \frac{1}{4.003}\left(\frac{1}{1.196} + \frac{1.443^{-2}}{1 - 1.443^{-1}} + \frac{1.443^{-99}}{(1 - 1/1.161)}\right) < 0.6,$$

which implies (3.19). This completes the proof of Lemma 3.3. □

Important in the study of $\mathscr{S}_L$ are both global bounds on the numbers x_i as well as a determination of where "most" of the volume lies.

Lemma 3.9. *Suppose* $\mathbf{x} \in \mathscr{S}_L$ *and let*

$$z_i = a_1 x_i + a_2 x_{i+1} + \cdots + a_{L-i+1} x_L \quad (1 \le i \le L).$$

Also set $x_0 = z_0 = 1$. *We have*

$$z_i \le \varrho z_{i-1} \quad (i \ge 1), \tag{3.21}$$

and

$$x_j \ge g_{i-j}^* x_i \quad (0 \le j \le i \le L). \tag{3.22}$$

Proof: First, define $d_i = g_{i+1} - g_i/\varrho$. From the definition of g_i, we have $d_0 = 1$, $d_1 = a_1 - 1/\varrho$ and

$$d_i = \sum_{j=1}^{i} a_j d_{i-j} \quad (i \ge 2).$$

Now replace parts of each multiple of x_j in the definition of z_{i-1} with z_{j+1}, since $x_j \ge z_{j+1}$. Lemma 3.6 gives $d_i > 0$ for $i \ge 2$, thus

$$z_{i-1} - z_i/\varrho \ge d_1 z_i + \sum_{j=0}^{L-i-1} (a_{j+2} - d_{j+2})x_{i+j} + \sum_{j=0}^{L-i} d_{j+2} z_{i+j+1} + d_{L-i+1} x_L$$

$$= \sum_{k=0}^{L-i} x_{i+k}\left(a_{k+2} - d_{k+2} + \sum_{j=0}^{k} a_{k-j+1} d_{j+1}\right) + d_{L-i+1} x_L$$

$$= d_{L-i+1} x_L.$$

To prove (3.22), fix i and note that the inequality is trivial for $j = i$ and $j = i - 1$. Assume it holds now for $j \ge k + 1$. Then by (I_k) and the induction hypothesis,

$$x_k \ge \sum_{h=1}^{L-k} a_h x_{k+h} \ge \sum_{h=1}^{i-k} a_h g_{i-k-h}^* x_i = g_{i-k}^* x_i.$$

□

Lemma 3.10. *If* $\mathbf{x} \in \mathscr{S}_L(\xi)$ *and* $H(\xi) \le 1.1$, *then* $x_j \le 3\varrho^{j-i} x_i$ *when* $i < j$ *and* $x_j < 3\varrho^j$ *for* $1 \le j \le L$.

Proof: Again set $x_0 = 1$. The maximum of ϱ^{-i}/g_i^* is $2.6211\ldots$, occurring at $i = 2$. When (3.2) holds, $\mathcal{S}_L(\xi) \in \mathcal{S}_L$, so Lemma 3.9 gives $x_j \leq 2.63\varrho^{j-i}x_i$. When (3.1) holds, by Lemmas 3.1 and 3.9,

$$x_j \leq 2.63(\xi_i \cdots \xi_{j-1})\varrho^{j-i}x_i \leq 3\varrho^{j-i}x_i. \qquad \square$$

Careful analysis of $\mathcal{S}_L$ reveals that most of the volume occurs with $x_i \approx \frac{L-i}{L}\varrho^i$ for each i, with the "standard deviation" from the mean increasing with i. This observation plays an important role in subsequent arguments. For now, we restrict our attention to the variable x_L, since results concerning the other variables will not be needed until Section 6. The next lemma shows that $x_L \approx \varrho^L/L$ for most of $\mathcal{S}_L$, a bound which is significantly smaller than the global upper bound given by Lemma 3.10.

Lemma 3.11. *Suppose $\beta < 1/g_L^*$. We have*

$$\mathrm{Vol}(\mathcal{S}_L \cap \{x_L \leq \beta\}) \ll T_L L\beta\varrho^{-L} \qquad (3.23)$$

and

$$\mathrm{Vol}(\mathcal{S}_L \cap \{x_L \geq \beta\}) \ll (1 - \beta g_L^*)^L T_L. \qquad (3.24)$$

Proof: Consider first $\mathbf{x} \in \mathcal{S}_L \cap \{x_L \leq \beta\}$. Since $(x_1, \ldots, x_{L-1}) \in \mathcal{S}_{L-1}$, the volume is $\leq \beta T_{L-1}$. Applying Lemma 3.3 now gives (3.23).

Next, suppose $\mathbf{x} \in \mathcal{S}_L \cap \{x_L \geq \beta\}$ and set $y_i = x_i - \beta g_{L-i}^*$ for each i. We have $y_{L-1} \geq y_L \geq 0$, $\mathbf{v}_j \cdot \mathbf{y} \leq 0$ for $1 \leq j \leq L-2$, and $\mathbf{v}_0 \cdot \mathbf{y} \leq 1 - \beta g_L^*$. The upper bound (3.24) follows from Lemmas 3.3 and 3.5. $\square$

We now return to the problem of relating $R_L(\xi; x)$ to $T_L(\xi)$. Recall the definition of $L_0(x)$ (1.7). For technical reasons, the lower bound for $V_k(x)$ will involve sums over a set which is slightly smaller than $\mathcal{R}_L(\xi; x)$.

Lemma 3.12. *If $1/(1000k^3) \leq \omega_{L_0-k} \leq 1/(10k^3)$ for $1 \leq k \leq L_0$, $\xi_i = 1 + \omega_i$ for each i, and $L \leq L_0$, then*

$$R_L(\xi; x) \ll (\log_2 x)^L T_L. \qquad (3.25)$$

If $1/(1000k^3) \leq \omega_{L_0-k} \leq 1/k^3$ and $\xi_i = 1 - \omega_i$ for each i, then there is an absolute constant M_1 so that whenever $M = [M_1 + 2C\log_3 P_0]$ and $L \leq L_0 - M$, we have

$$R_L^*(\xi; x) \gg (\log_2 x)^L T_L, \qquad (3.26)$$

where

$$R_L^*(\xi; x) = \sum_{n \in \mathcal{R}_L^*(\xi;x)} \frac{1}{\phi(n)},$$

and $\mathcal{R}_L^*(\xi; x)$ is the set of $n \in \mathcal{R}_L(\xi; x)$ with the additional restrictions

$$\log_2 q_{i-1}(n) \geq \frac{4}{\omega_i}(5/4)^{L-i} \quad (i \geq 1), \tag{3.27}$$

$$\log_2 q_i(n) \leq (1 - \omega_i)\log_2 q_{i-1}(n) \quad (i \geq 1), \tag{3.28}$$

$$q_{L-1}(n) \geq P_0. \tag{3.29}$$

Proof: Note that taken together, (3.25) and (3.26) imply that both $R_L(\xi; x)$ and $R_L^*(\xi; x)$ are $\asymp (\log_2 x)^L T_L$. The additional conditions defining $\mathcal{R}_L^*(\xi; x)$, which will be needed for the proof of the lower bound for $V_k(x)$, insure that none of the primes $q_i(n)$ are too small nor too close together.

The overall strategy is to bound $R_L(\xi; x)$ by upper and lower Riemann sums. For $1 \leq j \leq L$, let

$$\varepsilon_j = \frac{(5/4)^{L_0-j}}{\log_2 x}. \tag{3.30}$$

The constant $5/4$ was chosen because it is less than $\varrho^{-1/2}$. For a number n, write $q_i = q_{i-1}(n)$ and $x_i = x_{i-1}(n; x)$ for each i. Define $x_0 = 1$ and consider the inequality

$$a_1\varepsilon_{i+1} + \cdots + a_{L-i}\varepsilon_L \leq (\omega_i/8)x_i, \tag{3.31}$$

which basically says that x_i is not too small.

We attack (3.25) first. For $1 \leq j \leq L-2$, let P_j denote the set of $n \in \mathcal{R}_L(\xi; x)$ with (3.31) holding for $i < j$ but failing for $i = j$, and let U be the set of $n \in \mathcal{R}_L(\xi; x)$ with (3.31) satisfied for every $i \leq L-2$. Notice that (3.31) always holds for $i = 0$ if x is large enough. Evidently,

$$R_L(\xi; x) = \sum_{j=1}^{L-2}\sum_{n \in P_j}\frac{1}{\phi(n)} + \sum_{n \in U}\frac{1}{\phi(n)}. \tag{3.32}$$

If $n \in P_j$, then by (3.30) and (3.31)

$$x_j < y_j, \quad \text{where } y_j = 80\frac{\varepsilon_j}{\omega_j}. \tag{3.33}$$

Set $r = q_1 \cdots q_{j-1}$ and $s = q_j \cdots q_L$ (when $j = 1$ take $r = 1$). When $j \geq 3$, $r \in \mathcal{R}_L((\xi_0, \ldots, \xi_{j-3}); x)$ and each admissible $\mathbf{x} = (x_1, \ldots, x_{j-1})$ lies inside a box

$$B(\mathbf{m}) = \{m_i\varepsilon_i \leq x_i \leq (m_i + 1)\varepsilon_i, 1 \leq i \leq j-1\}, \tag{3.34}$$

where each m_i is a nonnegative integer. Since (3.31) holds for each $i < j$, it follows that

$$\cup B(\mathbf{m}) \subset \mathcal{S}_{j-1}^*(1 + 2\omega_0, \ldots, 1 + 2\omega_{j-3}),$$

where the * indicates that the inequalities $x_i \geq x_{i+1}$ are excluded from the definition (see (3.15)). By Corollary 3.2 and Lemma 3.3, the volume of this region is $\ll T_{j-1}^* \ll T_{j-1}$. Writing $E(z) = \exp\{(\log x)^z\}$, we have

$$\sum_r \frac{1}{\phi(r)} \leq \sum_{\mathbf{m}} \sum_{\mathbf{x} \in B(\mathbf{m})} \frac{1}{\phi(n)}$$

$$= \sum_{\mathbf{m}} \prod_{i=1}^{j-1} \left(\sum_{E(m_i \varepsilon_i) \leq p \leq E((m_i+1)\varepsilon_i)} \frac{1}{p-1} + O(p^{-2}) \right)$$

$$= \sum_{\mathbf{m}} (\varepsilon_1 \cdots \varepsilon_{j-1})(\log_2 x)^{j-1} \prod_{i=1}^{j-1} (1 + O((\varepsilon_i \log_2 x)^{-1}))$$

$$\ll (\log_2 x)^{j-1} \mathrm{Vol}(\mathscr{S}_{j-1}^*((1 + 2\omega_0, \ldots, 1 + 2\omega_{j-3})))$$

$$\ll (\log_2 x)^{j-1} T_{j-1}. \tag{3.35}$$

Also, if we define $T_0 = T_1 = 1$, inequality (3.35) also holds for $j = 1$ and $j = 2$. In addition, applying the same argument with $j - 1$ replaced with L gives

$$\sum_{n \in U} \frac{1}{\phi(n)} \ll (\log_2 x)^L T_L. \tag{3.36}$$

Crudely,

$$\sum_s \frac{1}{\phi(s)} \leq \left(\sum_{\frac{\log_2 p}{\log_2 x} \leq y_j} \frac{1}{p-1} \right)^{L-j+1} = (y_j \log_2 x + O(1))^{L-j+1}. \tag{3.37}$$

Let $h = L - j + 1$ and $k = L_0 - j$. By (3.35) and (3.37),

$$\sum_{n \in P_j} \frac{1}{\phi(n)} \leq \left(\sum_r \frac{1}{\phi(r)} \right) \left(\sum_s \frac{1}{\phi(s)} \right)$$

$$\ll (\log_2 x)^L T_L Q,$$

where

$$Q = \frac{T_{j-1}}{T_L} (\log_2 x)^{-h} (y_j \log_2 x + O(1))^h. \tag{3.38}$$

We claim that

$$Q \leq \exp\{h(-k/15 + 3\log k + O(1))\}. \tag{3.39}$$

From Lemma 3.3 and the definition of h and k, we have

$$\log Q = h \log(L y_j) + \frac{j(j-1) - L(L+1)}{2} \log \varrho + O(h)$$

$$= h(\log_4 x - \log_3 x + k \log(5/4) + 3 \log k) + \frac{hL - h^2/2}{2C} + O(h)$$

$$= hk(\log(5/4) - 1/4C) + 3h \log k + O(h),$$

from which (3.39) follows. We thus have

$$\sum_{n \in P_j} \frac{1}{\phi(n)} \ll (\log_2 x)^L T_L \exp\{h(-k/15 + O(\log k))\}.$$

The sum (on j) of the exponential factor is $O(1)$, and this completes the proof of (3.25).

Now suppose $\xi_i = 1 - \omega_i$ for each i and consider $\mathbf{x} \in \mathscr{S}'_L(\boldsymbol{\xi} - \boldsymbol{\omega})$, the subset of $\mathscr{S}_L(\boldsymbol{\xi} - \boldsymbol{\omega})$ with the additional restrictions $x_{i+1} \leq (1 - 2\omega_i)x_i$ for each i, (3.31) is satisfied for $1 \leq i \leq L - 2$ and $x_L \geq \varepsilon_L + \log_2 P_0 / \log_2 x$. By (3.31),

$$x_i \geq 5\varepsilon_i / \omega_i \quad (i \leq L - 2).$$

Consider any point $\mathbf{y}$ lying inside the box $B(\mathbf{m})$ containing $\mathbf{x}$ (see (3.34)). We have $y_L \geq \log_2 P_0 / \log_2 x$, $y_i \geq 4\varepsilon_i / \omega_i$ for each i, and

$$y_{i+1} \leq x_{i+1} + \varepsilon_{i+1} \leq (1 - 2\omega_i)x_i + \varepsilon_i \leq (1 - 2\omega_i)(y_i + \varepsilon_i) + \varepsilon_i$$

$$\leq (1 - \omega_i)y_i - \omega_i y_i + 2\varepsilon_i \leq (1 - \omega_i)y_i.$$

Furthermore, when $i \leq L - 2$ we have

$$a_1 y_{i+1} + \cdots + a_{L-i} y_L \leq (\xi_i - \omega_i + \omega_i/4)x_i$$

$$\leq \xi_i y_i.$$

Therefore, any n with $\mathbf{x}(n) \in B(\mathbf{m})$ satisfies $n \in \mathscr{R}^*_L(\boldsymbol{\xi}; x)$. Provided M_1 is large enough, (3.30) and (3.35) give

$$\sum_{\mathbf{x}(n) \in B(\mathbf{m})} \frac{1}{\phi(n)} = \prod_{i=1}^{L} (\varepsilon_i \log_2 x + O(1)) \gg (\varepsilon_1 \cdots \varepsilon_L)(\log_2 x)^L.$$

Therefore

$$R^*_L(\boldsymbol{\xi}; x) \gg (\log_2 x)^L \mathrm{Vol}(\cup B(\mathbf{m})) \geq (\log_2 x)^L \mathrm{Vol}(\mathscr{S}'_L(\boldsymbol{\xi} - \boldsymbol{\omega})). \qquad (3.40)$$

Let P_j denote the volume of the subset of $\mathscr{S}_L(\boldsymbol{\xi} - \boldsymbol{\omega})$ with (3.31) satisfied for $i < j$ but failing for $i = j$. Let V^*_j denote the volume of the subset of $\mathscr{S}_L(\boldsymbol{\xi} - \boldsymbol{\omega})$ with $x_{j+1} > (1 - 2\omega_j)x_j$, and let V_0 denote the volume with $x_L \leq \log_2 P_0 / \log_2 x$. With these definitions,

$$\mathrm{Vol}(\mathscr{S}'_L(\xi - \omega)) \geq T_L^*(\xi - \omega) - \sum_{j=1}^{L-2} P_j - \sum_{j=1}^{L-1} V_j^* - V_0.$$

Taking $M_1 \geq 10$ makes $\omega_L \leq 10^{-4}$ and $H(\xi - \omega) < 1.1$. Corollary 3.2 and the proof of Lemma 3.8 give

$$T_L^*(\xi - \omega) > 0.9 T_L^*$$

and

$$\sum_{j=1}^{L-1} V_j^* < 0.7 T_L^*.$$

By Lemmas 3.6 and 3.11,

$$V_0 \ll T_L \varrho^M (\log_2 P_0 + 1) \leq 0.05 T_L^*$$

if M_1 is large enough. Again set $h = L - j + 1$ and $k = L_0 - j$. If $\mathbf{x}(n)$ is counted in P_j, then $x_j < 80 \varepsilon_j / \omega_j$. A crude estimate combined with (3.38) and (3.39) gives

$$P_j \leq \left(\frac{80 \varepsilon_j}{\omega_j} \right)^{L-j+1} T_{j-1} = T_L \exp\{h(-k/15 + O(\log k))\}.$$

Since $k \geq M_1$,

$$\sum_{j=1}^{L-2} P_j \leq 0.05 T_L^*.$$

Thus, $\mathrm{Vol}(\mathscr{S}'_L(\xi - \omega)) \geq 0.1 T_L^*$. Combined with (3.40), this gives (3.26). □

Lemma 3.12 can be easily modified to relate sums with $\mathbf{x} \in \mathscr{R}$ for more general sets $\mathscr{R}$. For our applications, we will require the following generalization of Lemma 3.12. We omit the proof, as it requires only small changes to the proof of Lemma 3.12.

Lemma 3.13. *Suppose $\mathscr{R}$ is the intersection of $\mathscr{S}_L$ with one or two of the sets $\{x_i \leq \beta_1\}$, $\{x_i \geq \beta_2\}$, $\{\sum x_i \leq \beta_3\}$ or $\{\sum x_i \geq \beta_4\}$, where the β_i are positive real numbers. Then*

$$\sum \frac{1}{\phi(n)} \ll (\log_2 x)^L \mathrm{Vol}(\mathscr{R}),$$

where the sum is taken over $n \leq x$ with $(x_0(n; x), \ldots, x_{L-1}(n; x)) \in \mathscr{R}$.

Proof of Lemma 3.6: For $|z| < 1$, let

$$L(z) = \sum_{n=2}^{\infty} (\log n) z^n,$$

so that (cf. (1.3))

$$F(z) = (1 - z)L'(z) + 1 - \frac{1}{1 - z}. \tag{3.41}$$

The function $L(z)$ may be analytically continued to the split plane $\mathfrak{S} = \mathbb{C} \setminus [1, \infty)$ by means of the integral

$$L(z) = z \int_0^\infty \frac{1}{te^t(1 - z)} - \frac{1}{t(e^t - z)} \, dt. \tag{3.42}$$

This definition also provides an analytic continuation of $F(z)$ to $\mathfrak{S}$. Now let

$$G(z) = \sum_{n=0}^\infty g_n z^n \quad (|z| < \varrho). \tag{3.43}$$

Definition (1.9) implies

$$G(z) = \frac{1}{1 - F(z)} \tag{3.44}$$

when $|z| < \varrho$, which provides a meromorphic continuation of $G(z)$ to $\mathfrak{S}$. We need several analytic properties of $F(z)$.

Lemma 3.14. *We have*

$$F(z) = 1 - \frac{1}{\log |z|} + O(1/\log^2 |z|) \quad (|z| > 10)$$

and

$$F(z) = \frac{\log(1 - z)}{z - 1} + O(|z - 1|^{-1}) \quad (|z - 1| < 1/10).$$

Proof: For the first part, assume $|z| > 10$. Differentiating (3.42) and applying integration by parts gives

$$L'(z) = \int_0^\infty \frac{1}{te^t(1 - z)^2} - \frac{e^t}{t(e^t - z)^2} \, dt$$

$$= -\int_1^\infty \frac{e^t}{t(e^t - z)^2} \, dt + O(|z|^{-2})$$

$$= \frac{1}{z - e} \int_1^\infty \frac{dt}{t^2(e^t - z)} + \frac{1}{z - e} + O(|z|^{-2}).$$

When $1 \le t \le \log(|z|/2)$, $(e^t - z)^{-1} = -z^{-1}(1 + O(e^t/z))$, thus

$$\int_1^{\log(|z|/2)} \frac{dt}{t^2(e^t - z)} = -\frac{1}{z}\left(1 - \frac{1}{\log(|z|/2)}\right) + O\left(\frac{1}{|z|\log^2|z|}\right).$$

For the remaining t, deform the path of integration away from $\log z$, if necessary, so that $|e^t - z| \gg |e^t|$. It follows that

$$\int_{\log(|z|/2)}^{\infty} \frac{dt}{t^2(e^t - z)} \ll \frac{1}{|z|\log^2(|z|/2)}.$$

Thus

$$L'(z) = \frac{1}{z\log|z|} + O\left(\frac{1}{z\log^2|z|}\right),$$

and the first part of the lemma follows from (3.41). For z near 1 and z near the positive real axis we need to modify the integral representation of $L'(z)$. After applying integration by parts, adding various convergence factors and considerable manipulation, we obtain for $\Re z \ge \frac{1}{2}$

$$L'(z) = -\frac{\log(1 - z)}{z\log^2 z} + \frac{w(z) + 1}{(1 - z)^2}, \tag{3.45}$$

where

$$w(z) = \int_0^{\infty} \frac{(1 - z)(1 - e^t) - t^2(1 - z)^2(\log z)^{-2} + t(1 - ze^{-t})}{t^2(e^t - z)}\, dt. \tag{3.46}$$

Notice that the integrand defining $w(z)$ has no singularities provided $z \ne 1$. Also,

$$\lim_{z \to 1} w(z) = \int_0^{\infty} \frac{1 - t - e^{-t}}{t(e^t - 1)}\, dt = \kappa,$$

so that

$$F(z) = \frac{(z - 1)\log(1 - z)}{z\log^2 z} + O(|1 - z|^{-1}).$$

The second part of the lemma now follows from

$$\frac{z - 1}{z\log^2 z} = \frac{1}{z - 1} + O(z - 1). \qquad \square$$

The behavior of $F(z)$ near the branch cut is also required. When z is real and $z > 1$, define

$$F^*(z) = \lim_{\varepsilon \to 0^+} F(z + i\varepsilon), \qquad F_*(z) = \lim_{\varepsilon \to 0^+} F(z - i\varepsilon).$$

The integral (3.46) is continuous in a neighborhood of every real $z > 1$, and is real for real z. Taking imaginary parts in (3.45) and then applying (3.41) gives

Lemma 3.15. *We have*

$$\Im F_*(z) = \frac{\pi(z-1)}{z \log^2 z}, \quad \Im F^*(z) = \frac{-\pi(z-1)}{z \log^2 z}. \tag{3.47}$$

Lemmas 3.14 and 3.15 allow us to locate the zeros of $F(z) - 1$, which we need to apply contour integration with $G(z)$.

Lemma 3.16. *The equation $F(z) = 1$ has only the solution $z = \varrho$.*

Proof: Let $\mathscr{C} = \mathscr{C}(R, \varepsilon)$ be the positively oriented contour $C_1 \cup C_2 \cup C_3 \cup C_4$, where C_1 is the circle $|z| = R$, C_2 is the line segment from $z = R$ to $z = 1 + \varepsilon$ lying on the lower branch of $F(z)$, C_3 is the circle $|z - 1| = \varepsilon$, and C_4 is the line segment from $z = 1 + \varepsilon$ to $z = R$ lying on the upper branch of $F(z)$ (see figure 1).

By Lemmas 3.14 and 3.15, if R is sufficiently large and ε is sufficiently small, $F(z) \neq 1$ on or outside $\mathscr{C}(R, \varepsilon)$. For a contour C, let $w(f(z); C)$ denote the cumulative change in

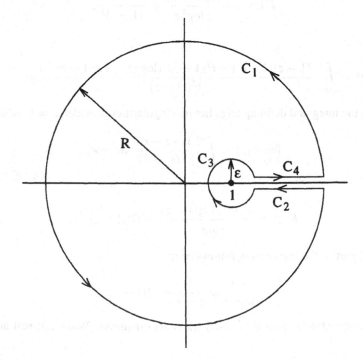

Figure 1. The contour $\mathscr{C}(R, \varepsilon)$.

$\arg f(z)$ as z traces C. By the argument principle, the number of zeros of $F(z) - 1$ is

$$\frac{1}{2\pi} w(F(z) - 1;\ \mathscr{C}(R, \varepsilon)) = \frac{1}{2\pi} \sum_{i=1}^{4} w(F(z) - 1; C_i).$$

By Lemma 3.14, when $|z| = R$, $F(z) - 1 = -1/\log R + O((\log R)^{-2})$, and therefore

$$w(F(z) - 1; C_1) \ll (\log R)^{-1}.$$

On C_2, by Lemma 3.15, $F(z) - 1$ stays in the upper half plane. Furthermore, by Lemma 3.14,

$$F(1 + \varepsilon) = \frac{\log \varepsilon}{\varepsilon} + O(1/\varepsilon),$$

and it follows that

$$w(F(z) - 1; C_2) \ll (\log R)^{-1} + |\log \varepsilon|^{-1}.$$

A similar argument shows that $w(F(z) - 1; C_4) \ll 1/\log R + 1/|\log \varepsilon|$. If $z = 1 - \varepsilon e^{i\theta}$, Lemma 3.14 gives

$$F(z) - 1 = \left(\frac{-i\theta - \log \varepsilon - \gamma}{\varepsilon} \right) e^{-i\theta} + O(1) = -\frac{\log \varepsilon}{\varepsilon} e^{-i\theta} + O(1/\varepsilon).$$

On C_3, θ goes from π down to $-\pi$, thus

$$w(F(z) - 1; C_3) = 2\pi + O(|\log \varepsilon|^{-1}).$$

Since $w(F(z) - 1;\ \mathscr{C}(R, \varepsilon))/2\pi$ is a positive integer, combining the estimates for $w(F(z) - 1; C_i)$ gives

$$w(F(z) - 1;\ \mathscr{C}(R, \varepsilon)) = 2\pi$$

provided R is sufficiently large and ε is sufficiently small. Therefore, $F(z) = 1$ has a unique solution z, namely $z = \varrho$. □

By Lemma 3.16, $G(z)$ is analytic in $\mathfrak{S}$ except for a simple pole at $z = \varrho$. Suppose $n \geq 2$, $\varepsilon > 0$ is small and R is large. Cauchy's integral formula gives

$$g_n = \frac{1}{2\pi i} \int_{\mathscr{C}(R, \varepsilon)} \frac{G(z)}{z^{n+1}} \, dz - \operatorname*{Res}_{z=\varrho} \left(\frac{G(z)}{z^{n+1}} \right). \tag{3.48}$$

By Lemma 3.14, when $|z| = R$, $1 - F(z) \gg (\log |z|)^{-1}$, so the integral over C_1 tends to zero as $R \to \infty$. Lemma 3.14 also implies that the integral over C_3 tends to zero as $\varepsilon \to 0$. On C_2 and C_4, Lemma 3.15 implies

$$|1 - F(z)| \geq \left(\frac{\pi(z - 1)}{z \log^2 z} \right),$$

and therefore

$$g_n = -\operatorname*{Res}_{z=\varrho}\left(\frac{G(z)}{z^{n+1}}\right) + \lim_{\varepsilon \to 0} \frac{1}{2\pi i} \int_{1+\varepsilon}^{\infty} \frac{F^*(z) - F_*(z)}{z^{n+1}|1 - F^*(z)|^2} \, dz$$

$$= \frac{1}{\varrho^{n+1} F'(\varrho)} - \int_1^{\infty} \frac{(z-1)}{z^{n+2}|1 - F^*(z)|^2 \log^2 z} \, dz.$$

The estimates in Lemma 3.14 guarantee that the above integral converges. This completes the proof of the first part of Lemma 3.6. For the second part, we have

$$\sum_{i=1}^{\infty} \gamma_i = \int_1^{\infty} \frac{dz}{z^2 |1 - F^*(z)|^2 \log^2 z} = O(1).$$

It is not difficult to show that in fact

$$\sum_{i=1}^{\infty} \gamma_i = 1 - \frac{1}{\varrho F'(\varrho)}. \qquad \square$$

4. An upper bound for $V(x)$

In this section, we prove that

$$V(x) \ll \frac{x}{\log x} \exp\{C(\log_3 x - \log_4 x)^2 + D(\log_3 x) - (D + 1/2 - 2C)\log_4 x\}. \quad (4.1)$$

We begin with the basic tools needed for the proof, which show immediately the significance of the set $\mathscr{S}_L(\xi; x)$. First, recall the definition of an S-normal prime (2.2)–(2.4).

Lemma 4.1. *Suppose y is sufficiently large, $k \geq 2$ and*

$$1 \geq \theta_1 \geq \cdots \geq \theta_{k+1} = \log_2 S / \log_2 y,$$

where $S \geq \exp\{(\log_2 y)^{30}\}$. Let $\log_2 E_j = \theta_j \log_2 y$ and $\delta_j = \delta(E_j, E_{j-1})$ for each j. Further suppose that for each $j \geq 2$, either $\theta_j = \theta_{j-1}$ or $\theta_{j-1} - \theta_j \geq 3\delta_j\theta_{j-1}$. The number of totients $v \leq y$ with a pre-image n divisible only by S-normal primes and satisfying

$$\log_2 q_j(n) \geq E_j \quad (1 \leq j \leq k)$$

is

$$\ll y(\log y)^{A+B} + y(\log y)^{-2},$$

where

$$A = -\sum_{j=1}^{k} a_j\theta_j, \qquad B = 2\sqrt{\frac{\log_2 S}{\log_2 y}} \sum_{j=2}^{k+1} \theta_{j-1}^{1/2} \, j \log j.$$

Proof: Suppose first that

$$E_1 \leq y^{1/20 \log_2 y}.$$

We may assume that $\Omega(v) \leq 10 \log_2 y$ and that neither n nor v is divisible by the square of a prime $\geq S$. By Lemmas 2.2 and 2.9, the number of exceptional v is $O(y/\log^2 y)$. Let m be the part of v composed of primes in $(S, E_1]$. Then

$$m \leq (E_1)^{\Omega(v)} \leq y^{1/2}.$$

By Lemma 2.4, the number of totients with a given m is

$$\ll \frac{y}{m} \frac{\log S}{\log E_1} = \frac{y}{m} (\log y)^{\theta_{k+1}-\theta_1}.$$

Since the primes $q_i(n)$ are S-normal, by (2.3)

$$\Omega(m, E_j, E_{j-1}) \geq j(\theta_{j-1} - \theta_j - \delta_j\theta_{j-1}) \log_2 y =: R_j \quad (2 \leq j \leq k+1).$$

Therefore, the total number, N, of totients counted satisfies

$$N \ll y(\log y)^{\theta_{k+1}-\theta_1} \prod_{j=2}^{k+1} \sum_{r \geq R_j} \frac{s_j^r}{r!},$$

where

$$s_j = \sum_{E_j < p \leq E_{j-1}} \frac{1}{p} \leq (\theta_{j-1} - \theta_j) \log_2 y + 1.$$

When $E_j = E_{j-1}, s_j = 0$. Otherwise

$$\frac{s_j}{R_j} \leq \frac{1}{j} \left(1 + \frac{2\delta_j\theta_{j-1}}{\theta_{j-1} - \theta_j}\right).$$

By hypothesis, $\delta_j\theta_{j-1} \leq \frac{1}{3}(\theta_{j-1} - \theta_j)$, so Lemma 2.1 implies

$$\sum_{r \geq R_j} \frac{s_j^r}{r!} \leq \left(\frac{es_j}{R_j}\right)^{R_j} \leq (\log y)^{j(\theta_{j-1}-\theta_j-\delta_j\theta_{j-1})(1-\log j+2\delta_j\theta_{j-1}/(\theta_{j-1}-\theta_j))}$$
$$\leq (\log y)^{(j-j \log j)(\theta_{j-1}-\theta_j)+(j \log j+j)\delta_j\theta_{j-1}}.$$

Therefore,

$$N \ll y(\log y)^{A^*+B^*},$$

where

$$A^* = -\sum_{j=1}^{k} a_j\theta_j + ((k+1)\log(k+1) - k)\theta_{k+1},$$

$$B^* = \sum_{j=2}^{k+1} (j \log j + j)\delta_j\theta_{j-1}.$$

The lemma follows by grouping the last term in A^* with B^* and using

$$\delta_j \theta_{j-1} \leq \sqrt{\theta_{j-1} \log_2 S / 2 \log_2 y}.$$

In the case where $E_1 > y^{1/20 \log_2 y}$, for each j set $E'_j = \min\{y^{1/20 \log_2 y}, E_j\}$ and then apply the above argument with E'_j in place of E_j. Since $\log_2 E_j - \log_2 E'_j < \log_3 y + 2$ for each j, the lemma follows in this case as well. $\square$

Recall (1.3) and (3.4) and let

$$A_j = a_1 + a_2 + \cdots + a_j. \tag{4.2}$$

For $\xi > 1$, let $N_k(\xi; y)$ be the number of totients $v \leq y$ with a pre-image n satisfying

$$a_1 x_1(n; y) + \cdots + a_k x_k(n; y) \geq \xi. \tag{4.3}$$

Lemma 4.2. *Suppose $k \geq 2$, $0 < \omega < 1/10$ and y is sufficiently large (say $y \geq y_0$). Then*

$$N_k(1 + \omega; y) \ll y(\log_2 y)^5 W(y)(\log y)^{-1-\omega^2/(150k^3 \log k)}.$$

Proof: Assume that

$$\omega^2 > 750 \frac{\log_3 y}{\log_2 y} k^3 \log k, \tag{4.4}$$

for otherwise the lemma is trivial. Define S by

$$\log_2 S = \frac{\omega^2}{25 k^3 \log k} \log_2 y, \tag{4.5}$$

so that $S \geq \exp\{(\log_2 y)^{30}\}$. Let v be a totient counted in $N_k(1 + \omega; y)$. We have

$$N_k(1 + \omega; y) \leq U_1(y) + U_2(y) + U_3(y) + U_4(y),$$

where

$U_1(y) = |\{\phi(n) \leq y : p \mid n \text{ for some prime } p \text{ which is not } S\text{-normal}\}|$,

$U_2(y) = |\{v : v < y/\log^2 y \text{ or } \Omega(v) \geq 10 \log_2 y\}|$,

$U_3(y) = |\{v : \text{ for some } d > S, d^2 \mid v \text{ or } d^2 \mid y \text{ for some } y \in \phi^{-1}(v)\}|$,

$U_4(y) = |\{v : v \text{ not counted in } U_1(y), U_2(y), \text{ or } U_3(y)\}|$.

By (4.5) and Lemmas 2.2, 2.9 and 2.10 we have

$$U_1(y) + U_2(y) + U_3(y) \ll \frac{y(\log_2 y)^5 W(y)}{\log y}(\log S)^{-1/6} + \frac{y \log_2 y}{S}$$

$$\ll y(\log_2 y)^5 W(y)(\log y)^{-1-\omega^2/(150k^3 \log k)}. \tag{4.6}$$

Let $\varepsilon = \omega/10$, and suppose n is a pre-image of a totient counted in $U_4(y)$ which satisfies (4.3). Let $x_i = x_i(n; y)$ for $1 \le i \le k$. Then there are numbers $\theta_1, \ldots, \theta_k$ so that $\theta_i \le x_i$ for each i, each θ_i is an integral multiple of ε/A_k, $\theta_1 \ge \cdots \ge \theta_k$, and

$$1 + \omega - \varepsilon \le a_1\theta_1 + \cdots + a_k\theta_k \le 1 + \omega. \tag{4.7}$$

For each admissible k-tuple $\boldsymbol{\theta}$, let $T(\boldsymbol{\theta}; y)$ denote the number of totients counted in $U_4(y)$ which have some pre-image n satisfying $x_i(n; y) \ge \theta_i$ for $1 \le i \le k$. Let j be the largest index with $\theta_j \ge \log_2 S/\log_2 y$. Clearly

$$\sum_{i>j} a_i\theta_i \le A_k \frac{\log_2 S}{\log_2 y}. \tag{4.8}$$

By (4.5), for each i either $\theta_{i-1} = \theta_i$ or

$$\theta_{i-1} - \theta_i \ge \frac{\varepsilon}{A_k} = \frac{\omega}{10A_k} > 3\sqrt{\frac{\log_2 S}{\log_2 y}} > 3\delta_i\theta_{i-1},$$

so the hypotheses of Lemma 4.1 are satisfied. By Lemma 4.1,

$$T(\boldsymbol{\theta}; y) \ll y(\log y)^{A+B} + y(\log y)^{-2},$$

where, by (4.7) and (4.8),

$$A = -\sum_{i=1}^{j} a_i\theta_i \le -(1 + 0.9\omega) + A_k \frac{\log_2 S}{\log_2 y}$$

and, by (4.5), (4.7) and the Cauchy-Schwarz inequality,

$$B \le \left(\frac{\log_2 S}{\log_2 y}(1 + \omega) \sum_{j=2}^{k+1} \frac{j^2 \log^2 j}{a_{j-1}} \right)^{1/2}$$

$$\le 4 \left(\frac{k^3 \log k \log_2 S}{\log_2 y} \right)^{1/2} \le \frac{4}{5}\omega.$$

Also

$$A_k \frac{\log_2 S}{\log_2 y} \le \frac{\omega^2}{50k^2} \le \frac{\omega}{2000}.$$

Using (4.4), the number of vectors $\boldsymbol{\theta}$ is trivially at most

$$\left(\frac{A_k}{\varepsilon} \right)^k \le \left(\frac{10k \log k}{\omega} \right)^k \ll \left(\frac{\log_2 y}{\log_3 y} \right)^{k/2} \le e^{k^3 \log k \log_3 y}$$

$$\le (\log y)^{\omega^2/750} \le (\log y)^{\frac{\omega}{7500}}.$$

Therefore,

$$U_4(y) \le \sum_\theta T(\theta; y) \ll y(\log y)^{-1-\omega/20},$$

which is smaller than the right side of (4.6).   □

Before proceeding with the main argument, we require a crude upper bound for $V(x)$ to get things started. We use the method of Pomerance [26] to show

$$W(x) \ll \exp\{(\log_3 x)^2\}. \tag{4.9}$$

(We could simply cite the result (1.2), but include a short argument below for the sake of completeness). Suppose x is large, and let $v \le x$ be a totient with pre-image n. By Lemmas 2.9 and 2.3, the number of v with an n satisfying either $p^2 \mid n$ for some prime $p > \log^3 x$, or $m \mid n$ for some m with $P^+(m) < m^{1/\log_2 x}$ is $O(x/\log^2 x)$. Let $\beta = 0.6$ and $\beta' = 0.599$. By Lemma 4.2, the number of v with

$$\sum_{i=1}^4 a_i x_i(n; x) > 1.02$$

is $O(xW(x)(\log x)^{-1.00000003})$. Since $\sum_{i=1}^4 a_i(\beta')^i > 1.02$, $x_i(n; x) \le (\beta')^i$ for some $i \le 4$ for each remaining totient. For $1 \le i \le 4$, let $U_i(x)$ denote the number of remaining totients with a pre-image satisfying $x_i(n; x) \le (\beta')^i$ and $x_j(n; x) > (\beta')^j$ for $1 \le j < i$. Write $v = \phi(q_0 \cdots q_{i-1})m$, so that $m \le \exp\{(\log x)^{\beta^i}\}$. Each prime q_j, $0 \le j \le i-1$ occurs to the first power and the number of possible totients for a fixed m is

$$\ll \frac{x}{\log x} \frac{1}{\phi(q_1 \cdots q_{i-1})m}.$$

By partial summation,

$$U_i(x) \ll \frac{x}{\log x}(\log_2 x)^i W\left(\exp\left\{(\log x)^{\beta^i}\right\}\right),$$

from which it follows that for some $i \le 4$ we have

$$W(x) \ll (\log_2 x)^i W\left(\exp\left\{(\log x)^{\beta^i}\right\}\right).$$

Iterating the above expression as in [26] gives (4.9).

Lemma 4.3. *We have*

$$\sum_{\substack{v \in \mathcal{V} \\ P^+(v) \le Z}} \frac{1}{v} \ll W(Z \exp \log_2 Z) \log_2 Z \ll \exp\left\{2(\log_3 Z)^2\right\}.$$

Proof: Let $f(y)$ denote the number of totients $v \leq y$ with $P^+(v) \leq Z$, and set $Z' = Z^{\log_2 Z}$. First suppose $y > Z'$. If $v > y^{1/2}$, then $P^+(v) < v^{2/\log_2 Z}$, so Lemma 2.3 gives $f(y) \ll y/\log^2 y$. For $y < Z'$, we use the trivial bound $f(y) \leq V(y)$. The lemma now follows from (4.9), $\log_2 Z' = \log_2 Z + \log_3 Z$ and partial summation.   □

Let $L = L_0(x)$ and for $0 \leq i \leq L - 2$, let

$$\omega_i = \frac{1}{10(L-i)^3}, \quad \xi_i = 1 + \omega_i. \tag{4.10}$$

Then $H(\xi) \leq 1.1$. Let v be a generic totient $\leq x$ with a pre-image n, and set $x_i = x_i(n; x)$ for $i \geq 0$. Evidently

$$V(x) \leq \sum_{j=0}^{L-2} M_j(x) + N(x),$$

where $M_j(x)$ denotes the number of totients $\leq x$ with a pre-image satisfying inequality (I_i) for $i < j$ but not satisfying inequality (I_j), and $N(x)$ denotes the number of totients with every pre-image satisfying $\mathbf{x} \in \mathscr{S}_L(\xi)$. By Lemma 4.2 and (4.9),

$$M_0(x) \ll x(\log_2 x)^5 W(x)(\log x)^{-1-1/(15000L^9 \log L)} \ll x/\log x.$$

Now suppose $1 \leq j \leq L - 2$, and set $k = L - j$. Let n be a pre-image of a totient counted in $M_j(x)$, and set

$$w = q_j(n)q_{j+1}(n) \cdots, \quad m = \phi(w).$$

Since (I_0) holds, $x_2 \leq \xi_0/(a_1 + a_2) < 0.9$. It follows that $q_0(n) > x^{1/3}$, whence $m < x^{2/3}$. By the definition of $M_j(x)$ and (4.10),

$$x_j \leq \xi_j^{-1}(a_1 x_{j+1} + a_2 x_{j+2} + \cdots) < \xi_{j-1}^{-1}(a_1 x_j + a_2 x_{j+1} + \cdots) \leq x_{j-1},$$

whence $q_{j-1} > q_j$ and $\phi(n) = \phi(q_0 \cdots q_{j-1})m$. For a given m, the number of choices for $q_0, \ldots, q_{j-1}$ is

$$\ll \frac{x}{m \log x} R_{j-1}((\xi_0, \ldots, \xi_{j-3}); x) \quad (j \geq 4)$$

and

$$\ll \frac{x}{m \log x} (\log_2 x)^{j-1} \quad (j \leq 3).$$

Let $F(y)$ be the number of $m \leq y$. Define Y_j by the relation

$$\frac{\log_2 Y_j}{(\log_3 Y_j)^2} = \frac{300k^3 \log k}{\omega_j^2}, \tag{4.11}$$

so that

$$\log_2 Y_j \ll k^{10}. \tag{4.12}$$

Since m is a totient, we have $F(y) \le V(y)$, but when $y > Y_j$ we can do much better. First note that $w \ll y \log_2 y$. By Lemma 2.3, the number of such w with $P^+(w) < y^{1/\log_2 y}$ is $O(y/(\log y)^3)$. Otherwise, we have $q_j(n) = P^+(w) \ge y^{1/\log_2 y}$ and

$$x_j \ge \frac{\log_2 y - \log_3 y}{\log_2 x}.$$

For $0 \le i \le k$, let

$$z_i = x_i(w; y) = \frac{\log_2 x}{\log_2 y} x_{i+j}.$$

Since (I_j) fails and $y > Y_j$, it follows that

$$a_1 z_1 + \cdots + a_k z_k \ge \frac{\log_2 x}{\log_2 y} (1 + \omega_j) x_j$$
$$\ge (1 + \omega_j/2).$$

By Lemma 4.2, (4.9) and (4.11), when $y \ge \max(y_0, Y_j)$ we have

$$F(y) \ll \frac{y}{\log y} \exp\left\{ 5 \log_3 y + \log W(y) - \frac{\omega_j^2}{150k^3 \log k} \log_2 y \right\}$$
$$\ll \frac{y}{\log y (\log_2 y)^2}.$$

Therefore, by partial summation,

$$\sum_m \frac{1}{m} \ll W(Y_j) \log_2 Y_j,$$

and thus

$$M_j(x) \ll \frac{x}{\log x} R_{j-1}((\xi_0, \ldots, \xi_{j-3}); x) W(Y_j) \log_2 Y_j. \tag{4.13}$$

By Corollary 3.4 and Lemma 3.12, we have

$$R_{j-1}((\xi_0, \ldots, \xi_{j-3}); x) \ll T_{j-1}(\log_2 x)^{j-1}$$
$$\ll \exp\{-k^2/4C - ((D+1)/2C - 1)k\} R_{L-1}(\xi'; x), \tag{4.14}$$

where $\xi' = (\xi_0, \ldots, \xi_{L-3})$. By (4.9) and (4.12), $W(Y_j) \log_2 Y_j = \exp\{O(\log^2 k)\}$, thus

$$\sum_{j=0}^{L-2} M_j(x) \ll \frac{x}{\log x} R_{L-1}(\xi'; x). \tag{4.15}$$

It remains to bound $N(x)$. Suppose n is a pre-image of a totient counted in $N(x)$. By Lemma 3.9,

$$x_L \le 3\varrho^L \le \frac{10 \log_3 x}{\log_2 x}.$$

If b is a nonnegative integer, let $N_b(x)$ be the number of totients counted in $N(x)$ with $b/\log_2 x \le x_L \le (b+1)/\log_2 x$. Consider now $n > x/\log^2 x$, a pre-image of a totient counted in $N_b(x)$, and set $Z_b = E((b+1)/\log_2 x) = \exp\exp(b+1)$. Let $w = q_L(n)q_{L+1}(n)\cdots$ and $q = q_1(n)\cdots q_{L-1}(n)w$. Since $x_2 < 0.9$ we have $q < x^{2/3}$ and for a fixed q, the number of possibilities for $q_0(n)$ is

$$\ll \frac{x}{\log x}\frac{1}{\phi(q)}.$$

First, by Corollary 3.2, Lemma 3.11 and the fact that

$$x_{L-1} \ge x_L \ge \frac{b/5}{(L-1)g_{L-1}^*},$$

we have

$$\sum \frac{1}{\phi(q_1 \cdots q_{L-1})} \ll R_{L-1}(\xi'; x)e^{-b/5}.$$

By Lemma 4.3 and (4.9),

$$\sum \frac{1}{w} \ll \exp\{2\log^2 b\}.$$

Combining these estimates gives

$$N_b(x) \ll \frac{x}{\log x}R_{L-1}(\xi'; x)\exp\{-b/5 + 2\log^2 b\}.$$

Summing on b gives $N(x) \ll \frac{x}{\log x}R_{L-1}(\xi'; x)$, which together with (4.15) gives

$$V(x) \ll \frac{x}{\log x}R_{L-1}(\xi'; x).$$

The upper bound (4.1) now follows from Lemmas 3.3 and 3.12.

5. The lower bound for $V_k(x)$

Our lower bound for $V_k(x)$ is obtained by constructing a set of numbers with multiplicative structure similar to the numbers counted by $N(x)$ in the upper bound argument. Suppose $A(d) = \kappa$ and $\phi(d_i) = d$ ($1 \le i \le \kappa$). Assume throughout that $x \ge x_0(d)$. The variable k is reserved as an index for certain variables below.

Define

$$M = M_2 + [(\log d)^{1/9}], \; M_2 \text{ is a sufficiently large absolute constant} \tag{5.1}$$

$$L = L_0(x) - M, \tag{5.2}$$

$$\xi_i = 1 - \omega_i, \quad \omega_i = (L_0 - i)^{-3} \quad (0 \le i \le L). \tag{5.3}$$

Notice that

$$\varrho^L \asymp \varrho^{-M} \frac{L}{\log_2 x} \tag{5.4}$$

and choose M_2 large enough so that

$$\frac{\omega_i}{\omega_{i+1}} > 1 - \frac{1}{1000} \quad (i \le L - 1). \tag{5.5}$$

Let B denote the set of integers $n = p_0 p_1 \cdots p_L$ with each p_i prime and

$$\phi(n) \le x/d, \tag{5.6}$$

$$(x_1(n; x), \ldots, x_L(n; x)) \in \mathscr{S}_L(\xi), \tag{5.7}$$

$$\log_2 p_i \ge (1 + \omega_i) \log_2 p_{i+1} \quad (0 \le i \le L - 1), \tag{5.8}$$

$$\log_2 p_i \ge \frac{4}{\omega_i} (5/4)^{L-i} \quad (0 \le i \le L), \tag{5.9}$$

$$p_L \ge \max(d + 1, 16). \tag{5.10}$$

We show that for most numbers $n \in B$, $A(d\phi(n)) = A(d)$, and thus $V_\kappa(x) \gg |B|$. By Corollary 3.4, Lemma 3.12, (4.1) and (5.1), we have

$$|B| \gg \frac{x}{d \log(x/d)} (\log_2(x/d))^L T_L$$

$$\gg \frac{1}{d} \exp\{-(\log d)^{2/9}/4C + O((\log d)^{1/9})\} V(x). \tag{5.11}$$

Consider the equation

$$d\phi(n) = \phi(n_1), \tag{5.12}$$

where $n \in B$. Let $q_0 > q_1 > \cdots$ be the prime factors of n_1. If $n \mid n_1$, then none of the primes q_i $(0 \le i \le L)$ occur to a power greater than 1. Otherwise, since $q_i = p_i$ for each i, (5.10) gives $\phi(n_1) \ge \phi(n) p_L > \phi(n) d$. Therefore

$$\phi(n_1) = \phi(n_1/n)\phi(n) = \phi(n)d,$$

which implies $n_1 = nd_i$ for some i. These we will call the trivial solutions to (5.12). We then have $A(d\phi(n)) = \kappa$ for each $n \in B$ for which (5.12) has no non-trivial solutions. The numbers n which give rise to non-trivial solutions may be grouped as follows. Let B_j be the set of $n \in B$ such that (5.12) holds for some n_1 with

$$q_i = p_i \; (0 \le i \le j-1), \quad p_j \ne q_j.$$

We then have

$$V_\kappa(x) \ge |B| - \sum_{j=0}^{L} |B_j|. \tag{5.13}$$

For n counted in B_j with $j \ge 1$, write $n = p_0 n_2 n_3$, where $n_2 = p_1 \cdots p_{j-1}$ (when $j = 1$ put $n_2 = 1$) and $n_3 = p_j \cdots p_L$. Also put

$$h = L - j + 1 = \omega(n_3). \tag{5.14}$$

For x large, (5.7) and (5.8) imply $p_0 > x^{1/2}$, so for each fixed $n_2 n_3$, the number of choices for p_0 is $O(x/(d\phi(n_2 n_3) \log x))$. Hence

$$|B_j| \ll \frac{x}{d \log x} \sum_{n_2} \frac{1}{\phi(n_2)} \sum_{n_3} \frac{1}{\phi(n_3)}. \tag{5.15}$$

Since $n_2 \in W_{j-1}((\omega_1, \ldots, \omega_{j-3}); x)$ when $j \ge 4$, Lemma 3.12 gives

$$\sum_{n_2} \frac{1}{\phi(n_2)} \ll (\log_2 x)^{j-1} T_{j-1}. \tag{5.16}$$

When $j \le 3$, we use the trivial bound $\sum 1/\phi(n_2) \ll (\log_2 x)^{j-1}$.

To attack the sum on n_3, let $B_j(y)$ denote the number of possible n_3 with $\phi(n_3) \le y$. In particular, $|B_0| = B_0(x/d)$. When $j \ge 1$, by (5.15), (5.16) and partial summation,

$$|B_j| \ll \frac{x(\log_2 x)^{j-1} T_{j-1}}{d \log x} \sum_y \frac{B_j(y)}{y^2}. \tag{5.17}$$

Each n_3 counted in $B_j(y)$ satisfies

$$d\phi(n_3) = \phi(n_4) \tag{5.18}$$

for some n_4 with $P^+(n_4) \ne p_j$.

If $\log_2 y < M^{18}$ we use a trivial upper bound for $B_j(y)$. By (5.14) and a Theorem of Hardy and Ramanujan [19], for some absolute constant c_5 we have

$$B_j(y) \ll \frac{y}{\log y} \frac{(\log_2 y + c_5)^{h-1}}{(h-1)!}. \tag{5.19}$$

For larger y we can do much better.

Lemma 5.1. *When*

$$\log_2 y \geq M^{18},$$
(5.20)

then

$$B_j(y) \ll \frac{y}{\log y (\log_2 y)^2}.$$

The reason we chose M^{18} will become evident in inequalities (5.39) and (5.40) below. Combining Lemma 5.1 with (5.19) gives

$$\sum_{n_3} \frac{1}{n_3} \ll \sum_y \frac{B_j(y)}{y^2} \ll 1 + (M^{18} + c_5)^h \leq M^{19h}.$$

Together with (5.1) and (5.17), this gives

$$|B_0| = B_0(x/d) \ll \frac{x}{d \log x} \exp\{19L \log M\}$$

and, using Lemma 3.3 when $j \geq 1$,

$$|B_j| \ll \frac{x}{d \log x} (\log_2 x)^{j-1} T_{j-1} \exp\{19h \log M\}$$

$$\ll \frac{x}{d \log x} (\log_2 x)^L T_L \exp\{h(19 \log M - M/2C - h^2/4C)\}.$$

If M_2 is sufficiently large, summing over j and using (5.11) gives

$$\sum_{j=0}^{L} |B_j| \leq |B|/2.$$

It follows from (5.11) and (5.13) that

$$V_\kappa(x) \gg |B| \gg_\varepsilon d^{-1-\varepsilon} V(x),$$

which completes the proof of Theorems 1 and 2.

Proof of Lemma 5.1:   Notice that $p_j \leq y$ means (for $j \leq L - 2$) that

$$\left(\frac{\log_2 p_{j+1}}{\log_2 y}, \ldots, \frac{\log_2 p_L}{\log_2 y} \right) \in \mathscr{S}_{L-j}((\xi_j, \ldots, \xi_{L-2})).$$

Thus, by Lemma 3.10 and (5.10) (and simple calculations when $j \geq L - 1$),

$$1 \leq \log_2 p_L \leq 3\varrho^{L-j} \log_2 y,$$

which implies

$$h \leq 2C \log_3 y + 3. \qquad (5.21)$$

The definition (5.1) and inequality (5.20) imply

$$d \leq \exp\{(\log_2 y)^{1/2}\}. \qquad (5.22)$$

We first remove from consideration those n_3 with $n_3 \leq y/\log^2 y$, $\Omega(\phi(n_4)) \geq 10 \log_2 y$, or those with $b^2 \mid \phi(n_3)$ or $b^2 \mid n_4$ for some $b > \log^3 y$. By Lemmas 2.2 and 2.9, the number of such n_3 is

$$O(y/\log^2 y).$$

Next define

$$S = \exp \exp\{(\log_3 y)^3\},$$
$$R = \exp \exp\{(\log_2 y)^{1/2}\}. \qquad (5.23)$$

We next remove from consideration those n_3 with either n_3 or n_4 divisible by a prime which is not S-normal. By Lemma 2.10 and (4.1), the number of n_3 removed is

$$\ll \frac{y(\log_2 y)^5 W(y)}{\log y}(\log S)^{-1/6}$$
$$\ll \frac{y}{\log y(\log_2 y)^2}. \qquad (5.24)$$

Let $B_j^*(y)$ denote the number of remaining n_3, so that

$$B_j(y) \ll \frac{y}{\log y(\log_2 y)^2} + B_j^*(y). \qquad (5.25)$$

For $j \leq L - 1$, we have $p_{j+1} \cdots p_L \leq p_{j+1}^h$, so by (5.1), (5.8), (5.14), (5.20) and (5.21),

$$\log_2(n_3/p_j) \leq \frac{\log_2 p_j}{1 + (h + M - 1)^{-3}} + \log h$$
$$\leq \frac{\log_2 y}{1 + (2C \log_3 y + 2 + M)^{-3}} + \log(2C \log_3 y + 3)$$
$$\leq \log_2 y - \frac{1}{2}M^{18}(36C \log M + 2 + M)^{-3} + \log(36C \log M + 3)$$
$$\leq \log_2 y - 10.$$

In particular, since $n_3 > y/\log^2 y$, this shows that

$$p_j > y^{9/10}. \qquad (5.26)$$

When $j = L$, (5.26) holds because $n_3 > y/\log^2 y$. We now group the n_3 counted in $B_j^*(y)$ according to the sizes of the prime factors $p_j, \ldots, p_L$. As before, set

$$\varepsilon_i = \frac{(5/4)^{L-i}}{\log_2 y} \qquad (j \leq i \leq L). \qquad (5.27)$$

For each n_3, define a vector $\zeta = \zeta(n_3) = (\zeta_j, \dots, \zeta_L)$ by $\zeta_j = 1$ and

$$\zeta_i = \varepsilon_i \left[\frac{\log_2 p_i}{\varepsilon_i \log_2 y} + 1 \right] \quad (j+1 \le i \le L). \tag{5.28}$$

For a given ζ, let $B_j(\zeta; y)$ be the number of n_3 counted by $B_j^*(y)$ satisfying

$$y^{9/10} \le p_j \le y, \quad \zeta_i - \varepsilon_i \le \frac{\log_2 p_i}{\log_2 y} < \zeta_i \quad (j+1 \le i \le L).$$

We then have

$$B_j^*(y) = \sum_{\zeta} B_j(\zeta; y), \tag{5.29}$$

where $\zeta = (\zeta_j, \dots, \zeta_L)$, each ζ_i is an integral multiple of ε_i and the sum is over ζ corresponding to at least one n_3. Note that when $j = L$ there is only one summand in (5.29).

Fix ζ and suppose n_3 is counted in $B_j(\zeta; y)$. From (5.1), (5.14), (5.20) and (5.21),

$$\frac{\varepsilon_j}{\omega_j} \le \frac{(5/4)^{h-1}(M+h-1)^3}{\log_2 y}$$

$$\le (5/4)^2 (2C \log_3 y + M + 2)^3 (\log_2 y)^{-1+2C \log(5/4)}$$

$$\le 2(36C \log M + M + 2)^3 M^{-11} \le 1/20.$$

When $j \ge L - 2$, (5.7) gives

$$a_1 \zeta_{j+1} + \dots + a_{L-j} \zeta_L \le (1 - \omega_j) + (a_1 \varepsilon_{j+1} + \dots + a_{L-j} \varepsilon_L)$$

$$\le (1 - \omega_j) + 10\varepsilon_j \le 1 - \omega_j/2. \tag{5.30}$$

We also need information on the spacing of consecutive components of ζ. By (5.5), (5.8), and (5.9), for each i,

$$\zeta_{i-1} \ge (1 + \omega_{i-1})x_i$$

$$\ge \zeta_i - \varepsilon_i + \omega_{i-1}x_i \ge \zeta_i + 2\varepsilon_{i-1} \tag{5.31}$$

and

$$\zeta_{i+1} \le \frac{\zeta_i}{1 + \omega_i} + \varepsilon_{i+1}$$

$$\le \zeta_i (1 - (3/4)\omega_i). \tag{5.32}$$

In particular, (5.31) implies that the intervals $[\zeta_i - \varepsilon_i, \zeta_i]$ do not overlap. Recall that $\zeta_j = 1$ and define $\zeta_{L+1} = 0$. Let J be the unique number with $j + 1 \le J \le L + 1$ and

$$\zeta_J \le \frac{\log_2 R}{\log_2 y} < \zeta_{J-1}. \tag{5.33}$$

Set

$$\zeta^* = \min\left(\frac{\log_2 R}{\log_2 y}, \zeta_{J-1} - \varepsilon_{J-1} - \frac{1}{2}\omega_{J-1}\zeta_{J-1}\right). \tag{5.34}$$

Since

$$\varepsilon_{J-1} \leq \zeta_{J-1}\omega_{J-1}/4 \leq \zeta_{J-1}/40,$$

it follows that

$$\frac{1}{2}\log_2 R \leq \zeta^* \log_2 y \leq \log_2 R$$

and, by (5.32),

$$\zeta_J < \zeta^*.$$

Thus, when $j \leq L$, $\log_2 p_J \leq \zeta^* \log_2 x$.

Given the ranges of $p_j, \ldots, p_{J-1}$ and the fact that all the prime divisors of n_3 and n_4 are S-normal, we can deduce that the $J - j$ largest prime factors of n_4 must be roughly the same size as their counterparts in n_3. Recall that $q_j = q_0(n_4)$, $q_{j+1} = q_1(n_4)$, etc.

Lemma 5.2. *Redefine $\zeta_J = \zeta^*$ and, for $j \leq i \leq J$, set*

$$\mu_i = \zeta_i - \varepsilon_i - \theta_i, \quad \nu_i = \zeta_i + \theta_i, \quad \theta_i = (3(i - j) + 1)\sqrt{\zeta_i}\left(\frac{\log_2 S}{\log_2 y}\right)^{1/2}. \tag{5.35}$$

Then, for $j \leq i \leq J - 1$, q_i occurs to the first power in the prime factorization of n_4 and

$$\mu_i \leq \frac{\log_2 q_i}{\log_2 y} \leq \nu_i. \tag{5.36}$$

Furthermore, if $\omega(n_4) > J - j$ then $\log_2 q_J \leq \nu_J \log_2 y$.

Proof: First note that $\Omega(n_4) < 10 \log_2 y$ implies

$$q_j > y^{1/10\log_2 y} > E(\mu_j),$$

where for brevity we set $E(z) = \exp\{(\log y)^z\}$. We now proceed by induction, assuming that (5.36) has been proved for $j \leq i \leq j + k - 1$, where $1 \leq k \leq J - j$. Let $g = j + k$ and $\delta_g = \delta(E(\mu_g), E(\zeta_g - \varepsilon_g))$. From (2.4) and (5.35), we have

$$\theta_g > (3k + 1)\zeta_g\delta_g.$$

Since each prime p_i is S-normal, by (2.3) we have

$$\Omega(\phi(n_3), E(\mu_g), E(\zeta_g - \varepsilon_g)) \geq (k + 1)(\theta_g - \zeta_g\delta_g)\log_2 y$$
$$> k(\theta_g + \zeta_g\delta_g)\log_2 y$$
$$\geq \Omega(\phi(q_j \cdots q_{g-1}), E(\mu_g), E(\zeta_g - \varepsilon_g)).$$

It follows from (5.18) that $q_g \geq E(\mu_g)$ and hence $\omega(n_4) \geq k + 1$ and q_g is S-normal. We also have $\nu_g \leq 2\zeta_g$ and $\delta_g \geq \delta(E(\zeta_g), E(\nu_g))$, hence

$$\Omega(\phi(n_3), E(\zeta_g), E(\nu_g)) \leq k(\theta_g + 2\zeta_g \delta_g) \log_2 y$$
$$< (k + 1)(\theta_g - \zeta_g \delta_g) \log_2 y. \tag{5.37}$$

If $q_g \geq E(\nu_g)$, then

$$\Omega(\phi(n_4), E(\zeta_g), E(\nu_g)) \geq (k + 1)(\theta_g - \zeta_g \delta_g) \log_2 y,$$

contradicting (5.18). Thus (5.36) holds for $i = j + k$ and therefore for all i by induction.

If $\omega(n_4) \geq J - j + 1$ and $q_J > E(\nu_J)$, we similarly reach a contradiction. Therefore $q_J \leq E(\nu_J)$. □

In summary, what we shown so far is

$$E(\zeta_i - \varepsilon_i) \leq p_i \leq E(\zeta_i), \quad E(\mu_i) \leq q_i \leq E(\nu_i) \quad (j \leq i \leq J - 1),$$

where

$$(p_j - 1) \cdots (p_L - 1)d = (q_j - 1) \cdots (q_{J-1} - 1)e \tag{5.38}$$

with $p_j \neq q_j$, $p_J \leq E(\zeta_J)$ (if $J \leq L$) and $P^+(e) \leq E(\nu_J)$.

The remaining argument requires that the intervals $[\mu_i, \nu_i]$ be sufficiently far apart. First, by (5.1), (5.20) and (5.21),

$$\omega_i = (M + h - 1)^{-3}$$
$$\geq ((\log_2 y)^{1/18} + 2C \log_3 y + 2)^{-3}$$
$$\geq (1/8)(\log_2 y)^{-1/6}. \tag{5.39}$$

Also, $\zeta_i \geq 4(\log_2 y)^{-1/2}$ for $i \leq J - 1$. Thus, by (5.1), (5.35) and (5.39), we have

$$\theta_i \leq \zeta_i (6C \log_3 y + 10)\zeta_i^{-1/2}(\log_2 S)^{1/2}(\log_2 y)^{-1/2}$$
$$\leq \zeta_i \cdot 3(\log_3 y)^{5/2}(\log_2 y)^{-1/4}$$
$$\leq 24\zeta_i \omega_i (\log_3 y)^{5/2}(\log_2 y)^{-1/12}$$
$$\leq \zeta_i \omega_i /10. \tag{5.40}$$

When $i \leq J - 1$, combining (5.1), (5.9), (5.32), (5.34), (5.35), (5.39) and (5.40) gives

$$\mu_i - \nu_{i+1} = \zeta_i - \zeta_{i+1} - \varepsilon_i - \theta_i - \theta_{i+1}$$
$$\geq \omega_i \zeta_i /4$$
$$\geq (1/8)(\log_2 y)^{-2/3}$$
$$\geq \frac{2 + \log_3 y}{\log_2 y}. \tag{5.41}$$

In addition

$$v_i - v_{i+1} \geq 5\theta_i \geq 5v_i\delta(S, E(v_i)). \qquad (5.42)$$

Now that each of p_i, q_i lies in a short interval and none of the intervals are too close together, we may apply the lower bound argument from [23], with great care taken to insure that the estimates are uniform in all of the parameters.

We make a further subdivision of the numbers n_3, counting separately those with $(p_j \cdots p_{J-1}, q_j \cdots q_{J-1}) = m_1$ and $p_J \cdots p_L = m_2$ (if $J = L + 1$ set $m_2 = 1$), the number of which we denote by $B_j(\zeta; m_1, m_2; y)$. Let

$$j = j_0 < j_1 < \cdots < j_{K-1} \leq J - 1$$

with $K \geq 1$ be the indices with $p_{j_i} \neq q_{j_i}$. Also, define $j_K = J$. For brevity, for $0 \leq k \leq K$ set

$$w_k = E\left(\zeta_{j_k} - \varepsilon_{j_k}\right), \quad u_k = E\left(\mu_{j_k}\right), \quad z_k = E\left(\zeta_{j_k}\right), \quad v_k = E\left(v_{j_k}\right)$$

and

$$\delta_k = \delta(v_k, v_{k-1}).$$

Then $B_j(\zeta; m_1, m_2; y)$ is at most the number of solutions of

$$\left(p_{j_0} - 1\right) \cdots \left(p_{j_{K-1}} - 1\right)\phi(m_2)d = \left(q_{j_0} - 1\right) \cdots \left(q_{j_{K-1}} - 1\right)e \leq y/(d\phi(m_1)) \qquad (5.43)$$

where e is a totient satisfying $P^+(e) \leq v_K$ and p_{j_i} and q_{j_i} are S-normal primes satisfying

$$w_k \leq p_{j_k} \leq z_k, \quad u_k \leq q_{j_k} \leq v_k. \quad (0 \leq k \leq K - 1) \qquad (5.44)$$

For each k, $0 \leq k \leq K$, let

$$s_k(n) = \prod_{\substack{p^a \| n \\ p \leq v_{j_k}}} p^a, \quad t_k(n) = \prod_{\substack{p^a \| n \\ v_{j_k} < p \leq v_{j_{k-1}}}} p^a. \qquad (5.45)$$

For each $2K$-tuple

$$\mathscr{A} = \left(p_{j_0}, \ldots, p_{j_{K-1}}; q_{j_0}, \ldots, q_{j_{K-1}}\right)$$

giving rise to a solution of (5.43), let $b = \phi(p_{j_0} \cdots p_{j_{K-1}} m_2)$ and define

$$\sigma_k(\mathscr{A}) = \left\{s_k(b); s_k\left(p_{j_0} - 1\right), \ldots, s_k\left(p_{j_{K-1}} - 1\right); s_k\left(q_{j_0} - 1\right), \ldots, s_k\left(q_{j_{K-1}} - 1\right)\right\},$$
$$\tau_k(\mathscr{A}) = \left\{t_k(b); t_k\left(p_{j_0} - 1\right), \ldots, t_k\left(p_{j_{K-1}} - 1\right); t_k\left(q_{j_0} - 1\right), \ldots, t_k\left(q_{j_{K-1}} - 1\right)\right\}.$$

Defining multiplication of $(2K + 1)$-tuples by componentwise multiplication, we have from (5.45),

$$\sigma_{k-1}(\mathscr{A}) = \sigma_k(\mathscr{A})\tau_k(\mathscr{A}). \qquad (5.46)$$

Let $\mathfrak{S}_k$ denote the set of $\sigma_k(\mathscr{A})$ arising from solutions $\mathscr{A}$ of (5.43) and $\mathfrak{T}_k$ the corresponding set of $\tau_k(\mathscr{A})$. By (5.46),

$$B_j(\zeta; m_1, m_2; y) \leq |\mathfrak{S}_0| = \sum_{\sigma \in \mathfrak{S}_1} \sum_{\substack{\tau \in \mathfrak{T}_1 \\ \sigma\tau \in \mathfrak{S}_0}} 1. \tag{5.47}$$

Let $\sigma^{(j)}$ denote the jth coordinate of a $(2K + 1)$-tuple σ. We will apply an iterative procedure based on the identity

$$\sum_{\sigma_{k-1} \in \mathfrak{S}_{k-1}} \frac{1}{\sigma_{k-1}^{(1)}} = \sum_{\sigma_k \in \mathfrak{S}_k} \frac{1}{\sigma_k^{(1)}} \sum_{\substack{\tau_k \in \mathfrak{T}_k \\ \sigma_k \tau_k \in \mathfrak{S}_{k-1}}} \frac{1}{\tau_k^{(1)}}. \tag{5.48}$$

The following two lemmas provide the necessary estimates for the application of (5.48).

Lemma 5.3. *For each* $\sigma \in \mathfrak{S}_1$, *we have*

$$\sum_{\substack{\tau \in \mathfrak{T}_1 \\ \sigma\tau \in \mathfrak{S}_0}} 1 \ll \frac{y(\log_2 y)^3}{\phi(m_1)\sigma^{(1)}(\log y)^{2+\nu_{j_1}}}.$$

Proof: Let $b = \sigma^{(1)}, r = \sigma^{(2)}, s = \sigma^{(T+J-j+2)}, t = \tau^{(1)} = \tau^{(2)} = \tau^{(T+J-j+2)}$. We have $t \leq y/(b\phi(m_1))$ and, by (5.1) and (5.8), $bm_1 \leq y^{1/10}$. Note that t is composed only of prime factors $\geq \nu_{j_1}$, and $rt + 1$ and $st + 1$ are unequal primes. Write $t = t'Q, Q = p^+(t)$. Given t', we use Lemma 2.6 to bound the number of Q, then Lemma 2.4 to bound the number of $t \leq z$. This gives

$$\sum_{\substack{\tau \in \mathfrak{T}_1 \\ \sigma\tau \in \mathfrak{S}_0}} 1 \ll \frac{y(\log_2 y)^3}{\phi(m_1)b(\log y)^{2+\nu_{j_1}}}. \qquad \square$$

Lemma 5.4. *If* $2 \leq k \leq K$ *and* $\sigma \in \mathfrak{S}_k$, *then*

$$\sum_{\substack{\tau \in \mathfrak{T}_k \\ \sigma\tau \in \mathfrak{S}_{k-1}}} \frac{1}{\tau^{(1)}} \ll (\log_2 y)^5 (\log y)^{-(\mu_{j_{k-1}} + \zeta_{j_{k-1}} - \varepsilon_{j_{k-1}}) + (k \log k + k)(\nu_{j_{k-1}} - \nu_{j_k} + \delta_k \nu_{j_{k-1}})}.$$

Proof: By (5.44), τ must have the form

$$\tau = \{w; f_0, \ldots, f_{k-1}, 1, \ldots, 1; b_0, \ldots, b_{k-1}, 1, \ldots, 1\},$$

where $w = f_0 \cdots f_{k-1} = b_0 \cdots b_{k-1}$. Write $f = f_{k-1}$ and $b = b_{k-1}$ for short, and let

$$r = \sigma^{(k+1)} = s_k(p_{j_{k-1}} - 1), \quad s = \sigma^{(T+k+1)} = s_k(q_{j_{k-1}} - 1).$$

Since $\sigma\tau \in \mathfrak{S}_{k-1}$, $rf+1$ and $sb+1$ are unequal primes. Let $Q_1 = P^+(f)$, $Q_2 = P^+(b)$, $f' = f/Q_1$ and $b' = b/Q_2$. Trivially

$$\Omega(f) = \Omega\big(p_{j_{k-1}} - 1, v_k, v_{k-1}\big) \le \Omega(n_3) \le 10\log_2 y,$$

so by (5.41),

$$f = \frac{p_{j_{k-1}} - 1}{r} \ge w_{k-1}v_k^{-10\log_2 y} \ge w_{k-1}^{1/2}.$$

This implies

$$Q_1 \ge w_{k-1}^{1/(20\log_2 y)}. \tag{5.49}$$

Similarly, we obtain

$$Q_2 \ge u_{k-1}^{1/(20\log_2 y)}. \tag{5.50}$$

We sum separately over $\mathfrak{T}_{k,1}$, the set of τ with $Q_1 = Q_2$ and $\mathfrak{T}_{k,2}$, the set of τ with $Q_1 \ne Q_2$. For the sum over $\mathfrak{T}_{k,1}$, let $w = tQ_1$. Since b_0 is uniquely determined by the $(2k-1)$-tuple

$$\mathbf{v} = (b_1, \ldots, b_{k-2}, b', f_0, \ldots, f_{k-2}, f'),$$

we have

$$\Sigma_1 := \sum_{\substack{\tau \in \mathfrak{T}_{k,1} \\ \sigma\tau \in \mathfrak{S}_{k-1}}} \frac{1}{\tau^{(1)}} = \sum_t \frac{h(t)}{t} \sum_{Q_1} \frac{1}{Q_1}.$$

Here $h(t)$ denotes the number of $\mathbf{v}$ satisfying

$$f_0 \cdots f_{k-2}f' = t, \quad b_1 \cdots b_{k-2}b' \mid t,$$

and $f'Q_1 + 1$, $b'Q_1 + 1$ are unequal primes. By sieve methods (Lemma 2.6), the number of $Q_1 \le z$ is $\ll z(\log z)^{-3}(\log_2 y)^3$ uniformly in f', b'. By (5.49) and partial summation, we have

$$\sum_{Q_1} \frac{1}{Q_1} \ll (\log_2 y)^4 (\log y)^{-2(\zeta_{j_{k-1}} - \varepsilon_{j_{k-1}})}. \tag{5.51}$$

Also, $h(t)$ is at most the number of dual factorizations of t into k factors each, i.e., $h(t) \le k^{2\Omega(t)}$. By (2.3),

$$\Omega(t) \le k\big(v_{j_{k-1}} - v_{j_k} + \delta_k v_{j_{k-1}}\big) \log_2 y =: I.$$

Thus

$$\sum_t \frac{h(t)}{t} \le \sum_{i \le I} \frac{k^{2i}H^i}{i!},$$

where

$$\sum_{v_{j_k} < p \leq v_{j_{k-1}}} \frac{1}{p} \leq (v_{j_{k-1}} - v_{j_k}) \log_2 y + 1 =: H.$$

By (5.42) and the fact that $v_{J-1} \geq \log_2 R_2 / \log_2 y$, we have

$$v_{j_{k-1}} - v_{j_k} > 2\delta_k v_{j_{k-1}},$$

hence $I < 3kH/2 \leq (3/4)k^2 H$. Applying Lemma 2.1 (with $\alpha \leq 3/4$) yields

$$\sum_t \frac{h(t)}{t} \leq \left(\frac{eHk^2}{I}\right)^I \leq (ek)^I$$

$$\leq (\log y)^{(k+k\log k)(v_{j_{k-1}} - v_{j_k} + \delta_k v_{j_{k-1}})}. \tag{5.52}$$

Combined with (5.51) this gives

$$\Sigma_1 \ll (\log_2 y)^4 (\log y)^{-2(\zeta_{j_{k-1}} - \varepsilon_{j_{k-1}}) + (k+k\log k)(v_{j_{k-1}} - v_{j_k} + \delta_k v_{j_{k-1}})}. \tag{5.53}$$

For the sum over $\mathfrak{T}_{k,2}$, set $w = tQ_1Q_2$. Then

$$\Sigma_2 := \sum_{\substack{\tau \in \mathfrak{T}_{k,2} \\ \sigma\tau \in \mathfrak{S}_{k-1}}} \frac{1}{\tau^{(1)}} = \sum_t \frac{h(t)}{t} \sum_{Q_1, Q_2} \frac{1}{Q_1 Q_2},$$

where $h(t)$ is the number of $\mathbf{v}$ corresponding to a given t, and $f'Q_1 + 1$, $b'Q_2 + 1$ are unequal primes. Note that

$$tQ_2 = f_0 \cdots f_{k-2} f', \quad tQ_1 = b_0 \cdots b_{k-2} b',$$

so Q_1 (respectively Q_2) divides one of the factors on the right side. If we fix the factors divisible by Q_1 and by Q_2, then the number of possible $\mathbf{v}$ is $\leq k^{2\Omega(t)}$ as before. Therefore, $h(t) \leq k^{2\Omega(t)+2}$. By sieve methods (Lemma 2.6), the number of $Q_1 \leq z$ (respectively $Q_2 \leq z$) is $\ll z(\log z)^{-2}(\log_2 y)$. Thus, by (5.49), (5.50) and partial summation, we have

$$\sum_{Q_1, Q_2} \frac{1}{Q_1 Q_2} = \sum_{Q_1} \frac{1}{Q_1} \sum_{Q_2} \frac{1}{Q_2} \ll (\log_2 y)^4 (\log y)^{-(\zeta_{j_{k-1}} - \varepsilon_{j_{k-1}} + \mu_{j_{k-1}})}.$$

Combined with (5.52) this gives

$$\Sigma_2 \ll k^2 (\log_2 y)^4 (\log y)^{-(\zeta_{j_{k-1}} - \varepsilon_{j_{k-1}} + \mu_{j_{k-1}}) + (k+k\log k)(v_{j_{k-1}} - v_{j_k} + \delta_k v_{j_{k-1}})}.$$

By (5.21), $k \leq K < 2C \log_3 y + 3$, so $k^2 \ll \log_2 y$. Adding Σ_1 and Σ_2 establishes the lemma. $\qquad\square$

Combining Lemmas 5.3 and 5.4 with (5.48) gives

$$B_j(\zeta; m_1, m_2; y) \ll \frac{y}{\phi(m_1)} (c_6 \log_2 y)^{5K} (\log y)^{-2-\nu_{j_1}+\sum_{k=2}^{K} b_k} \sum_{\sigma \in \mathfrak{S}_K} \frac{1}{\sigma^{(1)}}, \qquad (5.54)$$

where

$$b_k = \left(\nu_{j_{k-1}} - \nu_{j_k} + \delta_k \nu_{j_{k-1}}\right)(k \log k + k) - \left(\mu_{j_{k-1}} + \zeta_{j_{k-1}} - \varepsilon_{j_{k-1}}\right)$$
$$\leq \left(\zeta_{j_{k-1}} - \zeta_{j_k}\right)(k \log k + k) - 2\zeta_{j_{k-1}} + (k \log k + k + 2)\left(\theta_{j_{k-1}} + \delta_k \zeta_{j_{k-1}}\right).$$

From the definition of a_k, (5.1), (5.21), (5.30) and (5.35), the exponent of $\log y$ in (5.54) is at most

$$-2 + \sum_{k=1}^{K-1} a_k \zeta_{j_k} + (\log_3 y)^5 (\log_2 y)^{-1/2} \leq -1 - \omega_j/2 + (\log_3 y)^5 (\log_2 y)^{-1/2}. \qquad (5.55)$$

It remains to treat the sum on σ. Suppose $f = \sigma^{(1)}$ and consider (5.43). For a given f, the number of possible pairs (σ, m_2) which give $\sigma^{(1)} = f$ is at most the number of factorizations of f into $K + L - J + 1$ factors times the number of factorizations of df into $K + 1$ factors, which is at most

$$(K + 1)^{\Omega(df)}(K + L - J + 1)^{\Omega(f)} \leq h^{2\Omega(f)+\Omega(d)}.$$

First, by (5.1), (5.21) and (5.22),

$$h^{\Omega(d)} \leq h^{2 \log d} \leq \exp\{3 \log_4 y (\log_2 y)^{1/2}\}. \qquad (5.56)$$

By (2.3), (5.14), (5.21) and (5.23),

$$\Omega(f) \leq 2h \log_2 E(\zeta_J) \leq 2(2C \log_3 y + 3)(\log_2 y)^{1/2}.$$

Combined with (5.1), (5.21) and (5.56), this gives

$$h^{2\Omega(f)+\Omega(d)} \ll \exp\{7 \log_3 y (\log_2 y)^{1/2}\}. \qquad (5.57)$$

Also,

$$\sum_{P^+(f) \leq E(\zeta_J)} \frac{1}{f} \ll \log E(\zeta_J) \leq \exp\{(\log_2 y)^{1/2}\}.$$

Together with (5.57), this produces

$$\sum_{m_2} \sum_{\sigma \in \mathfrak{S}_K} \frac{1}{\sigma^{(1)}} \ll \exp\{8 \log_3 y (\log_2 y)^{1/2}\}. \qquad (5.58)$$

Therefore, by (5.54), (5.55) and (5.58),

$$\sum_{m_2} B_j(\zeta; m_1, m_2; y) \ll \frac{y}{\phi(m_1)} (\log y)^{-1-\omega_j/2+2(\log_3 y)^5 (\log_2 y)^{-1/2}}. \qquad (5.59)$$

Now m_1 has at most one prime factor in each interval $[w_i, z_i]$ $(j + 1 \le i \le J - 1)$ and no other prime factors. Therefore,

$$\sum_{m_1} \frac{1}{\phi(m_1)} \ll \prod_{i=j+1}^{J-1} (\varepsilon_i \log_2 y) \ll \exp\{(\log_3 y)^2\}.$$

Also, the number of possibilities for ζ is at most $(\varepsilon_j \cdots \varepsilon_L)^{-1} \le \exp\{2(\log_3 y)^2\}$. Summing over all possible m_1 and ζ and using (5.20) and (5.21), we have

$$B_j^*(y) \ll \frac{y}{\log y} (\log y)^{-\omega_j/2 + 3(\log_3 y)^5 (\log_2 y)^{-1/2}}$$

$$\ll \frac{y}{\log y} \exp\left\{ \frac{-\log_2 y}{2(2C \log_3 y + M + 3)^3} + 3(\log_3 y)^5 (\log_2 y)^{-1/2} \right\}$$

$$\ll \frac{y}{\log y} \exp\{-\log_2 y (\log_3 y)^{-4}\}.$$

Combining this with (5.25) completes the proof of Lemma 5.1. $\square$

6. The normal structure of totients

The proofs of Theorems 1 and 2 suggest that for most totients $m \le x$, all the pre-images n of m satisfy $(x_1, x_2, \ldots, x_L) \in S_L(\xi)$ with L near L_0 and ξ defined as in Section 4. We prove such a result below in Theorem 15, which is an easy consequence of Theorem 1 and the machinery created for its proof. From this, we deduce the normal size of $\Omega(m)$ (Theorem 12, Corollary 13) from the normal size of $x_1 + \cdots + x_L$ for $\mathbf{x} \in \mathscr{S}_L$. Likewise, the normal sizes of the primes $q_i(n)$ are deduced from results on the normal sizes of x_i for $\mathbf{x} \in \mathscr{S}_L$ (Theorems 10 and 11).

Theorem 15.   *Suppose* $0 \le \Psi < L_0(x)$, $L = L_0 - \Psi$ *and let*

$$\xi_i = \xi_i(x) = 1 + \frac{1}{10(L_0 - i)^3} \quad (0 \le i \le L - 2). \tag{6.1}$$

The number of totients $m \le x$ *with a pre-image* n *with*

$$(x_1(n; x), \ldots, x_L(n; x)) \notin \mathscr{S}_L(\xi) \tag{6.2}$$

is

$$\ll V(x) \exp\{-\Psi^2/4C\}.$$

Proof: As in Section 4, define $M_j(x)$ to be the number of totients $m \le x$ with a pre-image satisfying (I_i) for $i < j$, but not satisfying (I_j), where

$$\mathbf{x} = (x_1(n; x), \ldots, x_{L_0}(n; x)).$$

By Theorem 1, Corollary 3.4, (4.13) and (4.14), the number of totients $m \leq x$ with a pre-image n satisfying (6.2) is at most

$$\sum_{i \leq j} M_i(x) \ll \frac{x}{\log x} R_{L_0}(\xi; x) \exp\{-\Psi^2/4C - (D/2C + 1/2C - 1)\Psi + O(\log^2 \Psi)\}$$

$$\ll V(x) \exp\{-\Psi^2/4C\}. \qquad \qquad \Box$$

We show below that most of the contribution to $\mathscr{S}_L$ comes from $\mathbf{x}$ with

$$x_j \approx \varrho^j(1 - j/L) \quad (1 \leq j \leq L - 2).$$

Thus, for most totients $m \leq x$, we should have $\Omega(m) \approx 1/(1 - \varrho) \log_2 x$. Proving this requires accurate upper bounds on $T_L(\mathscr{R})$ for various regions $\mathscr{R}$, where for brevity we write

$$T_L(\mathscr{R}) = \text{Vol}(\mathscr{S}_L \cap \mathscr{R}).$$

For short, we write Σ for $\sum_{i=1}^{L} x_i$ and set $\beta_0 = \varrho/(1 - \varrho)$. As with the bounding of $N(x)$ in Section 4, we need to group totients by the size of x_L.

Lemma 6.1.   *Suppose $\alpha \leq 1/(3g_L^*)$ and $L \geq 100$. For $\beta_0 \leq \beta \leq 2$, we have*

$$T_L(\Sigma \geq \beta, x_L \geq \alpha) \ll \left(\frac{1 - \alpha g_L^*}{1 + a_1(\beta - \beta_0)/(1 - a_1\beta_0)}\right)^L T_L. \qquad (6.3)$$

When $0 \leq \beta \leq \beta_0$, we have

$$T_L(\Sigma \leq \beta, x_L \geq \alpha) \ll \left(\frac{\beta(1 - \alpha g_L^*)}{\beta_0}\right)^L T_L. \qquad (6.4)$$

Proof: For $j \geq 0$, let

$$f_j = \sum_{i=0}^{j} g_i, \quad f_j^* = \sum_{i=0}^{j} g_i^*.$$

Suppose $\mathbf{x} \in \mathscr{S}_L$ and let $\mathbf{1} = \{1, 1, \ldots, 1\}$. By (3.12),

$$-1 = \sum_{i=1}^{L-1} f_{i-1}\mathbf{v}_i + f_{L-1}^*\mathbf{v}_L. \qquad (6.5)$$

First suppose $x_1 + \cdots + x_L \geq \beta$ and $x_L \geq \alpha$. Set $y_i = x_i - \alpha g_{L-i}^*$ for each i. By (1.9), (3.9), (3.11) and (6.5),

$$\mathbf{v}_i \cdot \mathbf{y} \leq 0 \quad (1 \leq i \leq L),$$
$$\mathbf{v}_0 \cdot \mathbf{y} \leq 1 - \alpha g_L^* =: \alpha',$$
$$\mathbf{1} \cdot \mathbf{y} \geq \beta - \alpha f_{L-1}^* =: \beta'.$$

Thus, $\mathbf{u} \cdot \mathbf{y} \leq 0$, where $\mathbf{u} = \beta' \mathbf{v}_0 - \alpha' \mathbf{1}$. This relation will replace the inequality $\mathbf{v}_1 \cdot \mathbf{y} \leq 0$. By (3.13) and (6.5),

$$\mathbf{v}_0 + A\mathbf{u} + \sum_{j=2}^{L-1}(g_j + A(\beta' g_j - \alpha' f_{j-1}))\mathbf{v}_j + (g_L^* + A(\beta' g_L^* - \alpha' f_{L-1}^*))\mathbf{v}_L = \mathbf{0}, \quad (6.6)$$

where $A = a_1/(\alpha' - \beta' a_1)$. The restrictions on α and β force $a_1 \leq A \leq 22a_1$. To show each coefficient in (6.6) is positive, we first note that

$$\frac{f_{j-1}}{g_j} \leq \frac{f_1}{g_2} \leq 1.31,$$

which is deducible from Lemma 3.6. Therefore, the coefficient of $\mathbf{v}_j$ is at least

$$g_j(1 + A(\beta' - 1.31\alpha')) \geq g_j\left(1 + \frac{a_1(\beta - 1.36)}{0.819 - a_1\beta}\right) > 0.$$

Lemma 3.6 also gives

$$f_{j-1} = \beta_0 g_j + O(1), \quad f_{j-1}^* = \beta_0 g_j^* + O(1). \quad (6.7)$$

Applying Lemmas 3.3, 3.5 and 3.6, we deduce

$$T_L(\Sigma \geq \beta, x_L \geq \alpha) \ll \left(\frac{\alpha'}{1 + A(\beta' - \beta_0\alpha')}\right)^L T_L.$$

Some algebraic manipulation using (6.7) now gives (6.3).

Now suppose $x_1 + \cdots + x_L \leq \beta$ and $x_L \geq \alpha$. Adopting the same definitions for y_i, α' and β', we have

$$\mathbf{v}_i \cdot \mathbf{y} \leq 0 \quad (1 \leq i \leq L), \quad \mathbf{1} \cdot \mathbf{y} \leq \beta'.$$

Assume $\beta' > 0$, for otherwise the desired volume is zero by Lemma 3.9. By (6.5), (6.7) and Lemma 3.5,

$$T_L(\Sigma \leq \beta, x_L \geq \alpha) \ll (\beta'/\beta_0)^L T_L.$$

The identity (6.7) implies

$$\beta' = \beta - \beta_0\alpha(g_L^* + O(1)) \leq \beta(1 - \alpha g_L^*) + O(1/g_L^*),$$

which proves (6.4). $\qquad\qquad\qquad\qquad\qquad\qquad\qquad\qquad\qquad\qquad\qquad\qquad\qquad\qquad\quad$ □

Note that Lemma 6.1 gives non-trivial bounds only when $|\beta - \beta_0| \gg L^{-1}$.

Proof of Theorem 12:   Assume $\varepsilon \geq (\log_3 x)^{-1}$, for otherwise the theorem is trivial. Denote by $V^*(x)$ the number of totients $m \leq x$ satisfying

$$\left| \Omega(m) - \frac{\log_2 x}{1 - \varrho} \right| \geq \varepsilon \log_2 x.$$

Let $\Psi = \Psi(x) = [3\sqrt{\varepsilon \log_3 x}]$, $L = L_0(x) - \Psi$, and $S = \exp\{(\log_2 x)^{100}\}$. Let n be a generic pre-image of a totient $m \leq x$, and set $q_i = q_i(n)$ and $x_i = x_i(n; x)$ for $0 \leq i \leq L$. Also set $r = n/(q_0 \cdots q_L)$. Let $U(x)$ denote the number of totients $m \leq x$ with a pre-image n satisfying one of four conditions:

$$(x_1, x_2, \ldots, x_L) \notin \mathscr{S}_L(\xi), \tag{6.8}$$

$$p^2 \mid m \text{ for some prime } p \geq \log^2 x, \tag{6.9}$$

$$\text{some prime factor of } n \text{ is not } S\text{-normal} \tag{6.10}$$

$$\Omega(\phi(r)) \geq 2000 \varrho^{-2\Psi} (\log_3 x)^2. \tag{6.11}$$

By Theorem 15, Lemma 2.9 and Lemma 2.10, the number of totients $m \leq x$ with a pre-image satisfying either (6.8), (6.9) or (6.10) is

$$\ll \frac{x(\log_2 x)^6 W(x)}{\log x} \left\{ \exp\{-\Psi^2/4C\} + 1/\log x + (\log S)^{-1/6} \right\} \ll V(x)(\log_2 x)^{-2\varepsilon}.$$

Now suppose (6.10) and (6.11) hold, but (6.8) does not. By Lemma 3.10,

$$x_L \leq 3\varrho^L < \frac{10 \log_3 x}{\log_2 x} \varrho^{-\Psi} =: \frac{Z}{\log_2 x}.$$

Thus $\log_2 P^+(r) \leq Z$ and $\Omega(\phi(r)) \geq 20Z^2$. Given $q_1, \ldots, q_L$ and r, the number of possible q_0 is

$$\ll \frac{x}{\log x} \frac{1}{\phi(rq_1 \cdots q_L)}.$$

By Lemma 3.12,

$$\sum \frac{1}{\phi(q_1 \cdots q_L)} = R_L(\xi) \ll (\log_2 x)^L T_L.$$

Let $f(y)$ denote the number of possible $r \leq y$. By Lemma 2.3, the number of r with either $r < y/\log^2 y$ or $P^+(r) < y^{1/\log_2 y}$ is $O(y/\log^2 y)$. For remaining r,

$$Z \geq \log_2(P^+(r)) > \log_2(y^{1/\log_2 y}) > (9/10) \log_2 y$$

for large y. Since the prime factors of r are S-normal, $\Omega(\phi(r)) \leq 2Z\Omega(r)$. Thus $\Omega(r) \geq 10Z \geq 9 \log_2 y$, so Lemma 2.2 implies the number of such r is $O(y/\log^2 y)$. Therefore, $f(y) \ll y/\log^2 y$ and by partial summation,

$$\sum_r \frac{1}{\phi(r)} \ll \sum_r \frac{\log_2 r}{r} \ll 1.$$

By Corollary 3.4, the number of totients $m \leq x$ satisfying (6.10) and (6.11) but not (6.8) is

$$\ll \frac{x(\log_2 x)^L}{\log x} T_L \ll V(x)(\log_2 x)^{-2\varepsilon},$$

which implies

$$U(x) \ll V(x)(\log_2 x)^{-2\varepsilon}. \tag{6.12}$$

Denote by $U_1(x)$ and $U_2(x)$ the number of remaining totients $m \leq x$ with $\Omega(m) \geq (1 + \beta_0 + \varepsilon)\log_2 x$ and $\Omega(m) \leq (1 + \beta_0 - \varepsilon)\log_2 x$, respectively. By (2.2), (2.3) and (2.4), we have

$$\Omega(q_i - 1) = \log_2 q_i + O(\sqrt{\log_2 x \log_3 x}) \quad (1 \leq i \leq L).$$

Therefore, by (6.11),

$$\Omega(m) = (1 + x_1 + x_2 + \cdots + x_L)\log_2 x + O((\log_2 x)^{1/2}(\log_3 x)^{3/2}). \tag{6.13}$$

For $\mathbf{x} \in \mathscr{S}_L(\xi)$, let $y_i = (\xi_0\xi_1 \cdots \xi_{i-1})^{-1}x_i$ for each i. By Lemma 3.1, $\mathbf{y} \in \mathscr{S}_L$.

Suppose first that m is counted in $U_1(x)$, and set $J = [6C\log_4 x] + 1$. By Lemma 3.10, this implies

$$x_{J+1} + x_{J+2} + \cdots + x_L \leq 4\varrho^J \leq 4(\log_3 x)^{-3}.$$

Therefore, by (6.1) and (6.13),

$$\begin{aligned} y_1 + \cdots + y_L &\geq (\xi_0\xi_1 \cdots \xi_{J-1})^{-1}(x_1 + \cdots + x_J) \\ &\geq (1 + O((\log_3 x)^{-3}))(\beta_0 + \varepsilon - O((\log_3 x)^{-3})) \\ &\geq \beta_0 + \varepsilon - O((\log_3 x)^{-3}). \end{aligned} \tag{6.14}$$

We again write $n = q_0q_1 \cdots q_L r$, and divide the interval $[0, 1/g_L^*]$ into subintervals, considering separately the totients having a pre-image with y_L in a particular subinterval. We note that

$$\frac{1}{Lg_L^*} \ll \frac{\varrho^{-\Psi}}{\log_2 x}.$$

By Lemmas 3.13, 4.3 and 6.1, together with Theorem 1, the number of totients counted in $U_1(x)$ with $y_L \in [u/Lg_L^*, v/Lg_L^*]$ is

$$\begin{aligned} &\ll \frac{x}{\log x}(\log_2 x)^L \left(\frac{1 - u/L}{1 + a_1\varepsilon/(1 - a_1\beta_0)}\right)^L T_L \frac{v}{Lg_L^*}(\log_2 x)W(E(v/Lg_L^*)) \\ &\ll V(x)\exp\{-K\varepsilon\log_3 x - \Psi\log\Psi + O(\Psi)\}g(u, v), \end{aligned}$$

where $K = 2Ca_1/(1 - \beta_0a_1) = 1.166277\ldots$ and

$$g(u, v) = \exp\{-u + \Psi\log v + O(\log^2 v)\}.$$

The subintervals we use are $([L/3]/Lg_L^*, 1/g_L^*]$ and $(k/Lg_L^*, (k+1)/Lg_L^*]$ for $0 \le k \le [L/3] - 1$. For the first interval, we have

$$g([L/3], L) = \exp\{-[L/3] + O(\Psi \log_4 x + (\log_4 x)^2)\}$$
$$\le e^{-L/4}.$$

For the intervals with small k we obtain

$$\sum_{k \le \Psi^2} g(k, k+1) \le \exp\{O(\log^2 \Psi)\} \sum_{k=1}^{\infty} \frac{k^\Psi}{e^k}$$
$$= \exp\{\Psi \log \Psi - \Psi + O(\log^2 \Psi)\}$$

and for the intervals with large k we have

$$\sum_{k > \Psi^2} g(k, k+1) \le \exp\{-\Psi^2 + O(\Psi \log \Psi)\}.$$

Therefore,

$$U_1(x) \ll V(x) \exp\{-K\varepsilon \log_3 x + O(\Psi)\}.$$

When m is counted in $U_2(x)$ and x is large,

$$y_1 + \cdots + y_L \le x_1 + \cdots + x_L$$
$$\le \beta_0 - \varepsilon + O((\log_2 x)^{-1/2} (\log_3 x)^{3/2}).$$

By (6.4) and an argument similar to that used to bound $U_1(x)$, we have

$$U_2(x) \ll V(x) \exp\{-(2C/\beta_0)\varepsilon \log_3 x + O(\Psi)\}.$$

The first part of the theorem now follows, since $V^*(x) \le U(x) + U_1(x) + U_2(x)$ and $2C/\beta_0 > K$.

For the second part, consider again a totient m not counted in $U(x)$. Then

$$\Omega(m) - \omega(m) \le \sum_{i=0}^{L} \Omega(q_i - 1, 1, S) + \Omega(\phi(r))$$
$$\ll (\log_3 x)^2 + (\varrho^{-\Psi} \log_3 x)^2$$
$$\ll \exp\{O((\log_3 x)^{1/2})\},$$

and thus the theorem holds with $\Omega(m)$ replaced by $\omega(m)$.   □

There is a curious asymmetry between the bounds for $U_1(x)$ and $U_2(x)$, stemming from the asymmetry in the bounds (6.3) and (6.4). This is a real phenomenon, rather than a product of imprecise estimating and is a consequence of an asymmetry in the distribution of the numbers x_i when $\mathbf{x} \in \mathscr{S}_L$ when $i \ll \log L$. The details are found in Lemma 6.2 below.

Proof of Corollary 13: It suffices to prove the theorem with $g(m) = \Omega(m)$. Divide the totients $m \leq x$ into three sets, S_1, those with $\Omega(m) \geq 10 \log_2 x$, S_2, those not in S_1 but with $|\Omega(m) - \log_2 x/(1-\varrho)| \geq \frac{1}{2} \log_2 x$, and S_3, those not counted in S_1 or S_2. By Lemma 2.2,

$$|S_1| \ll \frac{x}{\log^2 x}$$

and by Theorem 12,

$$|S_2| \ll V(x) \exp\{-(K/2) \log_3 x + O(\sqrt{\log_3 x})\} \ll V(x)(\log_2 x)^{-1/2}.$$

Therefore

$$|S_3| = V(x)(1 - O((\log_2 x)^{-1/2})) \tag{6.15}$$

and also

$$\sum_{m \in S_1 \cup S_2} \Omega(m) \ll |S_1| \log x + |S_2| \log_2 x \ll V(x)(\log_2 x)^{1/2}. \tag{6.16}$$

For each $m \in S_3$, let

$$\varepsilon_m = \frac{\Omega(m)}{\log_2 x} - \frac{1}{1-\varrho}$$

and for each natural number N, let $S_{3,N}$ denote the set of $m \in S_3$ with $N \leq |\varepsilon_m| \log_3 x < N + 1$. By Theorem 12, (6.15) and (6.16),

$$\sum_{m \in \mathcal{V}(x)} \Omega(m) = O(V(x)\sqrt{\log_2 x}) + \sum_{0 \leq N \leq \frac{1}{2} \log_3 x} \sum_{m \in S_{3,N}} \Omega(m)$$

$$= \frac{\log_2 x}{1-\varrho} |S_3| + O\left(V(x)\frac{\log_2 x}{\log_3 x} \sum_N (N+1) e^{-KN + O(\sqrt{N})}\right)$$

$$= \frac{V(x) \log_2 x}{1-\varrho}\left(1 + O\left(\frac{1}{\log_3 x}\right)\right).$$

Due to the asymmetry in the estimates (6.3) and (6.4), there is probably a secondary term of order $V(x) \log_2 x/\log_3 x$.

We now turn our attention to examining the size of individual prime factors of a pre-image of a normal totient. As before, the key is estimating the volume of the corresponding subset of $\mathcal{S}_L$. Recall definition (1.9) and Lemma 3.6. Define

$$\lambda_i = \varrho^i g_i \quad (i \geq 0), \qquad \lambda = \lim_{i \to \infty} \lambda_i = \frac{1}{\varrho F'(\varrho)} < \frac{1}{3}. \tag{6.17}$$

Lemma 6.2. *Suppose $i \leq L - 2$, $\alpha < 1/g_L^*$ and $\alpha g_{L-i}^* < \beta < 1/g_i$. Define θ by*

$$\beta = \frac{\varrho^i(1 - i/L)}{1 + \theta}. \tag{6.18}$$

If $\theta > 0$, we have

$$T_L(x_i \le \beta, x_L \ge \alpha) \ll T_L \frac{i}{\theta L} \frac{(1 + \theta L/i)^i}{(1 + \theta)^L} e^{-Lag_L^*}. \tag{6.19}$$

If $\theta < 0$, we have

$$T_L(x_i \ge \beta, x_L \ge \alpha) \ll T_L \frac{(1 - \lambda)^i}{(1 + \theta)^L} \left(1 + \frac{\theta + i\lambda_i/L}{1 - \lambda_i} \right)^L e^{-\frac{2}{3}Lag_L^*}. \tag{6.20}$$

If $-i\lambda_i/L < \theta < 0$ and $\alpha \le -\theta(L - i)/(2ig_L^*)$, then

$$T_L(x_i \ge \beta, x_L \ge \alpha) \ll T_L \frac{i}{|\theta'|L} \frac{(1 + \theta'L/i)^i}{(1 + \theta')^L} e^{-Lag_L^*}, \tag{6.21}$$

where $\theta' = \theta + iag_L^*/(L - i)$.

Proof: For each inequality, we show that the region in question lies inside a simplex for which we may apply Lemma 3.5. The volume is then related to T_L via Lemma 3.3.

The basic strategy is similar to the proof of Lemma 6.1. Consider $\mathbf{x} \in \mathscr{S}_L$ with $x_L \ge \alpha$ and let $y_j = x_j - ag_{L-j}^*$ for each j. Then $\mathbf{v}_j \cdot \mathbf{y} \le 0$ $(1 \le j \le L)$ and $\mathbf{v}_0 \cdot \mathbf{y} \le 1 - ag_L^*$. Let $\alpha' = 1 - ag_L^*$ and $\beta' = \beta - ag_{L-i}^*$. We may assume that $\beta' > 0$, for $x_L \ge \alpha$ implies $x_i \ge ag_{L-i}^*$ (Lemma 3.10). We now apply a second linear transformation, setting $z_j = y_j - \beta'g_{i-j}$ for $j \le i$ and $z_j = y_j$ for $j > i$. We then have

$$\begin{aligned} \mathbf{v}_j \cdot \mathbf{z} &\le 0 \quad (1 \le j \le L, j \ne i), \\ \mathbf{v}_i \cdot \mathbf{z} &\le \beta', \\ \mathbf{v}_0 \cdot \mathbf{z} &\le \alpha' - \beta'g_i. \end{aligned} \tag{6.22}$$

With these definitions, $x_i \ge \beta$ is equivalent to $z_i \ge 0$ and $x_i \le \beta$ is the same as $z_i \le 0$. For any A satisfying

$$A > 0 \quad (\text{when } z_i < 0), \quad -g_i \le A < 0 \quad (\text{when } z_i \ge 0), \tag{6.23}$$

the desired volume is at most the volume of the simplex defined by

$$\begin{aligned} (\mathbf{v}_0 + (g_i + A)\mathbf{v}_i) \cdot \mathbf{z} &\le \alpha' + A\beta', \\ \mathbf{v}_j \cdot \mathbf{z} &\le 0, \\ \pm \mathbf{e}_i \cdot \mathbf{z} &\le 0. \end{aligned} \tag{6.24}$$

By (6.22),

$$(\mathbf{v}_0 + (g_i + A)\mathbf{v}_i) + \sum_{j<i} g_j \mathbf{v}_j + A\mathbf{e}_i + \sum_{j=i+1}^{L-1} (g_j + Ag_{j-i})\mathbf{v}_j + (g_L^* + Ag_{L-i}^*)\mathbf{v}_L = \mathbf{0}. \tag{6.25}$$

By (6.23), each vector on the left of (6.24) has a positive coefficient in the identity (6.25). Therefore, by (3.18) and Lemma 3.5,

$$T_L\left(x_i \lessgtr \beta, x_L \geq \alpha\right) \leq T_L^* \frac{g_i}{|A|} (\alpha' + A\beta')^L \prod_{j=i+1}^{L-1} \left(1 + A\frac{g_{j-i}}{g_j}\right)^{-1} \left(1 + A\frac{g_{L-i}^*}{g_L^*}\right)^{-1}$$

$$\ll T_L \frac{g_i}{|A|} \frac{(\alpha' + A\beta')^L}{(1 + A\varrho^i)^{L-i}}. \tag{6.26}$$

The last inequality follows from Lemma 3.6 and (6.23), which give

$$Ag_{j-i}/g_j \geq -1/(\varrho F'(\varrho))$$

and

$$1 + A\frac{g_{j-i}}{g_j} = (1 + A\varrho^i)(1 + O(\varrho^{j-i})).$$

Define η by

$$\frac{\beta'}{\alpha'} = \frac{\varrho^i(1 - i/L)}{1 + \eta}. \tag{6.27}$$

We work on (6.19) first. Assume that $x_i \leq \beta$ and $\theta > 0$. A convenient choice for A (which is optimal if the term $1/|A|$ in (6.26) is ignored) is

$$A = -\frac{L}{i\varrho^i} + \frac{(L - i)\alpha'}{i\beta'} = \frac{\eta L}{i\varrho^i}. \tag{6.28}$$

From (6.18) and (6.27),

$$1 + \eta = \frac{\beta\alpha'(1 + \theta)}{\beta'} = (1 + \theta)\frac{1 - \alpha g_L^*}{1 - \alpha g_{L-i}^*/\beta}. \tag{6.29}$$

By Lemma 3.6, for L sufficiently large and $i < L$ we have

$$g_{L-i}^*/\beta \geq \varrho^{-i}\frac{L}{L - i} g_{L-i}^* > g_L^*,$$

from which it follows that $\eta > \theta > 0$. Therefore, $A > 0$ and (6.26) gives

$$T_L(x_i \leq \beta, x_L \geq \alpha) \ll T_L \frac{i}{\eta L} \left(\frac{\alpha'}{1 + \eta}\right)^L (1 + \eta L/i)^i, \tag{6.30}$$

which implies (6.19), since the right side of (6.30) is a decreasing function of η. We do give up some accuracy with this replacement, but this is negligible in applications, since $x_L \approx 1/(Lg_L^*)$ for most of $\mathscr{S}_L$.

Now assume $x_i \geq \beta$ and $\theta < 0$. When $\beta \geq \varrho^i$, the inequality $\mathbf{v}_i \cdot \mathbf{x} \leq 0$ is superfluous by Lemma 3.9, so nothing is lost by ignoring this inequality. In any case, taking $A = -g_i$ in (6.26) gives

$$T_L(x_i \geq \beta, x_L \geq \alpha) \ll T_L \frac{(1 - \alpha g_L^* - g_i(\beta - \alpha g_{L-i}^*))^L}{(1 - \varrho^i g_i)^{L-i}}.$$

Now from Lemma 3.6,

$$\frac{g_i g_{L-i}^*}{g_L^*} < g_i \varrho^i < 1/3,$$

so

$$1 - \alpha g_L^* - g_i(\beta - \alpha g_{L-i}^*) < (1 - \beta g_i)(1 - (2/3)\alpha g_L^*)$$

and (6.20) follows. In many instances we can do better with another choice of A. To prove (6.21), assume $-i\lambda_i/L < \theta < 0$ and $\alpha \leq -\theta(L - i)/(2ig_L^*)$. Also define A as in (6.28). From (6.29) and Lemma 3.6,

$$\eta < (1 + \theta)\alpha' - 1 + \alpha g_{L-i}^*/\beta$$
$$< (1 + \theta)(1 - \alpha g_L^*) - 1 + \alpha g_L^*(1 + \theta)L/(L - i)$$
$$< \theta + i\alpha g_L^*/(L - i) = \theta'.$$

By the restrictions on α and θ, we have

$$-i\lambda_i/L < \theta < \eta < \theta' < \theta/20,$$

and therefore $-g_i \leq A < 0$. Inequality (6.26) now gives

$$T_L(x_i \geq \beta, x_L \geq \alpha) \ll T_L \frac{i}{|\eta| L} \left(\frac{\alpha'}{1 + \eta}\right)^L (1 + \eta L/i)^i. \qquad (6.31)$$

Since the right side of (6.31) is increasing with η, we may replace η with θ', which gives (6.21). $\qquad\qquad\qquad\qquad\qquad\qquad\qquad\qquad\qquad\qquad\qquad\qquad\qquad\qquad\qquad\square$

We now apply Lemma 6.2 to the problem of determining the size of $q_i(n)$ when n is a pre-image of a "normal" totient. For convenience, we define $V(x; \mathscr{C})$ to be the number of totients $m \leq x$ with a pre-image n satisfying condition $\mathscr{C}$.

Lemma 6.3. *Suppose x is sufficiently large, $L_0 = L_0(x)$ and $1 \leq i \leq L_0 - 2$. Write*

$$\beta = \frac{\varrho^i(1 - i/L_0)}{1 + \theta}. \qquad (6.32)$$

If $0 \leq \Psi \leq (\log_3 x)^{1/2}$, $i \leq L_0 - \Psi - 2$ *and* $\frac{(3/2)i\Psi}{(L_0-\Psi)(L_0-i)} < \theta \leq \frac{1}{2}$, *then*

$$V(x; q_i(n) \leq \beta \log_2 x) \ll V(x) \left(e^{-\Psi^2/4C} + \frac{i}{\eta L} \frac{(1 + \eta L/i)^i}{(1 + \eta)^L} e^{O(\Psi)} \right), \tag{6.33}$$

where $\eta = \theta - \frac{(3/2)i\Psi}{(L_0-\Psi)(L_0-i)}$ *and* $L = L_0 - \Psi$. *If* $-\frac{1}{2} \leq \theta + \frac{1}{5(L_0-i)^2} < 0$, *then*

$$V(x; q_i(n) \geq \beta \log_2 x)$$
$$\ll V(x) \exp\left\{ \frac{\lambda_i L_0 \theta}{1 - \lambda_i} + K_1 i + O(L_0\theta^2 + \sqrt{-L_0\theta} \log(-L_0\theta)) \right\}. \tag{6.34}$$

where $K_1 = \frac{\lambda}{1-\lambda} + \log(1 - \lambda) = 0.08734\ldots$ *If* $\frac{-i\lambda_i}{L_0} \leq \theta < 0$ *then*

$$V(x; q_i(n) \geq \beta \log_2 x)$$
$$\ll V(x) \exp\left\{ -\frac{1}{2} \frac{L_0(L_0 - i)}{i} \theta^2 + O\left(|\theta| \sqrt{\frac{L_0(L_0 - i)}{i}} \log(L_0 - i) \right) \right\}. \tag{6.35}$$

Proof: Set $L = L_0 - \Psi$ and define ξ_i as in Theorem 15. By Theorem 15, the number of totients $m \leq x$ with a pre-image n satisfying $(x_1(n; x), \ldots, x_L(n; x)) \notin \mathscr{S}_L(\xi)$ is

$$\ll V(x) \exp\{-\Psi^2/4C\}. \tag{6.36}$$

For remaining totients, let $x_i = x_i(n; x)$ and set $y_i = (\xi_0 \cdots \xi_{i-1})^{-1} x_i$ for each i, so that $\mathbf{y} \in \mathscr{S}_L$.

First, assume Ψ, i and θ satisfy the hypotheses of (6.33). Then $y_i \leq \beta$. To set up an application of Lemma 6.2 (6.19), define θ' by

$$\frac{\varrho^i(1 - i/L)}{1 + \theta'} = \beta.$$

Some algebraic manipulation gives

$$\theta' \geq \eta, \quad \eta = \theta - \frac{(3/2)i\Psi}{L(L_0 - i)}.$$

By the hypothesis on θ, $\eta > 0$. For each natural number k, we consider separately totients with y_L satisfying

$$\frac{k}{Lg_L^*} \leq y_L \leq \frac{(k + 1)}{Lg_L^*}. \tag{6.37}$$

Arguing as in the proof of Theorem 12, by Lemmas 3.13, 4.3 and 6.2, we have

$$V(x; q_i(n) \leq \beta \log_2 x) \ll \frac{x}{\log x} T_L(\log_2 x)^L \frac{i}{\eta L} \frac{(1 + \eta L/i)^i}{(1 + \eta)^L} \sum_{k \geq 0} e^{-k} W(Z_k) \log_2 Z_k,$$

where $\log_2 Z_k = (k+1)(\log_2 x)/(Lg_L^*)$. Estimating the sum on k as in the proof of Theorem 12 and applying Corollary 3.4 gives (6.33).

Now assume that $-\frac{1}{2} \le \theta + \frac{1}{5(L_0-i)^2} \le -\frac{i\lambda_i}{L_0}$. Then $y_i \ge (\xi_0 \cdots \xi_{i-1})^{-1}\beta$. From (6.1) and an argument similar to the case $\theta > 0$, we have

$$y_i \ge \frac{\varrho^i (1-i/L)}{1+\eta}, \quad \eta = \theta + \frac{1}{5(L_0-i)^2}.$$

Take $\Psi = [(2CL_0|\eta|)^{1/2}] + 1$. Using Lemma 6.2 (6.20) and the same subdivision of y_L, (6.34) follows in the same manner as the proof of (6.33). A couple of inequalities used are

$$\frac{i\lambda_i}{1-\lambda_i} = \frac{i\lambda}{1-\lambda} + O(1)$$

and

$$\left(1 + \frac{\eta + i\lambda_i/L}{1-\lambda_i}\right)^L \le \exp\left\{\frac{L\eta + i\lambda_i}{1-\lambda_i}\right\},$$

the former being a consequence of Lemma 3.6. The factor $2/3$ occurring in (6.20) accounts for the factor $e^{O(\Psi \log \Psi)}$ instead of the factor $e^{O(\Psi)}$.

For the last inequality, assume $-\frac{i\lambda_i}{L_0} \le \theta < 0$. We may also assume that

$$\theta \le -3\left(\frac{i}{L_0(L_0-i)}\right)^{1/2}, \tag{6.38}$$

for otherwise (6.35) is trivial. Define

$$\eta = \theta + \frac{1}{5(L_0-i)^2} \tag{6.39}$$

and

$$\Psi = \left[|\eta|\sqrt{\frac{2CL_0(L_0-i)}{i}}\right]. \tag{6.40}$$

Set $L = L_0 - \Psi$. By (6.38) and (6.39),

$$-\frac{i\lambda_i}{L} < -\frac{i\lambda_i}{L_0} \le \eta < -2\left(\frac{i}{L_0(L_0-i)}\right)^{1/2}. \tag{6.41}$$

Therefore, for x large,

$$2 \le \Psi \le \frac{1}{2}\sqrt{\frac{i(L_0-i)}{L_0}} \le \frac{1}{2}\sqrt{L_0-i}. \tag{6.42}$$

In particular, (6.42) implies $L_0 - i \geq 16$ and thus $\Psi \leq \frac{1}{8}(L_0 - i)$. Consider now a totient $m \leq x$, all of whose pre-images n satisfy $(x_1, \ldots, x_L) \in \mathscr{S}_L(\xi)$. As in the proof of (6.34), we have

$$y_i \geq \frac{\varrho^i(1 - i/L)}{1 + \eta}. \tag{6.43}$$

As before, we group such n as to the size of y_L (see (6.37)). However, the hypotheses of Lemma 6.2 require that

$$k \leq \frac{|\eta|}{2} \frac{L(L - i)}{i} =: k_0$$

in order to use (6.21). By (6.40),

$$k_0 < \frac{2C|\eta|L_0(L_0 - i)}{4Ci} \leq \frac{\Psi^2}{4C}.$$

For $k < \Psi^2/4C$, we shall use the bound (3.24) in place of (6.21). By (3.24) and previous estimates for the sum over k, the number of n counted with $k > \Psi^2/4C$ is

$$\ll V(x) \exp\{-\Psi^2/4C + O(\Psi \log \Psi)\}. \tag{6.44}$$

For $k \leq \Psi^2/4C$, define $\eta_k = \eta + \frac{ki}{L(L-i)}$. By (6.40) and (6.42),

$$\eta_k \leq \eta + \frac{|\eta|}{2} \frac{L_0(L_0 - i)}{i} \frac{i}{L(L - i)} < \eta/3 < 0.$$

Hence, by Lemmas 3.13, 4.3 and 6.2 (6.21), the number of n with $k \leq \Psi^2/4C$ is

$$\ll \frac{x}{\log x} T_L (\log_2 x)^L \frac{i}{\eta L} \sum_{k \leq \Psi^2} \frac{(1 + \eta_k L/i)^i}{(1 + \eta_k)^L} e^{-k} W(Z_k) \log_2 Z_k, \tag{6.45}$$

where again $\log_2 Z_k = (k + 1)(\log_2 x)/(Lg_L^*)$. To estimate the sum on k, we use

$$\frac{(1 + \eta_k L/i)^i}{(1 + \eta_k)^L} e^{-k} \leq \exp\left\{ -\frac{1}{2} \frac{L(L - i)}{i} \eta^2 - (1 + \eta)k \right\}$$

and

$$\sum_{k=0}^{\infty} \frac{k^\Psi}{e^{(1+\eta)k}} \ll \Gamma\left(\frac{\Psi}{1 + \eta}\right) \ll \exp\{(1 - 2\eta)\Psi \log \Psi\}.$$

Combining (6.36), (6.44) and (6.45) together with the bounds (6.38)–(6.42) readily gives (6.35).   □

Proof of Theorem 10: For the first part, we may assume $\varepsilon > 2(\frac{L_0(L_0-i)}{i})^{-1/2}$, for otherwise the result is trivial. Let $\beta = \frac{\beta_i}{1+\varepsilon} > \beta_i(1-\varepsilon)$ and take $\Psi = [\varepsilon\sqrt{L_0(L_0-i)/i}]$. Note that the bounds on ε give

$$2 \le \Psi < \frac{1}{6}(L_0 - i) \le L_0 - i - 2$$

and

$$\eta := \varepsilon - \frac{(3/2)i\Psi}{(L_0-\Psi)(L_0-i)} \ge \varepsilon\left(1 - 2\sqrt{\frac{i}{L_0(L_0-i)}}\right) > 0.$$

An application of Lemma 6.3 (6.33) now gives the first part of the theorem in the case $q_i(n)/\log_2 x \le \beta_i(1-\varepsilon)$. Now set $\beta = \beta_i(1+\varepsilon) = \frac{\beta_i}{1+\theta}$, so that

$$-\frac{i\lambda_i}{L_0} \le -\varepsilon \le \theta \le -\varepsilon + \varepsilon^2.$$

An application of Lemma 6.3 (6.35) completes the first part of the theorem.

The second part follows in a similar manner, except that now the bound for $V(x; \frac{q_i(n)}{\log_2 x} \ge \beta_i(1+\varepsilon))$ will be weaker than the bound for $V(x; \frac{q_i(n)}{\log_2 x} \le \beta_i(1-\varepsilon))$. For the former, Lemma 6.3 (6.34) gives the bound

$$V(x)\exp\left\{-\frac{\lambda_i}{1-\lambda_i}L_0\varepsilon + K_1 i + O(L_0\varepsilon^2 + (L_0\varepsilon)^{1/2}\log(L_0\varepsilon))\right\},$$

while for the latter, using $\Psi = [\sqrt{4CL_0\varepsilon}]$ in (6.33) we obtain the upper bound

$$V(x)\exp\{-L_0\varepsilon + i\log(1 + \varepsilon L_0/i) + O(L_0\varepsilon^2)\}. \qquad \square$$

Proof of Theorem 11: Assume first that $g(x) \ge G$, some large absolute constant, for otherwise the conclusion is trivial. For each i set

$$\varepsilon_i = \sqrt{\frac{i}{L_0(L_0-i)}\log(L_0-i)\log g(x)}.$$

When $1 \le i \le \log^3 L_0$, the second part of Theorem 10 gives

$$V\left(x; \left|\frac{q_i(n)}{\beta_i \log_2 x} - 1\right| \le \varepsilon_i\right)$$
$$\ll V(x)\exp\{-K_2\sqrt{i}\log_4 x \log g(x) + O(i^{1/4}\sqrt{\log_4 x \log g(x)})\},$$

where $K_2 = \lambda_1/(1-\lambda_1) \ge 0.265$. Now suppose $\log^3 L_0 \le i \le L_0 - g(x)$ and set $H = L_0 - i$. The first part of Theorem 10 gives

$$V\left(x; \left|\frac{q_i(n)}{\beta_i \log_2 x} - 1\right| \le \varepsilon_i\right) \ll V(x)\exp\left\{-\frac{1}{2}\log^2 H \log^2 g(x)\left(1 + O\left(\frac{1}{\log g(x)}\right)\right)\right\}$$
$$\ll V(x)\exp\{-(1/3)\log^2 H \log^2 g(x)\}$$

for x sufficiently large. Since $H \geq g(x)$, summing on i gives the desired upper bound on totients with an n not satisfying (1.10). Note that with n satisfying (1.10), we have $\Omega(n) \geq \omega(n) \geq L_0(x) - g(x)$. By Theorem 15, the number of totients $m \leq x$ with $\mathbf{x} \notin \mathscr{S}_L(\xi)$ is $O(V(x)e^{-\psi^2/4C})$ where $L = L_0 - \psi$, $\psi = \log g(x)$. Suppose $\mathbf{x}(n) \in \mathscr{S}_L(\xi)$, $n = q_0 \dots q_L r$ and $\Omega(r) > g(x)$. By Lemma 3.10,

$$\log_2 p^+(r) \leq \log_2 q_L \leq 3\psi \varrho^{-\psi} \leq 3 \log g(x) \cdot g(x)^{1/2C} \leq \frac{g(x)}{100}$$

if $g(x)$ is large. By Lemmas 2.2 and 2.3, $\sum \frac{1}{\phi(r)} \ll 1$. By Lemma 3.12 and Corollary 3.4, the number of totients $\leq x$ with such a pre-image n is $\ll V(x)e^{-(1/4C)\log^2 g(x)}$. $\square$

As a final remark, it is trivial that Theorems 10–12 and Corollary 13 hold with $V(x)$ and $V(x; \mathscr{C})$ replaced by the corresponding functions $V_k(x)$ and $V_k(x; \mathscr{C})$ (here the implied constants depend on k).

7. Conjectures of Sierpiński and Carmichael

7.1. Sierpiński's Conjecture

Prime k-tuples Conjecture (Dickson [6]). *Suppose $a_1, a_2, \dots a_k$ are positive integers and $b_1, b_2, \dots, b_k$ are integers so that no prime p divides $(a_1 n + b_1) \cdots (a_k n + b_k)$ for every integer n. Then the numbers $a_1 n + b_1, \dots, a_k n + b_k$ are simultaneously prime for infinitely many n.*

Schinzel's argument deducing Sierpiński's Conjecture from Hypothesis H requires $\gg k$ polynomials of degrees up to k to be simultaneously prime to show the existence of a number with multiplicity k. Below we follow a completely different approach, which is considerably simpler and requires only the simultaneous primality of three linear polynomials (the Prime 3-tuples Conjecture). The idea is to take a number m with multiplicity k and construct a multiple of it with multiplicity $k + 2$. This is motivated by the technique used in Section 5 where many numbers with multiplicity κ are constructed from a single example.

Lemma 7.1. *Suppose $A(m) = k$ and p is a prime satisfying*
   (i) $p > 2m + 1$,
   (ii) $2p + 1$ *and* $2mp + 1$ *are prime*,
   (iii) $dp + 1$ *is composite for all* $d \mid 2m$ *except* $d = 2$ *and* $d = 2m$.
Then $A(2mp) = k + 2$.

Proof: Suppose $\phi^{-1}(m) = \{x_1, \dots, x_k\}$ and $\phi(x) = 2mp$. Condition (i) implies $p \nmid x$, hence $p \mid (q - 1)$ for some prime q dividing x. Since $(q - 1) \mid 2mp$, we have $q = dp + 1$ for some divisor d of $2m$. We have $q > 2p$, so $q^2 \nmid x$ and $\phi(x) = (q - 1)\phi(x/q)$. By conditions (ii) and (iii), either $q = 2p + 1$ or $q = 2mp + 1$. In the former case, $\phi(x/q) = m$, which has solutions $x = (2p + 1)x_i$ ($1 \leq i \leq k$). In the latter case, $\phi(x/q) = 1$, which has solutions $x = q$ and $x = 2q$. $\square$

Now suppose $A(m) = k$ and let $d_1, \dots, d_j$ be the divisors of $2m$ with $3 \leq d_i < 2m$. Let $p_1, \dots, p_j$ be distinct primes satisfying $p_i > d_i$ for each i. Using the Chinese

Remainder Theorem, let $a \bmod b$ denote the intersection of the residue classes $-d_i^{-1} \bmod p_i$ ($1 \leq i \leq j$). Then for every h and i, $(a + bh)d_i + 1$ is divisible by p_i, hence composite for large enough h. The Prime k-tuples Conjecture implies that there are infinitely many numbers h so that $p = a + hb$, $2p + 1$ and $2mp + 1$ are simultaneously prime. By Lemma 7.1, $A(2mp) = k + 2$ for each prime p. Starting with $A(1) = 2$ and $A(2) = 3$, Sierpiński's Conjecture follows by induction on k.

Table 2 of [31] lists the smallest m for which $A(m) = k$ for $2 \leq k \leq 100$. In all cases, m is less than 100,000. A modest computer search revealed that for each k, $2 \leq k \leq 1000$, there is an $m < 23,000,000$ with $A(m) = k$. The smallest of these values (denoted m_k) are listed in Table 1.

7.2. Carmichael's Conjecture

The basis for computations of lower bounds for a counterexample to Carmichael's Conjecture is the following Lemma of Carmichael [4], as refined by Klee [21]. For short, let $s(n) = \prod_{p|n} p$ denote the square-free kernel of n.

Lemma 7.2. *Suppose* $\phi(x) = m$ *and* $A(m) = 1$. *If* $d \mid x$, e *divides* $\frac{x/d}{s(x/d)}$ *and* $P = 1 + e\phi(d)$ *is prime, then* $P^2 \mid x$.

From Lemma 7.2 it is easy to deduce $2^2 3^2 7^2 43^2 \mid x$. Here, following Carmichael, we break into two cases: (I) $3^2 \parallel x$ and (II) $3^3 \mid x$. In case (I) it is easy to show that $13^2 \mid x$. From this point onward Lemma 7.2 is used to generate a virtually unlimited set of primes P for which $P^2 \mid x$. In case (I) we search for P using $d = 1$, $e = 6k$ or $d = 9$, $e = 2k$, where k is a product of distinct primes (other than 2 or 3) whose squares we already know divide x. That is, if $6k + 1$ or $12k + 1$ is prime its square divides x. In case (II) we try $d = 9$, $e = 2k$ and $d = 27$, $e = k$, i.e., we test, whether or not $6k + 1$ and $18k + 1$ are primes.

As in [31], certifying that a number P is prime is accomplished with the following lemma of Lucas, Lehmer, Brillhart and Selfridge.

Lemma 7.3. *Suppose, for each prime* q *dividing* $n - 1$, *there is a number* a_q *satisfying* $a_q^{n-1} \equiv 1$ *and* $a_q^{(n-1)/q} \not\equiv 1 \pmod{n}$. *Then* n *is prime.*

The advantage of using Lemma 7.3 in our situation is that for a given P we are testing, we already know the prime factors of $P - 1$ (i.e., 2, 3 and the prime factors of k).

Our overall search strategy differs from [31]. In each case, we first find a set of 32 "small" primes P (from here on, P will represent a prime generated from Lemma 7.2 for which $P^2 \mid x$, other than 2 or 3). Applying Lemma 7.2, taking k to be all possible products of 1, 2, 3 or 4 of these 32 primes yields a set S of 1000 primes P, which we order $p_1 < \cdots < p_{1000}$. This set will be our base set. In particular, $p_{1000} = 796486033533776413$ in case (I) and $p_{1000} = 783994289507697435075\overline{19}$ in case (II). The calculations are then divided into "runs". For run #0, we take for k all possible combinations of 1, 2 or 3 of the primes in S. For $j \geq 1$, run #j tests every k which is the product of p_j and three larger primes in S.

Table 1. Smallest solution to $A(m) = k$.

k	m_k	k	m_k	k	m_k	k	m_k	k	m_k
2	1	40	1200	78	38640	116	7200	154	23760
3	2	41	15936	79	9360	117	8064	155	13440
4	4	42	3312	80	81216	118	54000	156	54720
5	8	43	3072	81	4032	119	6912	157	47040
6	12	44	3240	82	5280	120	43680	158	16128
7	32	45	864	83	4800	121	32400	159	48960
8	36	46	3120	84	4608	122	153120	160	139392
9	40	47	7344	85	16896	123	225280	161	44352
10	24	48	3888	86	3456	124	9600	162	25344
11	48	49	720	87	3840	125	15552	163	68544
12	160	50	1680	88	10800	126	4320	164	55440
13	396	51	4992	89	9504	127	91200	165	21120
14	2268	52	17640	90	18000	128	68640	166	46656
15	704	53	2016	91	23520	129	5760	167	15840
16	312	54	1152	92	39936	130	49680	168	266400
17	72	55	6000	93	5040	131	159744	169	92736
18	336	56	12288	94	26208	132	16800	170	130560
19	216	57	4752	95	27360	133	19008	171	88128
20	936	58	2688	96	6480	134	24000	172	123552
21	144	59	3024	97	9216	135	24960	173	20736
22	624	60	13680	98	2880	136	122400	174	14400
23	1056	61	9984	99	26496	137	22464	175	12960
24	1760	62	1728	100	34272	138	87120	176	8640
25	360	63	1920	101	23328	139	228960	177	270336
26	2560	64	2400	102	28080	140	78336	178	11520
27	384	65	7560	103	7680	141	25200	179	61440
28	288	66	2304	104	29568	142	84240	180	83520
29	1320	67	22848	105	91872	143	120000	181	114240
30	3696	68	8400	106	59040	144	183456	182	54432
31	240	69	29160	107	53280	145	410112	183	85536
32	768	70	5376	108	82560	146	88320	184	172224
33	9000	71	3360	109	12480	147	12096	185	136800
34	432	72	1440	110	26400	148	18720	186	44928
35	7128	73	13248	111	83160	149	29952	187	27648
36	4200	74	11040	112	10560	150	15120	188	182400
37	480	75	27720	113	29376	151	179200	189	139104
38	576	76	21840	114	6720	152	10080	190	48000
39	1296	77	9072	115	31200	153	13824	191	102816

(*Continued on next page.*)

Table 1. (*Continued.*)

k	m_k	k	m_k	k	m_k	k	m_k	k	m_k
192	33600	230	137088	268	89856	306	67200	344	192000
193	288288	231	73920	269	101376	307	133056	345	370944
194	286848	232	165600	270	347760	308	82944	346	57600
195	59904	233	184800	271	124800	309	114048	347	1181952
196	118800	234	267840	272	110592	310	48384	348	1932000
197	100224	235	99840	273	171360	311	43200	349	1782000
198	176400	236	174240	274	510720	312	1111968	350	734976
199	73440	237	104832	275	235200	313	1282176	351	473088
200	174960	238	23040	276	25920	314	239616	352	467712
201	494592	239	292320	277	96000	315	1135680	353	556800
202	38400	240	93600	278	464640	316	274560	354	2153088
203	133632	241	93312	279	200448	317	417600	355	195840
204	38016	242	900000	280	50400	318	441600	356	249600
205	50688	243	31680	281	30240	319	131040	357	274176
206	71280	244	20160	282	157248	320	168480	358	767232
207	36288	245	62208	283	277200	321	153600	359	40320
208	540672	246	37440	284	228480	322	168000	360	733824
209	112896	247	17280	285	357696	323	574080	361	576576
210	261120	248	119808	286	199584	324	430560	362	280800
211	24192	249	364800	287	350784	325	202752	363	63360
212	57024	250	79200	288	134784	326	707616	364	1351296
213	32256	251	676800	289	47520	327	611520	365	141120
214	75600	252	378000	290	238464	328	317952	366	399360
215	42240	253	898128	291	375840	329	624960	367	168960
216	619920	254	105600	292	236544	330	116640	368	194400
217	236160	255	257040	293	317520	331	34560	369	1067040
218	70560	256	97920	294	166320	332	912000	370	348480
219	291600	257	176256	295	312000	333	72576	371	147840
220	278400	258	264384	296	108864	334	480000	372	641520
221	261360	259	244800	297	511488	335	110880	373	929280
222	164736	260	235872	298	132480	336	1259712	374	1632000
223	66240	261	577920	299	354240	337	1350720	375	107520
224	447120	262	99360	300	84480	338	250560	376	352512
225	55296	263	64800	301	532800	339	124416	377	165888
226	420000	264	136080	302	218880	340	828000	378	436800
227	26880	265	213120	303	509184	341	408240	379	982080
228	323136	266	459360	304	860544	342	74880	380	324000
229	56160	267	381024	305	46080	343	1205280	381	307200

(*Continued on next page.*)

136

Table 1. (*Countinued.*)

k	m_k	k	m_k	k	m_k	k	m_k	k	m_k
382	496800	420	1879200	458	86400	496	2455488	534	368640
383	528768	421	1756800	459	1575936	497	499200	535	529920
384	1114560	422	90720	460	248832	498	834624	536	2036736
385	1609920	423	376320	461	151200	499	1254528	537	751680
386	485760	424	1461600	462	1176000	500	2363904	538	233280
387	1420800	425	349920	463	100800	501	583200	539	463680
388	864864	426	158400	464	601344	502	1029600	540	2042880
389	959616	427	513216	465	216000	503	2519424	541	3018240
390	1085760	428	715392	466	331776	504	852480	542	2311680
391	264960	429	876960	467	337920	505	1071360	543	1368000
392	470016	430	618240	468	95040	506	3961440	544	3120768
393	400896	431	772800	469	373248	507	293760	545	1723680
394	211200	432	198720	470	559872	508	1065600	546	1624320
395	404352	433	369600	471	228096	509	516096	547	262080
396	77760	434	584640	472	419328	510	616896	548	696960
397	112320	435	708480	473	762048	511	639360	549	1889280
398	1148160	436	522720	474	342720	512	4014720	550	734400
399	51840	437	884736	475	918720	513	266112	551	842400
400	152064	438	1421280	476	917280	514	2386944	552	874368
401	538560	439	505440	477	336000	515	126720	553	971520
402	252000	440	836352	488	547200	516	2469600	554	675840
403	269568	441	60480	479	548352	517	2819520	555	4306176
404	763776	442	1836000	480	129600	518	354816	556	1203840
405	405504	443	866880	481	701568	519	1599360	557	668160
406	96768	444	1537920	482	115200	520	295680	558	103680
407	1504800	445	1219680	483	1980000	521	1271808	559	2611200
408	476928	446	349440	484	1291680	522	304128	560	820800
409	944640	447	184320	485	1199520	523	3941280	561	663552
410	743040	448	492480	486	556416	524	422400	562	282240
411	144000	449	954720	487	359424	525	80640	563	3538944
412	528000	450	1435200	488	1378080	526	508032	564	861120
413	1155840	451	215040	489	2088000	527	2677248	565	221760
414	4093440	452	990720	490	399168	528	5634720	566	768000
415	134400	453	237600	491	145152	529	411840	567	2790720
416	258048	454	69120	492	2841600	530	2948400	568	953856
417	925344	455	384000	493	1622880	531	972000	569	7138368
418	211680	456	338688	494	1249920	532	2813184	570	655200
419	489600	457	741888	495	2152800	533	3975552	571	3395520

(*Continued on next page.*)

Table 1. (*Countinued.*)

k	m_k	k	m_k	k	m_k	k	m_k	k	m_k
572	3215520	610	1404000	648	1938816	686	712800	724	3556224
573	2605824	611	554400	649	316800	687	1059840	725	1943040
574	1057536	612	3120000	650	3075840	688	1733760	726	453600
575	1884960	613	1559040	651	860160	689	728640	727	4012800
576	3210240	614	1672704	652	746496	690	2673216	728	3182400
577	1159200	615	336960	653	2708640	691	1410048	729	4548960
578	4449600	616	2908224	654	5466240	692	120960	730	709632
579	272160	617	332640	655	635040	693	6292800	731	4736160
580	913920	618	1098240	656	4327680	694	3875040	732	2794176
581	393120	619	998400	657	1603584	695	684288	733	4018560
582	698880	620	230400	658	988416	696	423360	734	181440
583	2442240	621	1545600	659	1749600	697	3504384	735	699840
584	6914880	622	1056000	660	1096704	698	4915200	736	2069760
585	695520	623	4610304	661	2724480	699	5456640	737	285120
586	497664	624	1077120	662	6255360	700	6829056	738	993600
587	808704	625	3394560	663	456192	701	2946240	739	4134240
588	2146176	626	1188000	664	1915200	702	2864160	740	5298048
589	2634240	627	4331520	665	172800	703	2352000	741	6868800
590	4250400	628	299520	666	739200	704	2064384	742	561600
591	2336256	629	3219840	667	1491840	705	387072	743	943488
592	1516320	630	3590400	668	608256	706	1169280	744	1850688
593	268800	631	940800	669	1094400	707	984960	745	4490640
594	656640	632	677376	670	4078080	708	5193216	746	3516480
595	1032192	633	1742400	671	374400	709	2125440	747	504000
596	4743360	634	4589568	672	596160	710	6948864	748	4416000
597	4101120	635	3292800	673	3304800	711	253440	749	6438528
598	2410560	636	7104000	674	3052800	712	2970240	750	1278720
599	9922560	637	737280	675	756000	713	822528	751	532224
600	427680	638	1080000	676	1568160	714	4727808	752	3015936
601	662400	639	224640	677	1327104	715	3689280	753	2201472
602	1486080	640	2587200	678	2204928	716	844800	754	2845440
603	2227680	641	2370816	679	936000	717	6905088	755	3198720
604	1149120	642	2706048	680	190080	718	161280	756	6306048
605	138240	643	653184	681	1026432	719	201600	757	5428800
606	752640	644	1477440	682	1870848	720	2877120	758	2503872
607	397440	645	1848000	683	979200	721	435456	759	4623360
608	1880064	646	3446784	684	5520960	722	3252480	760	1336320
609	155520	647	792000	685	2066688	723	1403136	761	1975680

(*Continued on next page.*)

Table 1. (Countinued.)

k	m_k	k	m_k	k	m_k	k	m_k	k	m_k
762	2483712	800	5248800	838	2439360	876	1542240	914	15593472
763	6741504	801	460800	839	9767520	877	5280000	915	4868640
764	207360	802	2237760	840	3345408	878	19885824	916	3440640
765	4884480	803	16298496	841	564480	879	1722240	917	4919040
766	591360	804	1078272	842	1753920	880	10632960	918	443520
767	1907712	805	2527200	843	927360	881	1959552	919	3758400
768	432000	806	2550240	844	3407040	882	6884352	920	2545920
769	276480	807	1996800	845	4579200	883	6240000	921	2112000
770	1827840	808	5110560	846	2736000	884	2306304	922	345600
771	3302208	809	475200	847	5744640	885	5879808	923	3880800
772	259200	810	2115072	848	1321920	886	10964160	924	1468800
773	685440	811	2635776	849	3072000	887	2496000	925	22492800
774	3009600	812	3060288	850	524160	888	5144832	926	2756160
775	3346560	813	1393920	851	3245760	889	449280	927	8125920
776	5948640	814	1002240	852	4176000	890	16057440	928	3598560
777	8305920	815	2479680	853	1324800	891	4504320	929	4680000
778	3307392	816	2923200	854	7992000	892	2239488	930	4104000
779	12403200	817	4593600	855	8726400	893	11100672	931	2751840
780	5087232	818	1283040	856	1010880	894	881280	932	10178784
781	2703360	819	2426112	857	466560	895	2462400	933	2086560
782	6815232	820	1909440	858	7231680	896	9085440	934	11625120
783	19325952	821	5510400	859	10444800	897	2820096	935	552960
784	1016064	822	1918080	860	2861568	898	3897600	936	7683840
785	2842560	823	798336	861	7171200	899	6791040	937	7064064
786	2678400	824	14424480	862	302400	900	7689600	938	1905120
787	6138720	825	3172608	863	823680	901	10133760	939	5308416
788	7666560	826	1397760	864	3409920	902	774144	940	2099520
789	5400000	827	4245696	865	2119680	903	2405376	941	580608
790	3729600	828	6199200	866	3055104	904	1503360	942	5195520
791	1566720	829	1179360	867	576000	905	912384	943	12531456
792	2423520	830	1615680	868	2972160	906	1990656	944	506880
793	1853280	831	5679360	869	13167360	907	8458560	945	6439680
794	4039200	832	794880	870	2949120	908	3400320	946	3306240
795	1670400	833	1440000	871	4292352	909	4432320	947	4723200
796	1720320	834	3564288	872	311040	910	1689600	948	4663296
797	1437696	835	3840000	873	9803520	911	1641600	949	2963520
798	2407680	836	599040	874	6547968	912	6098400	950	3091200
799	2196480	837	3928320	875	13258080	913	8114400	951	1419264

(Continued on next page.)

Table 1. *(Countinued.)*

k	m_k	k	m_k	k	m_k	k	m_k	k	m_k
952	9914400	962	3196800	972	9561600	982	3815424	992	22958208
953	1788480	963	6105600	973	4821120	983	16524000	993	1344000
954	5107200	964	1958400	974	3010560	984	8631360	994	2059200
955	2203200	965	3886080	975	14276736	985	6716160	995	2449440
956	4133376	966	17860608	976	633600	986	4945920	996	7925760
957	2188800	967	10195200	977	748800	987	10386432	997	12109824
958	5581440	968	4953600	978	2672640	988	1166400	998	997920
959	9938880	969	1976832	979	13298688	989	4872960	999	12633600
960	5448960	970	7996800	980	2875392	990	1028160	1000	1360800
961	322560	971	4804800	981	3133440	991	1965600		

Table 2. Case I summary.

Run	# Primes	Sum $\log_{(10)} P$	CPU time
0	10,191,639	375,034,038.57	120.64
1	12,236,589	462,341,117.21	159.45
2	10,710,973	408,376,259.57	143.60
3	9,886,494	382,858,480.93	138.12
4	9,958,236	388,272,039.16	140.88
5	9,693,975	381,285,664.20	139.83
6	9,479,208	378,432,045.33	141.85
7	9,147,009	368,620,935.75	140.73
8	9,271,836	378,581,332.58	146.64
9	8,965,114	366,471,885.76	142.58
10	8,864,722	364,852,167.97	143.39
11	9,047,650	372,728,985.39	146.21
12	9,066,729	373,999,403.77	146.89
Total	126,520,174	5,001,854,356.19	1,850.81

Each candidate P is first tested for divisibility by small primes and must pass the strong pseudoprime test with bases 2, 3, 5, 7, 11 and 13 before attempting to certify that it is prime.

There are two advantages to this approach. First, the candidates P are relatively small (the numbers tested in case (I) had an average of 40 digits and the numbers tested in case (II) had an average of 52 digits). Second, $P - 1$ has at most 6 prime factors, simplifying the certification process.

To achieve $\prod P^2 > 10^{10^{10}}$, 13 runs (run #0 to run #12) were required in case (I) and 14 runs were required in case (II). Together these runs give Theorem 6. The number of primes found, $\log_{(10)} \prod P$ (logarithm to the base 10) and the CPU time (in hours) for each run are listed in Tables 2 and 3.

Table 3. Case II summary.

Run	# Primes	Sum $\log_{(10)} P$	CPU time
0	7,909,575	370,596,531.24	176.23
1	9,348,902	446,035,827.76	225.68
2	8,258,104	398,747,306.87	206.02
3	7,893,206	384,838,826.63	201.54
4	7,417,173	365,511,032.37	194.46
5	7,154,873	361,812,847.34	201.28
6	7,497,735	381,631,849.23	212.99
7	7,253,311	370,521,782.34	208.42
8	7,351,443	378,232,507.69	214.83
9	7,275,870	374,899,613.21	213.48
10	7,025,069	367,850,680.19	214.97
11	6,982,569	366,497,651.08	215.08
12	6,739,413	354,493,165.58	209.20
13	6,834,905	363,039,802.31	219.97
Total	104,942,148	5,284,709,423.84	2,914.15

The computer program was written in GNU C, utilizing Arjen Lenstra's Large Integer Package. Hardware consisted of a network of 200 MHz Pentium PCs running LINUX O/S. Each processor was given one "run" (with up to 14 runs executing concurrently) and the total CPU time used for all 27 runs was 4,765 hours.

Aside from Theorem 5, the only other known result concerning the behavior of $V_1(x)$ as $x \to \infty$ is the bound

$$\liminf_{x\to\infty} \frac{V_1(x)}{V(x)} \le \frac{1}{2}, \tag{7.1}$$

established by very elementary means in an unpublished note of Pomerance (see [27] and [22]). A modification of his argument, combined with the results of the above computations, yields the much stronger bound in Theorem 7. The following lemma is the key. Recall the definition of $V(x; k)$ given in the introduction.

Lemma 7.4. *We have*

$$V(x; a^2) \le V(x/a).$$

Proof: The lemma is trivial when $a = 1$ so assume $a \ge 2$. Let n be a totient with $x/a < n \le x$. First we show that for some integer $s \ge 0$, $a^{-s}n$ is a totient with an pre-image not divisible by a^2. Suppose $\phi(m) = n$. If $a^2 \nmid m$, take $s = 0$. Otherwise we can write $m = a^t r$, where $t \ge 2$ and $a \nmid r$. Clearly $\phi(ar) = a^{1-t}n$, so we take $s = t - 1$. Next, if n_1 and n_2 are two distinct totients in $(x/a, x]$, then $a^{-s_1}n_1 \neq a^{-s_2}n_2$ (since n_1/n_2 cannot be a power of a), so the mapping from totients in $(x/a, x]$ to totients $\leq x$ with a pre-image not divisible by a^2 is one-to-one. Thus

$$V(x) - V(x; a^2) \geq V(x) - V(x/a),$$

and the lemma follows.   □

The above computations show that if $\phi(x) = n$ and $A(n) = 1$, then x is divisible by either a^2 or b^2, where a and b are numbers greater than $10^{5,001,850,000}$.

Without loss of generality, suppose $a \leq b$. By Lemma 7.4, we have

$$V_1(x) \leq V(x/a) + V(x/b) \leq 2V(x/a). \tag{7.2}$$

Lemma 7.5. *Suppose $V_1(x) \leq bV(x/a)$, where $a > 1$ and $b > 0$. Then*

$$\liminf_{x \to \infty} \frac{V_1(x)}{V(x)} \leq \frac{b}{a}.$$

Proof: Let

$$c = \liminf_{x \to \infty} \frac{V_1(x)}{V(x)}$$

and suppose $c > 0$. For every $\varepsilon > 0$ there is a number x_0 such that $x \geq x_0$ implies $V_1(x)/V(x) \geq c - \varepsilon$. For large x, set $n = [\log(x/x_0)/\log a]$. Then

$$
\begin{aligned}
V(x) &= \frac{V(x)}{V(x/a)} \frac{V(x/a)}{V(x/a^2)} \cdots \frac{V(x/a^{n-1})}{V(x/a^n)} V(x/a^n) \\
&\leq b^n \frac{V(x)}{V_1(x)} \frac{V(x/a)}{V_1(x/a)} \cdots \frac{V(x/a^{n-1})}{V_1(x/a^{n-1})} V(ax_0) \\
&\leq b^n (c - \varepsilon)^{-n} (ax_0) = O\left(x^{-\log((c-\varepsilon)/b)/\log a}\right).
\end{aligned}
$$

This contradicts the trivial lower bound $V(x) \gg x/\log x$ if $c > b/a + \varepsilon$. Since ε is arbitrary, the lemma follows.   □

Theorem 7 follows immediately. Further improvements in the lower bound for a counterexample to Carmichael's Conjecture will produce corresponding upper bounds on $\liminf_{x \to \infty} V_1(x)/V(x)$. Explicit bounds for the $O(1)$ term appearing in Theorem 1 (which would involve considerable work to obtain) combined with (7.2) should give $\limsup_{x \to \infty} V_1(x)/V(x) \leq 10^{-5,000,000,000}$ as well.

Lemma 7.4 raises another interesting question. First, suppose d is a totient, all of whose pre-images m_i are divisible by k. The lower bound argument given in Section 5 shows that for at least half of the numbers $b \in B$, the totient $\phi(b)d$ has only the pre-images bm_i. In particular, all of the pre-images of such totients are divisible by k and Theorem 8 follows.

It is also natural to ask for which k do there exist totients, all of whose pre-images are divisible by k. A short search reveals examples for each $k \leq 11$ except $k = 6$ and $k = 10$. For $k = 2, 4$ and 8, take $d = 2^{18} \cdot 257$, for $k = 3$ or 9 take $d = 54 = 2 \cdot 3^3$, for $k = 5$ take $d = 12500 = 4 \cdot 5^5$, for $k = 7$, take $d = 294 = 6 \cdot 7^2$ and for $k = 11$, take $d = 110$. It appears that there might not be any totient, all of whose pre-images are divisible by 6, but I cannot prove this. Any totient with a unique pre-image must have that pre-image divisible by 6, so the non-existance of such numbers implies Carmichael's Conjecture.

I believe that obtaining the asymptotic formula for $V(x)$ will require simultaneously determining the asymptotics of $V_k(x)/V(x)$ (more will be said in Section 8) and $V(x; k)/V(x)$ for each k. It may even be necessary to classify totients more finely. For instance, taking $d = 4, k = 4$ in the proof of Theorem 2 (Section 5), the totients m constructed have $\phi^{-1}(m) = \{5n, 8n, 10n, 12n\}$ for some n. On the other hand, taking $d = 6, k = 4$ produces a different set of totients m, namely those with $\phi^{-1}(m) = \{7n, 9n, 14n, 18n\}$ for some n. Likewise, for any given d with $A(d) = k$, the construction of totients in Section 5 may miss whole classes of totients with multiplicity k. There is much further work to be done in this area.

8. The average of $A(m)$ and the ratio $V_k(x)/V(x)$

Proof of Theorem 3: Suppose x is sufficiently large and set $M = [\sqrt{4C \log N}]$, $L = L_0(x) - M$. Define ξ_i as in Theorem 15. By Theorem 15, the number of totients $m \leq x$ with a pre-image n satisfying $\mathbf{x}(n) \notin \mathscr{S}_L(\xi)$ is $O(V(x)/N)$ (here $\mathbf{x}(n) = (x_1(n; x), \ldots, x_L(n; x)))$. For $n \in \mathfrak{R}_L(\xi)$, set $q_i = q_i(n)$ and $x_i = x_i(n; x)$ for each i. For $0 \leq b < L$, let $N_b(x)$ denote the number of such n with

$$\frac{b}{Lg_L^*} \leq x_L < \frac{b+1}{Lg_L^*} =: \frac{Z_b}{\log_2 x}.$$

Write $n = q_0 \cdots q_{L-1} r$, so that $\log_2 P^+(r) \leq Z_b$. Also, $\log_2 Z_b \ll b\varrho^M$. By Corollary 3.4 and Lemmas 3.24 and 4.3,

$$N_b(x) \ll \frac{x}{\log x} R_{L-1}(\xi; x) e^{-b} (\log_2 Z_b) W(E(Z_b))$$

$$\ll V(x) \exp\{-b + M \log b - M \log M + O(M + \log^2 b)\}.$$

Summing on b (as in the proof of Theorem 12) gives

$$\sum_b N_b(x) \ll V(x) \exp\{O(M)\}.$$

Therefore, the number of totients $m \leq x$ with $A(m) \geq N$ is

$$O(V(x)N^{-1} \exp\{O(M)\}),$$

and the theorem follows. $\square$

In contrast, the average value of $A(m)$ over totients $m \le x$ is clearly $\ge x/V(x) = (\log x)^{1+o(1)}$. The vast differences between the "average" behavior and the "normal" behavior is a result of some totients having an enormous number of pre-images. In fact, Erdős [7] showed that there are infinitely many totients for which

$$A(m) \ge m^{c_4} \tag{8.1}$$

for some positive constant c_4. Recently, Friedlander [14] has shown that one may take

$$c_4 = 1 - (2\sqrt{e})^{-1} - \varepsilon > 0.696. \tag{8.2}$$

It is probable that $V_k(x)/V(x)$ approaches a limit for each k. Based on Theorem 9 and computations, we make the

Conjecture. *For $k \ge 2$,*

$$\lim_{x \to \infty} \frac{V_k(x)}{V(x)} = C_k.$$

Theorem 3 implies that $V_k(x)/V(x) \ll k^{-2+\varepsilon}$ on average. Table 4 lists values of $V(x)$ and the ratios $V_k(x)/V(x)$ for $2 \le k \le 7$.

Table 5 lists the value of $R_k := k^2 V_k(x)/V(x)$ for $2 \le k \le 200$ and $x = 500,000,000$. It seems to indicate that $C_k \asymp 1/k^2$ for large k. In fact, at $x = 500,000,000$ we have $1.75 \le R_k \le 2.05$ for $20 \le k \le 200$.

This data is very misleading, however. It is an immediate corollary of (8.1), (8.2) and Theorem 2 that for infinitely many k,

$$\frac{V_k(x)}{V(x)} \gg k^{-1/c_4+\varepsilon} \gg k^{-1.437} \quad (x > x_0(k)),$$

so $V_k(x)/V(x)$ is much larger than average for many k. Erdős has conjectured that every $c_4 < 1$ is admissible in (8.1), which implies that for every $\varepsilon > 0$,

$$\frac{V_k(x)}{V(x)} \gg k^{-1-\varepsilon} \quad (x > x_0(k))$$

for infinitely many k. The behavior of the constants C_k is probably quite complicated.

Table 4.

x	$V(x)$	V_2/V	V_3/V	V_4/V	V_5/V	V_6/V	V_7/V
1M	180,184	0.380727	0.140673	0.098988	0.042545	0.062730	0.020790
5M	840,178	0.379462	0.140350	0.102487	0.042687	0.063193	0.020373
10M	1,634,372	0.378719	0.140399	0.103927	0.042703	0.063216	0.020061
25M	3,946,809	0.378198	0.140233	0.105466	0.042602	0.063414	0.019819
125M	18,657,531	0.377218	0.140176	0.107873	0.042560	0.063742	0.019454
300M	43,525,579	0.376828	0.140170	0.108933	0.042517	0.063818	0.019284
500M	71,399,658	0.376690	0.140125	0.109509	0.042493	0.063851	0.019194

Table 5.

k	R_k	k	R_k	k	R_k	k	R_k	k	R_k	k	R_k	k	R_k
2	1.5068	31	1.9303	60	2.0054	89	1.9547	118	1.9689	147	1.8743	176	1.9202
3	1.2611	32	2.0016	61	1.9534	90	2.0111	119	1.9286	148	1.9293	177	1.8727
4	1.7521	33	1.9019	62	2.0099	91	1.9612	120	1.9652	149	1.8946	178	1.9130
5	1.0623	34	1.9958	63	1.9540	92	1.9681	121	1.9417	150	1.8864	179	1.9279
6	2.2986	35	1.8936	64	2.0303	93	1.9589	122	1.9276	151	1.8963	180	1.8632
7	0.9405	36	2.0179	65	1.9869	94	1.9856	123	1.9608	152	1.9030	181	1.8239
8	2.3641	37	1.9010	66	2.0354	95	1.9748	124	1.9552	153	1.9304	182	1.9202
9	1.2704	38	2.0453	67	1.9670	96	1.9541	125	1.9218	154	1.9235	183	1.8597
10	2.1228	39	1.9071	68	2.0175	97	1.9767	126	1.9643	155	1.9311	184	1.9067
11	1.5321	40	2.0368	69	1.9450	98	1.9785	127	1.9398	156	1.9206	185	1.9145
12	1.8610	41	1.9107	70	2.0194	99	1.9531	128	1.9023	157	1.8908	186	1.9256
13	1.6543	42	2.0341	71	1.9741	100	2.0034	129	1.9184	158	1.9097	187	1.8670
14	1.9224	43	1.9139	72	2.0292	101	1.9271	130	1.9721	159	1.9219	188	1.8415
15	1.6549	44	2.0483	73	1.9656	102	1.9935	131	1.9259	160	1.9372	189	1.8906
16	1.9389	45	1.9204	74	2.0048	103	1.9561	132	1.9677	161	1.9561	190	1.8970
17	1.7370	46	2.0269	75	1.9863	104	1.9876	133	1.9042	162	1.9415	191	1.8910
18	1.9301	47	1.9214	76	2.0125	105	1.9816	134	1.9465	163	1.8870	192	1.9263
19	1.7149	48	2.0430	77	1.9757	106	1.9712	135	1.9279	164	1.9468	193	1.8645
20	1.8759	49	1.9382	78	2.0217	107	1.9536	136	1.9405	165	1.9153	194	1.9213
21	1.8837	50	2.0118	79	1.9779	108	1.9942	137	1.9203	166	1.9011	195	1.8608
22	1.8775	51	1.9447	80	1.9924	109	1.9479	138	1.9719	167	1.9112	196	1.8794
23	1.8501	52	2.0100	81	1.9874	110	1.9779	139	1.9197	168	1.9124	197	1.8106
24	1.9370	53	1.9432	82	2.0199	111	1.9519	140	1.9619	169	1.9061	198	1.8444
25	1.7866	54	2.0345	83	1.9957	112	2.0025	141	1.9383	170	1.8894	199	1.8004
26	1.9580	55	1.9508	84	1.9793	113	1.9263	142	1.9289	171	1.8982	200	1.8588
27	1.9142	56	2.0222	85	1.9516	114	1.9962	143	1.9006	172	1.9424		
28	1.9519	57	1.9585	86	2.0193	115	1.9199	144	1.9040	173	1.8574		
29	1.8553	58	1.9922	87	1.9447	116	1.9901	145	1.9294	174	1.8751		
30	1.9599	59	1.9673	88	1.9749	117	1.9368	146	1.9107	175	1.8787		

9. Generalization to other multiplicative functions

The proofs of the theorems depend entirely on the definition of $\phi(n)$ as a multiplicative function, and not on the arithmetical interpretation of $\phi(n)$ as the number of positive integers less than n which are coprime to n. This suggests that corresponding results should hold for a wide class of similar multiplicative arithmetic functions, such as $\sigma(n)$, the sum of divisors function.

Suppose $f : \mathbb{N} \to \mathbb{N}$ is a multiplicative arithmetic function. We first define

$$
\begin{aligned}
\mathscr{V}_f &= \{f(n) : n \in \mathbb{N}\}, \\
V_f(x) &= |\mathscr{V}_f \cap [1, x]|, \\
f^{-1}(m) &= \{n : f(n) = m\}, \\
A_f(m) &= |f^{-1}(m)|, \\
V_{f,k}(x) &= |\{m \le x : A_f(m) = k\}|,
\end{aligned}
\tag{9.1}
$$

which are the analogous functions to the definitions (1.1). With only minor modifications to the arguments given in previous sections, we prove Theorem 14. By itself, condition (1.12) is enough to prove the lower bound for $V_f(x)$. Condition (1.13) is used only for the upper bound argument and the lower bound for $V_{f,k}(x)$.

The function $f(n) = n$, which takes all positive integer values, is an example of why zero must be excluded from the set in (1.12). Condition (1.13) insures that the values of $f(p^k)$ for $k \ge 2$ are not too small too often, and thus have little influence on the size of $V_f(x)$. It essentially forces $f(h)$ to be a bit larger than $h^{1/2}$ on average. It's probable that (1.13) can be relaxed, but not too much. For example, the multiplicative function defined by $f(p) = p - 1$ for prime p, and $f(p^k) = p^{k-1}$ for $k \ge 2$ clearly takes all integer values, while

$$
\sum_{h \text{ square-full}} \frac{1}{f(h)(\log_2 h)^2} \ll 1.
$$

Condition (1.13) also insures that $A(m)$ is finite for each f-value m. For example, a function satisfying $f(p^k) = 1$ for infinitely many prime powers p^k has the property that $A(m) = \infty$ for every f-value m.

Some functions appearing in the literature which satisfy the conditions of Theorem 14 are $\sigma(n)$, the sum of divisors function, $\phi^*(n)$, $\sigma^*(n)$ and $\psi(n)$. Here $\phi^*(n)$ and $\sigma^*(n)$ are the unitary analogs of $\phi(n)$ and $\sigma(n)$, defined by $\phi^*(p^k) = p^k - 1$ and $\sigma^*(p^k) = p^k + 1$ [5], and $\psi(n)$ is Dedekind's function, defined by $\psi(p^k) = p^k + p^{k-1}$.

In general, implied constants will depend on the function $f(n)$. One change that must be made throughout is to replace every occurrence of "$p - 1$" (when referring to $\phi(p)$) with "$f(p)$", for instance in the definition of S-normal primes in Section 2. Since the possible values of $f(p) - p$ is a finite set, Lemmas 2.7 and 2.8 follow easily with the new definitions. The most substantial change to be made in Section 2, however, is to Lemma 2.9, since we no longer have the bound $n/f(n) \ll \log_2 n$ at our disposal.

Lemma 2.9*. *The number of $m \in \mathscr{V}_f(x)$ for which either $d^2 \mid m$ or $d^2 \mid n$ for some $n \in f^{-1}(m)$ and $d > Y$ is $O(x(\log_2 x)^K / \varepsilon(Y^2))$, where $K = \max_p(p - f(p))$.*

Proof: The number of m with $d^2 \mid m$ for some $d > Y$ is $O(x/Y)$. Now suppose $d^2 \mid n$ for some $d > Y$, and let $h = h(n)$ denote the square-full part of n, that is

$$
h(n) = \prod_{\substack{p^a \| n \\ a \ge 2}} p^a.
$$

In particular, $h(n) > Y^2$. From the fact that $f(p) \geq p - K$ for all primes p, we have

$$f(n) = f(h)f(n/h) \gg \frac{f(h)n}{h}(\log_2 n/h)^{-K}.$$

Thus, if $f(n) \leq x$, then

$$\left(\log_2 \frac{n}{h}\right)^{-K}\frac{n}{h} \ll \frac{x}{f(h)}.$$

Therefore, the number of possible n with a given h is crudely

$$\ll \frac{x(\log_2 x)^K}{f(h)}.$$

By (1.13), the total number of n is at most

$$\ll x(\log_2 x)^K \sum_{h \geq Y^2} \frac{1}{f(h)}$$

$$\ll \frac{x(\log_2 x)^K}{\varepsilon(Y^2)} \sum_{h \geq 16} \frac{\varepsilon(h)}{f(h)}$$

$$\ll \frac{x(\log_2 x)^K}{\varepsilon(Y^2)}. \qquad \square$$

The final modification to Section 2 is to Lemma 2.10, which needs an additional term $O(x(\log_2 x)^K/\varepsilon(S^2))$ in the conclusion. This comes from the use of Lemma 2.9* to count the number of totients with a pre-image divisible by a square of a prime $p > S$.

In Section 3, only the proof of Lemma 3.12 requires modification. For the upper bound (3.25), we must take into account contributions of numbers which are not square-free using only (1.13). First change the definition of $R_L(\xi; x)$ to

$$R_L(\xi; x) = \sum_{n \in \mathcal{R}_L(\xi;x)} \frac{1}{f(n)},$$

and for each $\mathbf{m}$ (see (3.34) and (3.35)), consider separately the sum, $S_1(\mathbf{m})$, over square-free n and the sum, $S_2(\mathbf{m})$, over non-square-free n. The key is inequality (3.35). Note first that $\mathbf{x} \in \mathcal{S}_L(\xi)$ implies $x_i > x_{i+2}$ for each i, hence any $n \in \mathcal{R}_L(\xi; x)$ is cube-free. From (1.12), we have

$$S_1(\mathbf{m}) = \prod_{i=1}^{j-1}(\varepsilon_i \log_2 x + O(1)). \qquad (9.2)$$

For n counted in $S_2(\mathbf{m})$, let $I(n)$ denote the set of indices i ($1 \leq i \leq j - 1$) with $p_i \parallel n$, and let $\mathscr{I}$ denote the set of possible $I(n)$. Note that $I(n)$ contains at most $j - 3$ indices, since n is not square-free. Also, by (1.13),

$$\sum_{t=1}^{\infty} \frac{1}{f(t^2)} \ll 1.$$

Also define $g(n) = \prod q_i$, the product being over all $i \notin I(n)$. In particular, $g(n)$ is a square. In analogy to (3.35), we have

$$
\begin{aligned}
S_2(\mathbf{m}) &= \sum_{\substack{I \in \mathscr{I}}} \sum_{\substack{x(n) \in B(\mathbf{m}) \\ I(n)=I}} \frac{1}{g(n)} \prod_{i \in I} \frac{1}{f(p_i)} \\
&\leq \sum_{I \in \mathscr{I}} \sum_{i \in I} \left(\sum_{p_i} \frac{1}{f(p_i)} \right) \left(\sum_{t=1}^{\infty} \frac{1}{f(t^2)} \right) \\
&\ll \sum_{I \in \mathscr{I}} \prod_{i \in I} (\varepsilon_i \log_2 x) \qquad\qquad\qquad\qquad (9.3) \\
&= (\varepsilon_1 \cdots \varepsilon_{j-1})(\log_2 x)^{j-1} \sum_{I \in \mathscr{I}} \prod_{i \notin I} (4/5)^{L_0-i} \\
&\leq (\varepsilon_1 \cdots \varepsilon_{j-1})(\log_2 x)^{j-1} \sum_{k \geq 2} \frac{u^k}{k!}, \quad u = \sum_{i \geq 0} (4/5)^i = 5 \\
&\ll (\varepsilon_1 \cdots \varepsilon_{j-1})(\log_2 x)^{j-1},
\end{aligned}
$$

where the range of p_i is $E(m_i \varepsilon_i) \leq p_i \leq E((m_i + 1)\varepsilon_i)$. This takes care of the possibility of r containing a square factor. A possible square factor in s is dealt with more easily. In analogy to (3.37), we have

$$\sum_{s} \frac{1}{\phi(s)} \leq \left(\sum_{\frac{\log_2 P}{\log_2 x} \leq y_j} \sum_{k \geq 1} \frac{1}{f(p^k)} \right)^{L-j+1} = (y_j \log_2 x + O(1))^{L-j+1}.$$

If we fix $\omega_i = (10(L_0 - i)^3)^{-1}$ for each i and suppose $L \leq L_0 - 20$, we do not need to worry about $n \in P_j$ with $q_{j-1} = q_j$. If this were to occur, then (3.31) would imply $\omega_{j-1} > (5/4)\omega_j$, contradicting the definition of ω_i.

For the lower bound (3.26), we first modify the definitions of L and $\mathscr{R}_L^*(\xi; x)$. One complication is that the mapping $p \to f(p)$ may not be one-to-one. We say a prime p is "bad" if $f(p) = f(p')$ for some prime $p' \neq p$ and say p is "good" otherwise. By (1.12) and Lemma 2.5, the number of bad primes $\leq y$ is $O(y/\log^2 y)$, so $\sum_{p \, \text{bad}} 1/p$ converges. The other complication has to do with "small" values of $f(p^k)$ for some prime powers p^k with $k \geq 2$. For each prime p, define

$$Q(p) := \min_{k \geq 2} \frac{f(p^k)}{f(p)}. \qquad\qquad\qquad (9.4)$$

Introduce another parameter d (which will be the same d as in Theorem 2) and suppose $L \leq L_0 - M$ where M is a sufficienlty large constant depending on P_0 and d. If follows from (1.13) and (9.4) that

$$\sum_{Q(p) \leq d} \frac{1}{p} = O(d).$$

Specify in the definition of $\mathscr{R}_L^*(\xi; x)$ the additional restrictions that every p_i is "good" and satisfies $Q(p_i) > d$. We then have

$$\sum_{\substack{E(m\varepsilon) \leq p \leq E((m+1)\varepsilon) \\ p \text{ not bad} \\ Q(p) > d}} 1/p = \varepsilon \log_2 x + O(d).$$

Therefore, by the proof of Lemma 3.12 (3.26), we have

$$R_L^*(\xi; x) \gg V_f(x).$$

In Section 4, we need to add a term $O(y(\log_2 y)^K/\varepsilon(S^2))$ to the conclusion of Lemma 4.1 due to the use of Lemma 2.9*. A modified Lemma 4.2 is stated below.

Lemma 4.2*. *Suppose* $k \geq 300$, $\omega = 1/(10k^3)$, $y \geq y_0$ *and* $\log_3 y \geq k/3$. *Then*

$$N_k(1 + \omega; y) \ll y(\log_2 y)^5 W(y)(\log y)^{-1-1/(15000k^9 \log k)}.$$

The only difference in the proof is that we have $\log_2 S = \frac{\log_2 y}{2500k^9 \log k}$, so in $U_1(y)$, $U_3(y)$ and $T(\theta; x)$ we have

$$\frac{y(\log_2 y)^K}{\varepsilon(S^2)} \ll y \exp\left\{-\frac{\log_2 y}{2500k^9 \log k}(\log_3 y - 10\log k - 8)^{20} + K \log_3 y\right\}$$
$$\ll y/\log^2 y.$$

In the main upper bound argument, we first eliminate the possibility of large squares dividing n using Lemma 2.9*. Defining Y_j by $\log_3 Y_j = k/3$ (see (4.11)) insures that (4.13) and (4.14) imply (4.15).

The additional restrictions p_i "good" and $Q(p_i) > d$ introduced for the proof of Lemma 3.12 are needed for the lower bound argument in Section 5. First, we modify slightly the definition of the set B. In place of (5.6) use

$$f(n) \leq x/d$$

and in place of (5.10) put $Q(p_i) > d$ $(0 \leq i \leq L)$. Also add the condition that none of the primes p_i are bad. Fortunately, the numbers in B are square-free by definition. The Eq. (5.12) becomes

$$df(n) = f(n_1). \tag{9.5}$$

Since $Q(p_i) > d$ for each p_i, if $n \mid n_1$ and one of the primes q_i $(0 \le i \le L)$ occurs to a power grater than 1, then $\phi(n_1) > d\phi(n)$. Therefore, the $L + 1$ largest prime factors of n_1 occur to the first power only, which forces $n_1 = nm_i$ for some i (the trivial solutions). For nontrivial solutions, we have at least one index i for which $p_i \ne q_i$, and hence $f(p_i) \ne f(q_i)$ (since each p_i is "good"). Obvious changes are made to (5.43) and the definitions of $\sigma_k(\mathscr{A})$ and $\tau_k(\mathscr{A})$. In Lemma 5.3, the phrase "$rt + 1$ and $st + 1$ are unequal primes" is replace by "$rt + a$ and $st + a'$ are unequal primes for some pair of numbers (a, a') with $a, a' \in \mathscr{P}$." Here $\mathscr{P}$ denotes the set of possible values of $f(p) - p$. As the number of pairs (a, a') is finite, this poses no problem in the argument. Similar changes are made in several places in Lemma 5.4.

It is not possible to prove analogs of Theorems 5–9 for general f satisfying the hypotheses of Theorem 14. One reason is that there might not be any "Carmichael Conjecture" for f, e.g., $A_\sigma(3) = 1$, where σ is the sum of divisors function. Furthermore, the proof of Theorem 9 depends on the identity $\phi(p^2) = p\phi(p)$ for primes p. If, for some $a \ne 0$, $f(p) = p + a$ for all primes p, then the argument of [13] shows that if the multiplicity k is possible and r is a positive integer, then the multiplicity rk is possible. For functions such as $\sigma(n)$, for which the multiplicity 1 is possible, this completely solves the problem of the possible multiplicities. For other functions, it shows at least that a positive proportion of multiplicities are possible. If multiplicity 1 is not possible, and $f(p^2) = pf(p)$, the argument in [14] shows that all multiplicities beyond some point are possible.

We can, however, obtain information about the possible multiplicities for more general f by an induction argument utilizing the next lemma. Denote by $a_1, \ldots, a_K$ the possible values of $f(p) - p$ for prime p.

Lemma 7.1*. *Suppose $A_f(m) = k$. Let p, q, s be primes and $r \ge 2$ an integer so that*
  (i) *s and q are "good" primes,*
 (ii) *$mf(s) = f(q)$,*
(iii) *$f(s) = rp$,*
 (iv) *$p \nmid f(\pi^b)$ for every prime π, integer $b \ge 2$ with $f(\pi^b) \le mf(s)$,*
  (v) *$dp - a_i$ is composite for $1 \le i \le K$ and $d \mid rm$ except $d = r$ and $d = rm$.*
  Then $A_f(mrp) = k + A_f(1)$.

Proof: Let $f^{-1}(m) = \{x_1, \ldots, x_k\}$ and suppose $f(x) = mrp$. By condition (iv), $p \mid f(\pi)$ for some prime π which divides x to the first power. Therefore, $f(\pi) = dp$ for some divisor d of mr. Condition (v) implies that the only possibilities for d are $d = r$ or $d = rm$. If $d = r$, then $f(\pi) = rp = f(p)$ which forces $\pi = s$ by condition (i). By conditions (ii) and (iii), we have $f(x/s) = m$, which gives solutions $x = sx_i$ $(1 \le i \le k)$. Similarly, if $d = rm$, then $\pi = q$ and $f(x/q) = 1$, which has $A_f(1)$ solutions. $\square$

By the Chinese Remainder Theorem, there is an arithmetic progression $\mathscr{A}$ so that condition (v) is satisfied for each number $p \in \mathscr{A}$, while still allowing each $rp + a_i$ and $rmp + a_i$ to be prime. To eliminate primes failing condition (iv), we need the asymptotic form of the Prime k-tuples Conjecture due to Hardy and Littlewood [18] (actually only the case where $a_i = 1$ for each i is considered in [18]; the conjectured asymptotic for k arbitrary polynomials can be found in [2]).

Prime k-tuples Conjecture (asymptotic version).   *Suppose $a_1, \ldots, a_k$ are positive integers and $b_1, \ldots, b_k$ are integers so that no prime divides*

$$(a_1 n + b_1) \ldots (a_k n + b_k)$$

for every integer n. Then for some constant $C(\mathbf{a}, \mathbf{b})$, the number of $n \leq x$ for which $a_1 n + b_1, \ldots, a_k n + b_k$ are simultaneously prime is

$$\sim C(\mathbf{a}, \mathbf{b}) \frac{x}{\log^k x} \quad (x \geq x_0(\mathbf{a}, \mathbf{b})).$$

A straightforward calculation using (1.13) gives

$$|\{\pi^b : f(\pi^b) \leq y\}| \ll y/\varepsilon(y).$$

If s is taken large enough, the number of possible $p \leq x$ satisfying condition (iv) (assuming r and m are fixed and noting condition (iii)) is $o(x/\log^3 x)$. The procedure for determining the set of possible multiplicities with this lemma will depend on the behavior of the particular function. Complications can arise, for instance, if m is even and all of the a_i are even (which makes condition (ii) impossible) or if the number of "bad" primes is $\gg x/\log^3 x$.

Acknowledgment

Much of the early work for this paper was completed while the author was enjoying the hospitality of the Institute for Advanced Study, supported by National Science Foundation grant DMS 9304580.

References

1. R.C. Baker and G. Harman, "The difference between consecutive primes," *Proc. London Math. Soc.* **72**(3) (1996), 261–280.
2. P.T. Bateman and R.A. Horn, "A heuristic asymptotic formula concerning the distribution of prime numbers," *Math. Comp.* **16** (1962), 363–367.
3. R.D. Carmichael, "On Euler's ϕ-function," *Bull. Amer. Math. Soc.* **13** (1907), 241–243.
4. R.D. Carmichael, "Note on Euler's ϕ-function," *Bull. Amer. Math. Soc.* **28** (1922), 109–110.
5. E. Cohen, "Arithmetical functions associated with the unitary divisors of an integer," *Math. Z.* **74** (1960), 66–80.
6. L.E. Dickson, "A new extension of Dirichlet's theorem on prime numbers," *Messenger of Math.* **33** (1904), 155–161.
7. P. Erdős, "On the normal number of prime factors of $p - 1$ and some related problems concerning Euler's ϕ-function," *Quart. J. Math.* (Oxford) (1935), 205–213.
8. P. Erdős, "Some remarks on Euler's ϕ-function and some related problems," *Bull. Amer. Math. Soc.* **51** (1945), 540–544.
9. P. Erdős, "Some remarks on Euler's ϕ-function," *Acta Arith.* **4** (1958), 10–19.
10. P. Erdős and R.R. Hall, "On the values of Euler's ϕ-function," *Acta Arith.* **22** (1973), 201–206.
11. P. Erdős and R.R. Hall, "Distinct values of Euler's ϕ-function," *Mathematika* **23** (1976), 1–3.
12. P. Erdős and C. Pomerance, "On the normal number of prime factors of $\phi(n)$," *Rocky Mountain J. of Math.* **15** (1985), 343–352.

13. K. Ford and S. Konyagin, "On two conjectures of Sierpiński concerning the arithmetic functions σ and ϕ," *Proceedings of the Number Theory Conference dedicated to Andrzej Schinzel on his 60th birthday* (to appear).
14. K. Ford, "The number of solutions of $\phi(x) = m$" *Annals of Math.* (to appear).
15. J. Friedlander, "Shifted primes without large prime factors," *Number theory and applications* (Banff, AB, 1988), Kluwer Acad. Publ., Dorbrecht 1989, pp. 393–401.
16. H. Halberstam and H.-E. Richert, *Sieve Methods*, Academic Press, London, 1974.
17. R.R. Hall and G. Tenenbaum, *Divisors*, Cambridge University Press, 1988.
18. G.H. Hardy and J.E. Littlewood, "Some problems of 'Partitio Numerorum': III. On the representation of a number as a sum of primes," *Acta Math.* **44** (1923) 1–70.
19. G.H. Hardy and S. Ramanujan, "The normal number of prime factors of a number n," *Quart. J. Math.* **48** (1917), 76–92.
20. A. Hildebrand and G. Tenenbaum, "Integers without large prime factors," *J. Théor. Nombres Bordeaux* **5** (1993), 411–484.
21. V. Klee, "On a conjecture of Carmichael," *Bull. Amer. Math. Soc.* **53** (1947), 1183–1186.
22. L.E. Mattics, "A half step towards Carmichael's conjecture, solution to problem 6671," *Amer. Math. Monthly* **100** (1993), 694–695.
23. H. Maier and C. Pomerance, "On the number of distinct values of Euler's ϕ-function," *Acta Arith.* **49** (1988), 263–275.
24. P. Masai and A. Valette, "A lower bound for a counterexample to Carmichael's Conjecture," *Bollettino U.M.I.* (1982), 313–316.
25. S. Pillai, "On some functions connected with $\phi(n)$," *Bull. Amer. Math. Soc.* **35** (1929), 832–836.
26. C. Pomerance, "On the distribution of the values of Euler's function," *Acta Arith.* **47** (1986), 63–70.
27. C. Pomerance, "Problem 6671," *Amer. Math. Monthly* **98** (1991), 862.
28. A. Schinzel, "Sur l'equation $\phi(x) = m$," *Elem. Math.* **11** (1956), 75–78.
29. A. Schinzel, "Remarks on the paper 'Sur certaines hypothèses concernant les nombres premiers'", *Acta Arith.* **7** (1961/62), 1–8.
30. A. Schinzel and W. Sierpiński, "Sur certaines hypothèses concernant les nombres premiers," *Acta Arith.* **4** (1958), 185–208.
31. A. Schlafly and S. Wagon, "Carmichael's conjecture on the Euler function is valid below $10^{10,000,000}$," *Math. Comp.* **63** (1994), 415–419.

THE RAMANUJAN JOURNAL 2, 153–165 (1998)

A Mean-Value Theorem for Multiplicative Functions on the Set of Shifted Primes

KARL-HEINZ INDLEKOFER*   k-heinz@uni-paderborn.de
Faculty of Mathematics and Informatics, University of Paderborn, Warburger Straße 100, 33098 Paderborn, Germany

NIKOLAI M. TIMOFEEV   timofeev@vgpu.elcom.ru
pr. Stroitelei 11, Vladimir State Ped. University, 600024 Vladimir, Russia

Dedicated to the memory of Paul Erdős

Received April 4, 1997; Accepted March 16, 1998

Abstract. Let f be a complex-valued multiplicative function, let p denote a prime and let $\pi(x)$ be the number of primes not exceeding x. Further put

$$m_p(f) := \lim_{x \to \infty} \frac{1}{\pi(x)} \sum_{p \le x} f(p+1), \quad M(f) := \lim_{x \to \infty} \frac{1}{x} \sum_{n \le x} f(n)$$

and suppose that

$$\limsup_{x \to \infty} \frac{1}{x} \sum_{n \le x} |f(n)|^2 < \infty, \quad \sum_{p \le x} |f(p)|^2 \ll x(\ln x)^{-\varrho},$$

with some $\varrho > 0$. For such functions we prove:

If there is a Dirichlet character χ_d such that the mean-value $M(f\chi_d)$ exists and is different from zero, then the mean-value $m_p(f)$ exists.

If the mean-value $M(|f|)$ exists, then the same is true for the mean-value $m_p(|f|)$.

Key words: multiplicative functions, shifted primes, mean-value

1991 Mathematics Subject Classification: Primary 11K65, 11N64, 11N37

1. Introduction and results

Let $f : \mathbb{N} \to \mathbb{C}$ be a complex-valued multiplicative function, let p denote a prime and let $\pi(x)$ be the number of primes below x. The aim of this paper is to find general conditions under which the mean-value

$$m_p(f) := \lim_{x \to \infty} \frac{1}{\pi(x)} \sum_{p \le x} f(p+1)$$

of f on the set of shifted primes exists.

*Supported in part by a grant of the DFG (Deutsche Forschungsgemeinschaft).

It is natural to expect that, if the mean-value

$$M(f) := \lim_{x \to \infty} \frac{1}{x} \sum_{n \leq x} f(n)$$

of f on the set of natural numbers exists, then the same holds for $m_p(f)$. For each positive α we shall say that f belongs to the class $\mathcal{L}_\alpha$ if

$$\|f\|_\alpha := \limsup_{x \to \infty} \left(\frac{1}{x} \sum_{n \leq x} |f(n)|^\alpha \right)^{1/\alpha} < +\infty.$$

Using the results of our previous paper [3] we prove the following theorem.

Theorem. *Let f be a complex-valued multiplicative function. Let $f \in \mathcal{L}_2$ and satisfy the condition*

$$\sum_{p \leq x} |f(p)|^2 \leq A_1 x \ln^{-\varrho} x \tag{1}$$

for some $\varrho > 0$ and some positive constant A_1. If there is a Dirichlet character $\chi_d \bmod d$ such that the mean-value $M(f \chi_d)$ exists and is different from zero, then the mean-value $m_p(f)$ exists and

$$m_p(f) = \frac{\mu(\delta)}{\varphi(\delta)} \prod_p \left(1 + \sum_{r=1}^{\infty} \frac{1}{\varphi(p^r)} (\chi_\delta(p^r) f(p^r) - \chi_\delta(p^{r-1}) f(p^{r-1})) \right),$$

where $\chi_\delta \bmod \delta$ is a primitive character which generates χ_d.

If the mean-value $M(|f|)$ of $|f|$ exists, then the mean-value $m_p(|f|)$ of $|f|$ on the set $\{p + 1\}$ exists, too.

Remark 1. If $f \in \mathcal{L}_\alpha, \alpha > 2$, then $f \in \mathcal{L}_2$ and

$$\sum_{p \leq x} |f(p)|^2 \leq \left(\sum_{n \leq x} |f(n)|^\alpha \right)^{2/\alpha} \pi(x)^{1 - \frac{2}{\alpha}} \ll x \ln^{\frac{2}{\alpha} - 1} x.$$

Hence, f satisfies condition (1).

Remark 2. There is an example (see [6]) such that $M(f) = 0$ but $m_p(f) \neq 0$.

2. Some lemmata

The next result will play a key role in the proof of the theorem.

Lemma 1 (see Theorem 2 of [4]).   *Let f_i, $i = 1, \ldots, k$, be complex-valued multiplicative functions satisfying the conditions*

$$A(n) = \sum_{i=1}^{k} \alpha_i f_i(n) \geq 0, \quad n = 1, 2, \ldots,$$

where $\alpha_i \in \mathbb{C}$ and where $f_i \in \mathcal{L}_2$ and satisfy (1), $i = 1, \ldots, k$. Then, for some $\varrho > 0$,

$$\frac{1}{\pi(x)} \sum_{p \leq x} A(p+1) \ll \frac{\ln y}{x} \sum_{n \leq x} A(n) + \frac{1}{y^{\varrho}}$$

holds uniformly for $2 \leq y \leq \ln x$.
If f_i, $i = 1, \ldots, k$, satisfy (1) and the estimates

$$\sum_{n \leq x} |f_i(n)|^2 \leq A_2 x \tag{2}$$

with some positive constant A_2, then the constant implied in the symbol $\ll$ depends only on A_1 and A_2.

The next lemma was proved in [6] for the case $|f(n)| \leq 1$.

Lemma 2.   *Let f be a complex-valued multiplicative function, assume that $f \in \mathcal{L}_2$ and satisfies condition (1). Suppose further that there exists a primitive character χ_d^* to a modulus d such that*

$$\sum_{p \leq x} |1 - \chi_d^*(p) f(p)| \frac{\ln p}{p} = o(\ln x) \tag{3}$$

holds as $x \to \infty$. Then we have

$$\frac{1}{\pi(x)} \sum_{p \leq x} f(p+1) = \frac{\mu(d)}{\varphi(d)} \prod_{p \leq x} \left(1 + \sum_{r=1}^{\infty} \frac{1}{\varphi(p^r)} (\chi_d^*(p^r) f(p^r) - \chi_d^*(p^{r-1}) f(p^{r-1})) \right)$$

$$+ o \left(\exp \left(\sum_{p \leq x} \frac{|f(p)| - 1}{p} \right) \right) + o(1).$$

Proof: Given a parameter $t \geq 2$ that will be chosen later, we define a multiplicative function f_1 by

$$f_1(p^r) = \begin{cases} f(p^r) & \text{if } p \leq t, r \geq 1, \\ \overline{\chi_d^*}(p^r) & \text{if } p > t, r \geq 2 \end{cases}$$

and

$$f_1(p) = \begin{cases} f(p) & \text{if } p > t, |f(p)| < 3/2, \\ \overline{\chi_d^*}(p) & \text{if } p > t, |f(p)| \geq 3/2. \end{cases}$$

Next, for a multiplicative function $f(n)$ we introduce a function $g(f; n)$ which is multiplicative in n and defined by $g(f; p^r) = \sqrt{f(p^r)}$ for $r \geq 2$ and all primes p, $g(f; p) = \sqrt{f(p)}$ if $p \mid d$ and $g(f; p)\sqrt{\chi_d^*(p)} = \sqrt{f(p)\chi_d^*(pL)}$ if $p \nmid d$. Here $\sqrt{z} = \sqrt{|z|} \cdot \exp(\frac{1}{2}i \arg z)$ with $\arg z \in [-\pi, \pi)$. Then $g^2(f; n) = f(n)$, and if $p \nmid d$ the inequality $|g(f; p) + g(\chi_d^*; p)| = |\sqrt{f(p)\chi_d^*(p)} + 1| \geq 1$ holds. We now apply Lemma 1 for

$$A(n) = |g(f; n) - g(f_1; n)|^2$$
$$= g(f; n)\overline{g(f; n)} - g(f; n)\overline{g(f_1; n)} - \overline{g(f; n)}g(f_1; n) + g(f_1; n)\overline{g(f_1; n)}.$$

To do this we first prove that f_1 satisfies the conditions of the theorem. For $n > 1$ let $n = m \cdot u \cdot v$, where v is only divisible by primes $q \leq t$, $m \cdot u$ possesses only prime divisors $q > t$, m and u are coprime and $q^2 \mid m$ if $q \mid m$ (it is possible that one or two of the numbers m, u and v are equal to 1). We have $|f_1(n)| \leq |f(u \cdot v)|$ and therefore

$$\sum_{n \leq x} |f_1(n)|^2 \leq \sum_{\substack{m \leq x, \\ p|m \to p^2|m}} \sum_{n \leq x/m} |f^2(n)| \leq A_2 \cdot x \prod_p \left(1 + \frac{2}{p^2}\right).$$

Hence, f_1 satisfies (2) and obviously (1), too. Applying Cauchy's inequality shows that $g(f; n)\overline{g(f; n)}$, $g(f; n)\overline{g(f_1; n)}$, $\overline{g(f; n)}g(f_1; n)$, $g(f_1; n)\overline{g(f_1; n)}$ fulfill the conditions of Lemma 1 Using Lemma 1 we get

$$\frac{1}{\pi(x)} \sum_{p \leq x} |g(f; p+1) - g(f_1; p+1)|^2 \ll \frac{\ln y}{x} \sum_{n \leq x} |g(f; n) - g(f_1; n)|^2 + \frac{1}{y^\beta},$$

where $\beta > 0$ and $2 \leq y \leq \ln x$. Let us denote the last sum by $\sum$. Then obviously (cf. the definition of f_1) $\sum \leq \sum_1 + \sum_2$, where the sum in $\sum_1$ runs over the natural numbers $n \leq x$ with the property that n is divisible by the square p^2 of some prime $p > t$, and where in $\sum_2$ we sum over those $n \leq x$, which possess some prime divisor $p > t$ such that $p \parallel n$ and $|f(p)| \geq \frac{3}{2}$. For the first sum we obtain

$$\sum_1 \ll \sum_{p > t} \sum_{r \geq 2} \frac{x}{p^r}(|f(p^r)| + 1)$$

$$\ll x \left(\sum_n \frac{|f(n)|^2 + 1}{n^{\frac{4}{3}}}\right)^{1/2} \left(\sum_{p > t} \sum_{r \geq 2} \frac{1}{(p^r)^{\frac{2}{3}}}\right)^{1/2} \ll \frac{x}{\sqrt[6]{t}}.$$

Here we made use of the fact that $f \in \mathcal{L}_2$ which implies, by partial summation,

$$\sum_n \frac{|f(n)|^2}{n^{\frac{4}{3}}} < +\infty.$$

Concerning the second sum we write the relation (3) in the equivalent form $\sum_{p \leq x} |1 - \chi_d^*(p) f(p)| \frac{\ln p}{p} \leq \varepsilon(x) \ln x$ where $\varepsilon(x) \searrow 0$, and assume $\varepsilon(x) \ln x \to \infty$ as $x \to \infty$.

From this we conclude

$$\sum_{2} \ll x \sum_{\substack{t < p \leq x, \\ |f(p)| \geq 3/2}} \frac{|f(p)|}{p} \ll \frac{x}{\ln t} \sum_{p \leq x} |1 - \chi_d^*(p) f(p)| \frac{\ln p}{p} \ll \sqrt{\varepsilon(x)},$$

if $t \geq x^{\sqrt{\varepsilon(x)}}$. Choosing $t \geq \max(\ln^6 x, x^{\sqrt{\varepsilon(x)}})$ and collecting the estimates for $\sum_1$ and $\sum_2$, respectively, we arrive at

$$\frac{1}{\pi(x)} \sum_{p \leq x} |g(f; p+1) - g(f_1; p+1)|^2 \ll \sqrt{\varepsilon(x)} \ln y + \frac{1}{y^\beta}. \qquad (4)$$

Let us now define a multiplicative function f_2 by $f_2(p) = \overline{\chi_d^*(p)}$ if $p > t$, $|f(p)| < 3/2$ and $f_2(p^r) = f_1(p^r)$ otherwise. Then

$$\sum_{n \leq x} |f_1(n) - f_2(n)|^2 \leq \sum_{n_1 n_2 \leq x} |f_1(n_1)|^2 \left| \prod_{p|n_2} f(p) - \prod_{p|n_2} \overline{\chi_d^*(p)} \right|^2,$$

where for all prime divisors p of n_2 we have $p > t$, $|f(p)| < 3/2$, $\mu^2(n_2) = 1$ and if $p \mid n_1$ then either $p \leq t$ or $p^2 \mid n_1$ or $|f(p)| \geq 3/2$. Using the inequality

$$|1 - x_1 \ldots x_n|^2 = |1 - x_1 + x_1 - x_1 x_2 + \cdots - x_1 \cdots x_n|^2$$

$$\leq \left(1 + \max_{i=1,2,\ldots,n} |x_1 \cdots x_i|^2 \right) (|1 - x_1|^2 + |1 - x_2|^2 + \cdots + |1 - x_n|^2) n$$

we arrive at

$$\sum_{n \leq x} |f_1(n) - f_2(n)|^2 \leq 5 \left(\frac{3}{2} \right)^{2 \frac{\ln x}{\ln t}} \frac{\ln x}{\ln t} \sum_{n_1 \leq x/t} |f_1(n_1)|^2$$

$$\times \sum_{\substack{t < p \leq x/n_1, \\ |f(p)| \leq 3/2}} |f(p) - \overline{\chi_d^*(p)}| \sum_{n_2 \leq x/(n_1 p)} 1.$$

Applying Selberg's sieve (see [5]) yields

$$\sum_{n_2 \leq x/(n_1 p)} 1 \ll \frac{1}{\ln t} \frac{x}{n_1 p} + 1$$

and therefore

$$\frac{1}{x} \sum_{n \leq x} |f_1(n) - f_2(n)|^2 \ll \left(\frac{3}{2} \right)^{2 \frac{\ln x}{\ln t}} \frac{\ln x}{\ln t} \left(\frac{1}{\ln t} \sum_{n \leq x} \frac{|f_1(n)|^2}{n} \frac{1}{\ln t} \sum_{p \leq x} |f(p) - \overline{\chi_d^*(p)}| \frac{\ln p}{p} \right.$$

$$\left. + \frac{1}{\ln t} \sum_{p \leq x} |f(p) - \overline{\chi_d^*(p)}| \ln p \frac{1}{x} \sum_{n \leq x/p} |f_1(n)|^2 \right).$$

Since f_1 satisfies (2) we obtain

$$\frac{1}{x}\sum_{n\leq x}|f_1(n)-f_2(n)|^2 \ll \left(\frac{3}{2}\right)^{2\frac{\ln x}{\ln t}}\left(\frac{\ln x}{\ln t}\right)^3\frac{1}{\ln t}\sum_{p\leq x}|f(p)-\overline{\chi_d^*}(p)|\frac{\ln p}{p}\ll\sqrt{\varepsilon(x)},$$

if $(\frac{3}{2})^{2\frac{\ln x}{\ln t}}(\frac{\ln x}{\ln t})^3\leq\frac{1}{\sqrt{\varepsilon(x)}}$, for example $\frac{\ln x}{\ln t}=\frac{1}{20}\ln\frac{1}{\varepsilon(x)}$. Hence f_2 satisfies (2), too, and Lemma 1 shows

$$\frac{1}{\pi(x)}\sum_{p\leq x}|g(f_1;p+1)-g(f_2;p+1)|^2\ll\frac{\ln y}{x}\sum_{n\leq x}|g(f_1;n)-g(f_2;n)|^2+\frac{1}{y^\beta}.$$

Following the same arguments as before we obtain

$$\frac{1}{x}\sum_{n\leq x}|g(f_1;n)-g(f_2;n)|^2\ll\left(\frac{3}{2}\right)^{2\frac{\ln x}{\ln t}}\left(\frac{\ln x}{\ln t}\right)^3\frac{1}{\ln t}\sum_{p\leq x}|g(f;p)-g(\overline{\chi_d^*};p|\frac{\ln p}{p}.$$

If $p\nmid d$, we have $|g(f;p)+g(\overline{\chi_d^*};p|\geq 1$ and therefore

$$\frac{1}{\pi(x)}\sum_{p\leq x}|g(f_1;p+1)-g(f_2;p+1)|^2\ll\sqrt{\varepsilon(x)}\ln y+y^{-\beta}.$$

Put $y=(\varepsilon(x))^{-\frac{4}{\beta}}$. Using the above estimate and the corresponding inequality (4), we have

$$\frac{1}{\pi(x)}\sum_{p\leq x}|f(p+1)-f_2(p+1)|^2$$

$$\leq\left(\frac{1}{\pi(x)}\sum_{p\leq x}|g(f_2;p+1)-g(f;p+1)|^2\right)^{1/2}$$

$$\cdot\left(\frac{1}{\pi(x)}\sum_{p\leq x}(|g(f;p+1)|+|g(f_2;p+1)|)^2\right)^{1/2}\ll\sqrt[5]{\varepsilon(x)}.$$

Here, for the last sum, we made use of the fact that f, f_2 satisfy (2) and obtained, by Lemma 1, the estimate

$$\ll\frac{\ln v}{x}\sum_{n\leq x}(|f(n)|+|f_2(n)|)+\frac{1}{v^\beta}=O(1).$$

Thus, we proved

$$\frac{1}{\pi(x)}\sum_{p\leq x}f(p+1)=\frac{1}{\pi(x)}\sum_{p\leq x}f_2(p+1)+o(1),\qquad(5)$$

where $f_2(p^r)=f(p^r)$ if $p\leq t$, $f_2(p^r)=\overline{\chi_d^*}(p^r)$ if $p>t$ and $t\to\infty$, $\ln t/\ln x\to 0$ as $x\to\infty$.

It remains to investigate the asymptotic behaviour of $f_2(p+1)$. For this we put $p+1 = k_1 k_2$, where k_1 is only divisible by primes $q \leq t$ and k_2 possesses only prime divisors $q > t$, i.e., $(k_2, P(t)) = 1$, where $P(t) = \prod_{q \leq t} q$.

Then we have

$$\frac{1}{\pi(x)} \sum_{p \leq x} f_2(p+1) = \frac{1}{\pi(x)} \sum_{k_1 \leq x/t} f(k_1) \frac{1 + (-1)^{k_1}}{2} \sum_{p+1 = k_1 k_2 \leq x+1} \overline{\chi_d^*}(k_2)$$

$$+ O\left(\frac{1}{\pi(x)} \sum_{p+1 = k_1 \leq x} |f(k_1)| \right) + o(1).$$

Applying Lemma 1 we obtain

$$\frac{1}{\pi(x)} \sum_{p+1 = k_1 \leq x+1} |f(k_1)| \ll \frac{\ln y}{x} \sum_{k_1 \leq x+1} |f(k_1)| + y^{-\beta}$$

$$\ll \frac{\ln y}{x} \left(\sum_{n \leq x+1} |f(n)|^2 \sum_{k_1 \leq x+1} 1 \right)^{1/2} + y^{-\beta}.$$

Hence, using the inequality (see [7])

$$\sum_{k_1 \leq x} 1 \ll x \exp\left(-\frac{\ln x}{\ln t} \right) \tag{6}$$

for $t > \exp(\sqrt{\ln x})$ with $y = \ln x / \ln t$ gives

$$\frac{1}{\pi(x)} \sum_{p+1 = k_1 \leq x+1} |f(k_1)| \ll \ln y \exp\left(-\frac{1}{2} \frac{\ln x}{\ln t} \right) + y^{-\beta} = o(1).$$

Thus, we proved

$$\frac{1}{\pi(x)} \sum_{p \leq x} f_2(p+1) = \frac{1}{\pi(x)} \sum_{k_1 \leq x/t} f(k_1) \frac{(1 + (-1)^{k_1})}{2}$$

$$\times \sum_{\substack{b=1, \\ (b(bk_1-1),d)=1}}^{d} \overline{\chi_d^*}(b) I_{k_1}(x, t, b) + o(1),$$

where $I_{k_1}(x, t, b) := \text{card}\{p : p \leq x, p \equiv bk_1 - 1 \pmod{k_1 d}, (\frac{p+1}{k_1}, P(t)) = 1\}$. Observe that d is a fixed number, and therefore we may assume $d < t$.

Applying the Fundamental Lemma of sieve methods (see [5]) we obtain

$$I_{k_1}(x, t, b) = \frac{\pi(x)}{\varphi(k_1 d)} \prod_{\substack{q \leq t, \\ q \nmid d}} (1 - \varrho(q))(1 + \theta H) + O\left(\sum_{\substack{\delta \leq \xi^3, \\ \delta | P(t)}} 3^{\omega(\delta)} |\eta(x, \delta)| \right),$$

where $0 < \theta < 1$,

$$\eta(x, \delta) := \text{card}\{p \le x, p \equiv -1(\text{mod } k_1\delta), p \equiv bk_1 - 1(\text{mod } k_1 d)\} - \frac{\pi(x)}{\varphi(k_1 d)}\varrho(\delta),$$

$$H := \exp\left(-\frac{\ln\xi}{\ln t}\left\{\ln\frac{\ln\xi}{S} - \ln\ln\frac{\ln\xi}{S} - 2\frac{S}{\ln\xi}\right\}\right), \quad \varrho(\delta) = \frac{\varphi(dk_1)}{\varphi(\delta dk_1)},$$

$$S := \sum_{p \le t}\frac{\varrho(p)}{1 - \varrho(p)}\ln p, \quad \max(\ln t, S) \le \frac{1}{8}\ln\xi.$$

Since $p \equiv -1(\text{mod } k_1\delta)$ and $p \equiv bk_1 - 1(\text{mod } k_1 d)$ we deduce that (δ, d) must divide b. But $(b, d) = 1$ and therefore $(\delta, d) = 1$. Set $\xi^3 = \sqrt{x}/(k_1 d \ln^A x)$, where A is a large positive number, and let $\alpha = 1/100$. Then

$$\frac{1}{\pi(x)}\sum_{p \le x} f_2(p+1) = \sum_{k_1 \le x^\alpha} f(k_1)\frac{(1 + (-1)^{k_1})}{2} \sum_{\substack{b=1, \\ (b(bk_1-1),d)=1}}^{d} \overline{\chi_d^*}(b)\frac{1}{\varphi(k_1 d)}$$

$$\cdot \prod_{\substack{p \le t, \\ p \nmid d}}\left(1 - \frac{\varphi(dk_1)}{\varphi(pdk_1)}\right)\left(1 + O\left(\exp\left(-\frac{\ln x}{\ln t}\right)\right)\right) + R_1 + R_2,$$

where, by Vinogradov-Bombieri's theorem (see [1]),

$$R_1 = \frac{1}{\pi(x)}\sum_{k \le \sqrt{x}/\ln^A x} \tau_4(k)\max_{(l,k)=1}\left|\pi(x, k, l) - \frac{\pi(x)}{\varphi(k)}\right| \ll \ln^{-B(A)} x$$

and

$$R_2 = \frac{1}{\pi(x)}\sum_{x^\alpha \le k_1 \le x/t}|f(k_1)|\text{card}\{p : p \le x, p + 1 = k_1 k_2\}$$

$$\le \frac{1}{\pi(x)}\sum_{x^\alpha \le k_1 \le x/t}|f(k_1)|\text{card}\left\{n : n \le \frac{x}{k_1}, (n(nk_1 - 1), P(t)) = 1\right\}.$$

Using Selberg's sieve method (see [5]) we obtain

$$\text{card}\left\{n : n \le \frac{x}{k_1}, (n(nk_1 - 1), P(t)) = 1\right\} \ll \frac{x}{\varphi(k_1)\ln^2 t}.$$

Hence, by Cauchy's inequality,

$$R_2 \ll \frac{\ln x}{\ln^2 t}\left(\sum_{k_1 \le x}\frac{|f(k_1)|^2}{k_1}\sum_{x^\alpha \le k_1 \le x}\frac{k_1}{\varphi^2(k_1)}\right)^{1/2}.$$

We have

$$\sum_{x^\alpha \le k_1 \le x} \frac{k_1}{\varphi^2(k_1)} \le \sum_{\delta \le x} \frac{\tau(\delta)}{\delta^2} \sum_{x^\alpha \le k_1 \delta \le x} \frac{1}{k_1}$$

$$\ll \ln x \sum_{\delta > x^{\frac{\alpha}{2}}} \frac{\tau(\delta)}{\delta^2} + \sum_{\delta \le x^{\frac{\alpha}{2}}} \frac{\tau(\delta)}{\delta^2} \sum_{x^{\frac{\alpha}{2}} \le k_1 \le x} \frac{1}{k_1}.$$

Using the inequality (6) we see that

$$\sum_{x^{\frac{\alpha}{2}} \le k_1 \le x} \frac{1}{k_1} \le \max_{x^{\frac{\alpha}{2}} \le Q \le x} \frac{1}{Q} \operatorname{card}\{k_1 : k_1 \le 2Q\} \ll \ln x \exp\left(-\frac{1}{200} \frac{\ln x}{\ln t}\right).$$

Since $f \in \mathcal{L}_2$ and

$$\sum_{\delta > u} \frac{\tau(\delta)}{\delta^2} \ll \frac{\ln 2u}{u}$$

we obtain

$$R_2 \ll \left(\frac{\ln x}{\ln t}\right)^2 \exp\left(-\frac{1}{400} \frac{\ln x}{\ln t}\right) = o(1)$$

if $\frac{\ln x}{\ln t} \to \infty$ as $x \to \infty$. Thus we proved

$$\frac{1}{\pi(x)} \sum_{p \le x} f_2(p+1) = \Gamma(x)\left(1 + O\left(\exp\left(-\frac{\ln x}{\ln t}\right)\right)\right) + o(1), \tag{7}$$

where

$$\Gamma(x) := \sum_{k_1 \le x^\alpha} f(k_1) \frac{(1 + (-1)^{k_1})}{2} \sum_{\substack{b=1, \\ (b(bk_1 - 1), d) = 1}}^d \overline{\chi_d^*}(b) \frac{1}{\varphi(k_1 d)} \prod_{\substack{p \le t, \\ p \nmid d}} \left(1 - \frac{\varphi(dk_1)}{\varphi(pdk_1)}\right).$$

The main term equals

$$\sum_{k_1} f(k_1) \frac{(1 + (-1)^{k_1})}{2} \sum_{\substack{b=1, \\ (b(bk_1 - 1), d) = 1}}^d \overline{\chi_d^*}(b) \frac{1}{\varphi(k_1 d)} \cdot \prod_{\substack{p \le t, \\ p \nmid d}} \left(1 - \frac{\varphi(dk_1)}{\varphi(pdk_1)}\right)$$

$$+ O\left(\sum_{k_1 > x^\alpha} \frac{|f(k_1)|}{\varphi(k_1 d)} \prod_{2 < p \le t} \left(1 - \frac{1}{p-1}\right)\right) = \Gamma_1(x) + R_3.$$

By the inequality $\varphi(dk_1) \ge \varphi(d)\varphi(k_1)$ we obtain

$$R_3 \ll \frac{1}{\ln t \ln x^\alpha} \sum_{k_1 > x^\alpha} \frac{|f(k_1)|}{\varphi(k_1)} \ln k_1 \ll \frac{1}{\ln t \ln x} \sum_{p^r, p \le t} \frac{|f(p^r)| \ln p^r}{p^r} \sum_{k_1} \frac{f(k_1)}{\varphi(k_1)}.$$

Since $f \in \mathcal{L}_2$ it follows

$$\sum_p \sum_{r \geq 2} \frac{|f(p^r)| \ln p^r}{p^r} \leq \left(\sum_n \frac{|f(n)|^2}{n^{\frac{3}{4}}} \right)^{1/2} \left(\sum_p \sum_{r \geq 2} \frac{r^2 \ln^2 p}{(p^r)^{\frac{2}{3}}} \right) < +\infty$$

and therefore by (3)

$$R_3 \ll \frac{1}{\ln t \ln x} \left(\sum_{p \leq t} \frac{|f(p) - \overline{\chi_d^*}(p)|}{p} \ln p + \ln t \right) \exp \left(\sum_{p \leq t} \frac{|f(p)|}{p} \right)$$

$$\ll \frac{\ln t}{\ln x} \exp \left(\sum_{p \leq t} \frac{|f(p)| - 1}{p} \right).$$

Using the condition (3) in the above mentioned form and the inequality $|a - b| \geq \|a| - |b\|$ we get

$$\sum_{p \leq t} \frac{|f(p)| - 1}{p} \leq \sum_{p \leq x} \frac{|f(p)| - 1}{p} + \sum_{t \leq p \leq x} \frac{|f(p) - \overline{\chi_d^*}(p)|}{p} \frac{\ln p}{\ln t}$$

$$\leq \sum_{p \leq x} \frac{|f(p)| - 1}{p} + \varepsilon(x) \frac{\ln x}{\ln t}.$$

Now, let $\frac{\ln t}{\ln x} \to 0$ and $\varepsilon(x) \frac{\ln x}{\ln t} \to 0$ as $x \to \infty$. Then by (5) and (7) we arrive at

$$\frac{1}{\pi(x)} \sum_{p \leq x} f(p + 1) = \Gamma_1(x)(1 + o(1)) + o \left(\exp \left(\sum_{p \leq x} \frac{|f(p)| - 1}{p} \right) + 1 \right).$$

The main term $\Gamma_1(x)$ equals

$$\sum_{b=1}^{d} \frac{\overline{\chi_d^*}(b)}{\varphi(d)} \sum_{\delta | d} \mu(\delta) \sum_{\substack{k_1, \\ bk_1 \equiv 1 (\mathrm{mod}\, \delta)}} \frac{f(k_1)}{\varphi(k_1 d)} \varphi(d) \frac{(1 + (-1)^{k_1})}{2} \prod_{\substack{p \leq t, \\ p \nmid d}} \left(1 - \frac{\varphi(dk_1)}{\varphi(pdk_1)} \right)$$

$$= \prod_{\substack{p \leq t \\ p \nmid 2d}} \left(1 - \frac{1}{p - 1} \right) \sum_{b=1}^{d} \frac{\overline{\chi_d^*}(b)}{\varphi(d)} \times \sum_{\delta | d} \frac{\mu(\delta)}{\varphi(\delta)} \sum_{\chi_\delta} \chi_\delta(b)$$

$$\times \sum_{k_1} \frac{\chi_\delta(k_1) f(k_1)}{\varphi(k_1 d)} \varphi(d) \frac{(1 + (-1)^{k_1})}{2} \cdot \left(1 - \frac{1}{2} \right) \prod_{\substack{p | k_1, \\ p \nmid 2d}} \left(1 - \frac{1}{p} \right) \left(1 - \frac{1}{p - 1} \right)^{-1}.$$

The sum over k_1 can be represented as a product. Hence

$$\Gamma_1(x) = \sum_{b=1}^{d} \frac{\overline{\chi_d^*}(b)}{\varphi(d)} \sum_{\delta|d} \frac{\mu(\delta)}{\varphi(\delta)} \sum_{\chi_\delta} \chi_\delta(b) \prod_{\substack{p|d, \\ p \neq 2}} \left(1 + \sum_{r=1}^{\infty} \frac{\chi_\delta(p^r)f(p^r)}{p^r}\right)$$

$$\cdot \prod_{\substack{p \leq t, \\ p \nmid d}} \left(1 + \sum_{r=1}^{\infty} \frac{1}{\varphi(p^r)}(\chi_\delta(p^r)f(p^r) - \chi_\delta(p^{r-1})f(p^{r-1}))\right).$$

If $\delta \neq d$ or if $\delta = d$ but $\chi_\delta \neq \chi_d^*$, where χ_d^* is the primitive character mod d which occurs in the condition (3), we have

$$\prod_{\substack{p \leq t, \\ p \nmid d}} \left(1 + \sum_{r=1}^{\infty} \frac{1}{\varphi(p^r)}(\chi_\delta(p^r)f(p^r) - \chi_\delta(p^{r-1})f(p^{r-1}))\right)$$

$$\ll \exp\left(-\sum_{p \leq t} \frac{1 - \operatorname{Re}\chi_\delta(p)f(p)}{p - 1}\right)$$

$$\ll \exp\left(-\sum_{z < p \leq t} \frac{1 - \operatorname{Re}\chi_\delta(p)\overline{\chi_d^*}(p)}{p} + \sum_{z < p \leq t} \frac{|\overline{\chi_d^*}(p) - f(p)|}{p}\right).$$

Using (3) in the above mentioned form again we have

$$\sum_{z < p \leq t} \frac{|\overline{\chi_d^*}(p) - f(p)|}{p} \leq \frac{1}{\ln z} \sum_{z < p \leq t} \frac{|1 - \chi_d^* f(p)|}{p} \ln p \ll \varepsilon(t)\frac{\ln t}{\ln z} \ll 1$$

if $z = t^{\varepsilon(t)}$. It is easy to see that $\overline{\chi_d^*}(n)\chi_\delta(n) \neq \chi_0$ and $\delta \mid d$. Hence, there exists an arithmetic progression to a modulus d such that $1 - \operatorname{Re}\chi_\delta(p)\overline{\chi_d^*}(p) = \sigma > 0$ for $p \equiv l(\mathrm{mod}\ d)$. This implies

$$\sum_{z < p \leq t} \frac{1 - \operatorname{Re}\chi_\delta(p)\overline{\chi_d^*}(p)}{p} \geq \sigma \sum_{\substack{z < p \leq t, \\ p \equiv l(\mathrm{mod}\ d)}} \frac{1}{p} \geq \sigma_1 \ln \frac{\ln t}{\ln z},$$

where $\sigma_1 = \frac{\sigma}{2\varphi(d)} > 0$. Thus, if $\chi_\delta \neq \chi_d^*$, we have

$$\prod_{\substack{p \leq t, \\ p \nmid d}} \left(1 + \sum_{r=1}^{\infty} \frac{1}{\varphi(p^r)}(\chi_\delta(p^r)f(p^r) - \chi_\delta(p^{r-1})f(p^{r-1}))\right) \ll \left(\frac{\ln z}{\ln t}\right)^{\sigma_1} = (\varepsilon(t))^{\sigma_1}$$

and therefore

$$\frac{1}{\pi(x)} \sum_{p \leq x} f(p+1) = \frac{\mu(d)}{\varphi(d)} \prod_{p \leq t} \left(1 + \sum_{r=1}^{\infty} \frac{1}{\varphi(p^r)}(\chi_d^*(p^r)f(p^r) - \chi_d(p^{r-1})f(p^{r-1}))\right)$$

$$+ o\left(\exp\left(\sum_{p \leq x} \frac{|f(p)| - 1}{p}\right) + 1\right).$$

Using (3) again we see that

$$\prod_{t<p\le x}\left(1+\sum_{r=1}^{\infty}\frac{1}{\varphi(p^r)}(\chi_d^*(p^r)f(p^r)-\chi_d(p^{r-1})f(p^{r-1}))\right)-1$$

$$\ll\sum_{t<p\le x}\frac{|1-\chi_d^*(p)f(p)|}{p}\ll\varepsilon(x)\frac{\ln x}{\ln t}\to 0$$

as $x\to\infty$, and this ends the proof of Lemma 2. $\square$

3. Proof of theorem.

If $\|f\|_1=0$ then (see Corollary 5 of [4]) $m_p(|f|)=m_p(f)=0$. If $\|f\|_1>0$ we have (see Theorem 1 of [3])

$$\sum_{p\le x}\frac{|f(p)|-1}{p}\le c<\infty\quad(x\ge 2). \tag{8}$$

If the mean value $M(f\chi_d)$ of $f\chi_d$ exists and is different from zero and if $f\in\mathcal{L}_2$ then (see Theorem 2 of [3]) the series

$$\sum_p\frac{\chi_d(p)f(p)-1}{p},\qquad\sum_{|f(p)|\le 3/2}\frac{|\chi_d(p)f(p)-1|^2}{p},$$

$$\sum_{|f(p)|\ge 3/2}\frac{|f(p)|^2}{p},\qquad\sum_p\sum_{r\ge 2}\frac{|f(p^r)|}{p^r} \tag{9}$$

converge.

Now, let χ_δ be a primitive character which generates χ_d. Then the series (9) with χ_δ replaced by χ_d converge, too. Using Cauchy's inequality we have

$$\frac{1}{\ln x}\sum_{p\le x}|\chi_\delta(p)f(p)-1|\frac{\ln p}{p}\le 2\frac{\ln y}{\ln x}\sum_{\substack{p\le y,\\|f(p)|\ge 3/2}}\frac{|f(p)|^2}{p}+2\sum_{\substack{p>y,\\|f(p)|\ge 3/2}}\frac{|f(p)|^2}{p}$$

$$+\left(\sum_{|f(p)|\le 3/2}\frac{|\chi_\delta(p)f(p)-1|^2}{p}\frac{1}{\ln^2 x}\sum_{p\le y}\frac{\ln^2 p}{p}\right)^{1/2}$$

$$+\left(\sum_{\substack{p>y,\\|f(p)|\le 3/2}}\frac{|\chi_\delta(p)f(p)-1|^2}{p}\frac{1}{\ln^2 x}\sum_{p\le x}\frac{\ln^2 p}{p}\right)^{1/2}.$$

Since the series (9) converge, we obtain letting $y \to \infty$ such that $\ln y / \ln x \to 0$ as $x \to \infty$,

$$\frac{1}{\ln x} \sum_{p \leq x} |\chi_\delta(p) f(p) - 1| \frac{\ln p}{p} = o(1)$$

as $x \to \infty$. Thus f satisfies the conditions of Lemma 2. Hence, by Lemma 2 and (8) we conclude that

$$\frac{1}{\pi(x)} \sum_{p \leq x} f(p+1) = \frac{\mu(\delta)}{\varphi(\delta)} \prod_{p \leq x} \left(1 + \sum_{r=1}^{\infty} \frac{1}{\varphi(p^r)} (\chi_\delta(p^r) f(p^r) \right.$$
$$\left. - \chi_\delta(p^{r-1}) f(p^{r-1})) \right) + o(1).$$

Since the series (9) converge the Theorem is proved.

References

1. E. Bombieri, "On the large sieve," *Mathematika* **12** (1965), 201–225.
2. P.D.T.A. Elliott, "Probabilistic Number Theory I," *Grundlehren der math. Wissenschaften* **239** (1979).
3. K.-H. Indlekofer, "Properties of uniformly summable multiplicative functions," *Periodica Math. Hungarica* **17**(2) (1986), 143–161.
4. K.-H. Indlekofer, N.M. Timofeev, "Estimates for multiplicative functions on the set of shifted primes," to appear in *Acta Math. Hung.*
5. K. Prachar, *Primzahlverteilung*, Springer-Verlag, Berlin, 1979.
6. N.M. Timofeev, "Multiplicative functions on the set of shifted prime numbers," *Math. USSR Izvestija* **39**(3) (1992), 1189–1207.
7. A.I. Vinogradov, "On numbers with small prime divisors," *Dokl. Akad. Nauk SSSR* **109** (1956), 683–686. (Russian).

Since the series (7) converge, we obtain letting $y \to \infty$ so such that $\ln y/\ln x \to 0$ as $x \to \infty$,

$$\frac{1}{\ln x} \sum_{p \le x} \frac{|x(p)f(p) - 1|}{p} \ln p = o(1)$$

as $x \to \infty$, that f satisfies the conditions of Lemma 2. Hence, by Lemma 2 and (8) we conclude that

$$\frac{1}{\pi(x)} \sum_{p \le x} f(p+1) = \frac{f(2)}{\phi(2)} \prod_{p \ge 2} \frac{f(p)}{\phi(p)} \left(1 + \sum_{j=1}^{\infty} \frac{1}{\phi(p^j)} (x_p f(p^j) - \right.$$

$$\left. - x_p(p^{j-1}))(f(p^{j-1})) \right) + o(1)_{x}.$$

Since the series (9) converge, the Theorem is proved.

References

1. E. Bombieri, "On the large sieve," Mathematika 12 (1965), 201–225.
2. H.D.T.A. Elliott, *Probabilistic Number Theory I*, Grundlehren der math. Wissenschaften 239 (1979).
3. A.R. Daboussi, "Properties of uniformly summable multiplicative functions," Period. Math. Hungar. 12 (1981), 143–161.
4. P.D.T.A. Elliott, K.H. Indlekofer, "Estimates for multiplicative functions on the set of shifted primes," to appear in Acta Math. Hung.
5. K. Prachar, *Primzahlverteilung*, Springer-Verlag, Berlin 1979.
6. N.M. Timofeev, "Multiplicative functions on the set of shifted prime numbers," Mat. USSR Izvestija 39:3 (1992), 1189–1207.
7. A.I. Vinogradov, "On numbers with small prime divisors," Dokl. Akad. Nauk SSSR 109 (1956), 683–686. (Russian).

THE RAMANUJAN JOURNAL 2, 167–184 (1998)

Entiers Lexicographiques

ANDRÉ STEF   Andre.Stef@iecn.u-nancy.fr
GÉRALD TENENBAUM   Gerald.Tenenbaum@ciril.fr
Institut Élie Cartan, Université Henri Poincaré–Nancy 1, BP 239, 54506 Vandœuvre Cedex, France

À la mémoire de Paul Erdős, pour la grâce et la lumière

Received April 11, 1997; Accepted January 14, 1998

Abstract.   An integer n is called *lexicographic* if the increasing sequence of its divisors, regarded as words on the (finite) alphabet of the prime factors (arranged in increasing size), is ordered lexicographically. This concept easily yields to a new type of multiplicative structure for the exceptional set in the Maier-Tenenbaum theorem on the propinquity of divisors, which settled a well-known conjecture of Erdős. We provide asymptotic formulae for the number of lexicographic integers not exceeding a given limit, as well as for certain arithmetically weighted sums over the same set. These results are subsequently applied to establishing an Erdős-Kac theorem with remainder for the distribution of the number of prime factors over lexicographic integers. This provides quantitative estimates for lexicographical exceptions to Erdos' conjecture that also satisfy the Hardy-Ramanujan theorem.

Key words: distribution of divisors, Erdős-Kac theorem, Erdős conjecture, distribution of prime factors, distribution of arithmetic functions, sieve

1991 Mathematics Subject Classification: 11N25, 11N37, 11N64

1. Introduction

Désignons par $\{p_j(n)\}_{j=1}^{\omega(n)}$ la suite croissante des facteurs premiers d'un entier n, par $\nu_j(n)$ l'unique entier ν tel que $p_j(n)^{\nu} \parallel n \, (1 \leqslant j \leqslant \omega(n))$ et posons

$$n_j := \prod_{i<j} p_i(n)^{\nu_i(n)} \quad (1 \leqslant j \leqslant \omega(n)).$$

De nombreux problèmes de crible s'énoncent en termes de comparaison des tailles respectives de $p_j(n)$ et n_j. Un important résultat d'Erdős [3], précisé ultérieurement par Bovey [2], énonce que

$$\max_{1 \leqslant j \leqslant \omega(n)} \frac{\log n_j}{\log p_j(n)} \sim \frac{\log_3 n}{\log_4 n}$$

lorsque n tend vers l'infini en restant dans une suite convenable de densité unité. Si $\{d_j(n)\}_{j=1}^{\tau(n)}$ désigne la suite croissante des diviseurs de n, Tenenbaum a montré dans [9] que l'on a pour tout n

$$\max_{1 \leqslant j < \tau(n)} d_{j+1}(n)/d_j(n) = \max_{1 \leqslant j \leqslant \omega(n)} p_j(n)/n_j. \tag{1.1}$$

L'étude de la fonction de répartition de cette fonction arithmétique est susceptible de sur-prenantes applications à des problèmes apparemment disjoints de théorie des nombres, cf. [9, 11].

Nous nous proposons ici d'entreprendre l'étude duale, c'est-à-dire celle de la répartition de la fonction

$$W(n) := \min_{1 \leqslant j \leqslant \omega(n)} p_j(n)/n_j. \tag{1.2}$$

Nous convenons que $W(1) = +\infty$ et nous désignons par $\mathcal{L}(y)$ l'ensemble des entiers $n \geqslant 1$ tels que $W(n) > y$.

On sait que l'une des questions centrales dans l'étude de la structure fine des diviseurs d'un entier normal est le conflit entre les deux relations d'ordre naturelles dont on peut munir l'ensemble des $d_j(n)$ (voir en particulier [4, 5, 10]). La structure d'ordre additif est la trace de l'ordre de $\mathbb{Z}$. La structure d'ordre multiplicatif est induite par l'ordre lexicographique, obtenu en associant à chaque diviseur

$$d = \prod_{1 \leqslant j \leqslant \omega(n)} p_j(n)^{\alpha_j} \quad (0 \leqslant \alpha_j \leqslant \nu_j(n))$$

le mot $\alpha_{\omega(n)} \cdots \alpha_1$ de longueur $\omega(n)$.

Notons σ_n la permutation de $\{j : 1 \leqslant j \leqslant \tau(n)\}$ qui transforme l'ordre additif dans l'ordre lexicographique. Il est facile de constater qu'*une condition nécessaire et suffisante pour que σ_n coïncide avec l'identité est que $n \in \mathcal{L}(1)$*. En effet, si $n \notin \mathcal{L}(1)$, il existe un j tel que $p_j = p_j(n) \leqslant n_j$. Puisque $p_j \nmid n_j$, on a en fait $p_j < n_j$. On peut donc écrire $p_j = d_s(n)$, $n_j = d_t(n)$ avec $s < t$. Or $\sigma_n(s) > \sigma_n(t)$ par définition de n_j. Il s'ensuit que σ_n n'est pas l'identité. Réciproquement, pour $n \in \mathcal{L}(1)$, considérons $d_s(n)$ et $d_t(n)$, avec $s < t$. Désignant par j le plus grand indice tel que $d_s(n)$ et $d_t(n)$ aient des valuations $p_j(n)$-adiques distinctes. On peut alors décomposer $d_s(n) = a_s p_j(n)^\beta b$, et $d_t(n) = a_t p_j(n)^\gamma b$ avec $a_s \mid n_j, a_t \mid n_j$, et $b \mid n/n_{j+1}$. Comme $n_j < p_j$, on a nécessairement $\beta < \gamma$. Cela implique $\sigma_n(s) < \sigma_n(t)$. Comme s et t sont quelconques, on obtient bien que σ_n coïncide avec l'identité.

Ce critère nous a conduit à désigner $\mathcal{L}(1)$ comme l'*ensemble des entiers lexicographiques*. Une seconde propriété attrayante de cet ensemble d'entiers fait l'objet de l'énoncé suivant.

Théorème 1. *Soit n un nombre entier. On a*

$$W(n) = \min_{1 \leqslant j \leqslant \tau(n)} d_{j+1}(n)/d_j(n) \tag{1.3}$$

si, et seulement si, $n \in \mathcal{L}(1)$.

Démonstration. La condition est clairement nécessaire puisque le membre de droite de (1.3) est au moins 1. Montrons qu'elle est suffisante. Soit $j = k$ un indice réalisant le minimum du membre de droite de (1.3). On a $d_{k+1}(n) = p_\ell^{\alpha_\ell} \ldots p_1^{\alpha_1}$ pour un certain indice ℓ, $1 \leqslant \ell \leqslant \omega(n)$, avec $\alpha_\ell \geqslant 1$ et $\alpha_j \geqslant 0$ pour $1 \leqslant j < \ell$. Alors $d_{k+1}(n) \mid n_{\ell+1}$ et puisque $p_{\ell+1} > n_{\ell+1}$, on a aussi $d_k(n) \mid n_{\ell+1}$, d'où, pour des exposants $\beta_j \geqslant 0$ convenables, $d_k(n) = p_\ell^{\beta_\ell} \ldots p_1^{\beta_1}$. Soit m le plus petit indice tel que $\alpha_j = \beta_j$ pour $m < j \leqslant \ell$. Comme $d_k(n) \neq d_{k+1}(n)$, on a $m \geqslant 1$. Si $\alpha_m < \beta_m$, alors

$$d_{k+1}(n)/d_k(n) \leqslant n_m/p_m < 1,$$

ce qui est manifestement impossible. Donc $\alpha_m > \beta_m$, et il suit

$$d_{k+1}(n)/d_k(n) \geqslant p_m/n_m. \tag{1.4}$$

Comme $n_m < p_m$, on a en fait égalité dans (1.4). Par définition de k, cela implique

$$p_m/n_m = \min p_j/n_j,$$

ce qu'il fallait démontrer. □

Soit $\mathcal{E}$ l'ensemble des entiers n tels que

$$\min d_{j+1}(n)/d_j(n) \geqslant 2. \tag{1.5}$$

On a donc $\mathcal{L}(2) \subset \mathcal{E}$, d'après le Théorème 1. Une célèbre conjecture d'Erdős, établie par Maier et Tenenbaum [6], affirme que $\mathcal{E}$ est de densité nulle, autrement dit

$$E(x) := |\mathcal{E} \cap [1, x]| = o(x) \quad (x \to \infty).$$

Hall et Tenenbaum ont donné dans [5] l'estimation effective

$$E(x) \ll x (\log_3 x)^{4\beta}/(\log_2 x)^\beta,$$

avec $\beta = 1 - (1 + \log_2 3)/\log 3 \approx 0,00415$. Stef a amélioré considérablement cette majoration [7] et a obtenu une minoration (peut-être optimale) de $E(x)$. Il montre en fait que l'on a pour une constante positive convenable c

$$x (\log x)^{-\beta+o(1)} \ll E(x) \ll x e^{-c\sqrt{\log_2 x}}. \tag{1.6}$$

Cet encadrement reflète bien l'état actuel de nos connaissances sur la structure multiplicative des entiers. La majoration est obtenue en montrant que les facteurs premiers satisfont statistiquement à certaines propriétés relatives d'équirépartition, et la valeur explicite de la fonction de x résultante indique le degré d'uniformité auquel les méthodes disponibles permettent de parvenir. La minoration découle d'un principe heuristique plus simple : si le nombre total des quantités distinctes de la forme $\log(d_j(n)/d_k(n))$, qui est normalement « voisin » de $3^{\omega(n)}$, est inférieur à $\log n$, on peut s'attendre à ce que le minimum ne soit pas trop proche de 0. Cette condition est statistiquement réalisée lorsque le nombre total $\omega(n)$ des facteurs premiers de n n'excède pas $(\log_2 n)/\log 3$, ce qui correspond à l'exposant β.

Le fait que ce principe puisse être rendu effectif nous donne un premier renseignement sur la structure de l'ensemble exceptionnel $\mathcal{E}$ pour la conjecture d'Erdős. Il contient en particulier, pour chaque $\varepsilon > 0$, presque tous les entiers n sans facteur premier $\leqslant \exp(\log x)^\varepsilon$ et tels que $\omega(n) < (1 - \varepsilon)(\log_2 n)/\log 3$. Cela induit *ipso facto* la question d'exhiber d'autres structures multiplicatives représentatives des entiers de $\mathcal{E}$. Quelle minoration peut-on donner, par exemple, pour le nombre des n de $\mathcal{E}$ n'excédant pas x et possédant le nombre « normal » de facteurs premiers, soit $\omega(n) \sim \log_2 n$? L'étude de la répartition de la fonction $\omega(n)$ sur les entiers lexicographiques fournit une première réponse.

Il est bien connu que la fonction arithmétique $n \mapsto t^{\omega(n)}$ permet souvent d'opérer un « recentrage exponentiel » (selon l'expression utilisée en théorie des probabilités) d'une somme arithmétique autour d'un ensemble d'entiers ayant une quantité prescrite de facteurs premiers. Nous allons montrer que cette technique fonctionne encore, quoique sous une forme plus sophistiquée, dans l'étude des nombres lexicographiques.

Posons $\mathcal{L}(x, y) := \mathcal{L}(y) \cap [1, x]$. Les motivations exposées plus haut nous conduisent à étudier le comportement asymptotique de la quantité

$$L(x, y; t) := \sum_{n \in \mathcal{L}(x, y)} t^{\omega(n)}. \tag{1.7}$$

Introduisons quelques notations pour énoncer plus commodément notre résultat. Nous posons pour $\delta \in \,]-\infty, 1[$

$$J(\delta) := \int_0^{1/2} v^{-\delta} \frac{dv}{1 - v}. \tag{1.8}$$

Il est clair que J définit une bijection strictement croissante de $]-\infty, 1[$ sur $]0, +\infty[$. Nous désignons, pour chaque $t > 0$, par $\delta(t)$ l'unique solution de l'équation

$$t J(\delta) = 1. \tag{1.9}$$

On note que $\delta(t) \geqslant 0$ si, et seulement si, $t \leqslant 1/\log 2$.

Enfin, nous définissons $\varrho(t) := 1 - \delta(t)$ et nous introduisons la fonction positive

$$\eta(t) := \min(\varrho(t), 1/\varrho(t)^2) \quad (t > 0). \tag{1.10}$$

Théorème 2. *Pour chaque $y \geqslant 1$, il existe une fonction $K_y: \,]0, \infty[\to \,]0, \infty[$, de classe $\mathcal{C}^1$, telle que l'on ait*

$$L(x, y; t) = \left\{ K_y(t) + O_T\left(\frac{(\log_2 x)^2}{(\log x)^{\eta(t)}} \right) \right\} \frac{x}{(\log x)^{\delta(t)}} \quad (x \geqslant 3) \tag{1.11}$$

pour tout $T > 1$ et uniformément dans le domaine $1/T \leqslant t \leqslant T, 1 \leqslant y \leqslant T$.

Lorsque $t = 1$, on obtient une formule asymptotique pour la fonction de comptage des entiers lexicographiques. On a $\delta(1) \approx 0,2228$.

Posons $L(x, y) := L(x, y; 1)$ et

$$G_x(z) := L(x, y)^{-1} \sum_{\substack{n \leqslant x \\ \omega(n) \leqslant z \log_2 x}} 1.$$

Pour $t = e^{-u/\log_2 x}$, la quantité $L(x, y; t)/L(x, y)$ est la valeur au point u de la transformée de Laplace bilatérale de $G_x(z)$, soit

$$L(x, y; e^{-u/\log_2 x})/L(x, y) = \widehat{G}_x(u) := \int_{-\infty}^{\infty} e^{-uz} \, dG_x(z).$$

En reportant dans (1.11), on obtient que

$$\lim_{x \to \infty} \widehat{F}_x(u) = e^{u^2/2} \quad (u \in \mathbb{R}) \tag{1.12}$$

avec $F_x(z) := G_x(a + z\sqrt{b/\log_2 x})$, $a := -\delta'(1)$, $b := -\delta'(1) - \delta''(1)$. Cela implique immédiatement une version du théorème d'Erdős-Kac pour les entiers lexicographiques.

En fait, le Théorème 2 fournit une forme effective de (1.12). Pour exploiter ce renseignement supplémentaire, il est nécessaire de disposer d'un théorème d'inversion de Laplace quantitative, par exemple analogue au théorème de Berry-Esseen pour la transformée de Fourier. Le seul résultat de ce type disponible dans la littérature est, à notre connaissance, celui de Alladi [1]. Le théorème suivant, qui affine l'estimation de Alladi, est un cas particulier d'un résultat récent, obtenu par les auteurs [8], d'inversion quantitative pour la transformation de Laplace réelle. Ici et dans tout l'article nous posons

$$\Phi(z) := \frac{1}{\sqrt{2\pi}} \int_{-\infty}^{z} e^{-u^2/2} du.$$

Théorème A. *Soit F une fonction de répartition satisfaisant, pour $\varepsilon \in \,]0, 1[$, $\kappa > 0$ et $R > 1$, à*

(i) $\displaystyle \widehat{F}(u) := \int_{-\infty}^{\infty} e^{-uz} \, dF(z) = e^{u^2/2}\{1 + O(\varepsilon)\} \quad (0 \leqslant u \leqslant \kappa)$,

(ii) $\displaystyle \widehat{F}(u) \ll e^{u^2/2} \quad (-R \leqslant u \leqslant R)$.

Alors on a

$$\sup_{z \in \mathbb{R}} |F(z) - \Phi(z)| \ll_\kappa \frac{\log R}{R} + \frac{\log_2(1/\varepsilon)}{\sqrt{\log(1/\varepsilon)}}. \tag{1.13}$$

Le résultat de Alladi, prouvé par une méthode différente, fournissait la borne sensiblement moins précise $\ll_\kappa 1/\sqrt{R} + (\log |\log \varepsilon|/|\log \varepsilon|)^{1/8}$. Il est à noter, ainsi qu'il est établi dans [8], que la dépendance en ε de (1.13) est optimale, au facteur $\log_2(1/\varepsilon)$ près, dans les conditions de généralité indiquées. La conjonction des Théorèmes 2 et A fournit immédiatement le résultat suivant de type Erdős–Kac effectif, autrement dit de convergence vers la loi de Gauss avec estimation de l'erreur.

Théorème 3. *Soit* $T > 1$. *On a uniformément pour* $1/T \leqslant t \leqslant T$, $1 \leqslant y \leqslant T$, $z \in \mathbb{R}$,

$$\frac{1}{L(x, y; t)} \sum_{\substack{n \in \mathcal{L}(x, y) \\ \omega(n) \leqslant a(t) \log_2 x + z \sqrt{b(t) \log_2 x}}} t^{\omega(n)} = \Phi(z) + O\left(\frac{\log_4 x}{\sqrt{\log_3 x}}\right), \qquad (1.14)$$

avec $a(t) = -t\delta'(t)$, $b(t) = -t\delta'(t) - t^2\delta''(t) > 0$.

Compte tenu des restrictions, signalées plus haut, inhérentes à une estimation de type (1.13), le terme d'erreur de (1.14) peut être considéré, au facteur $\log_4 x$ près, comme la limite naturelle d'une méthode fondée sur l'estimation des sommes $L(x; y, t)$ avec t réel.

Il est facile de voir que pour tout n de $\mathcal{L}(1)$, on a $\omega(n) \leqslant \{1/\log 2 + o(1)\} \log_2 n$. En choisissant $t = 1$, on obtient un théorème de Hardy–Ramanujan pour les entiers lexicographiques.

Corollaire 1. *L'ordre normal de la fonction* $\omega(n)$ *dans la suite* $\mathcal{L}(y)$ *est* $a \log_2 n$ *avec* $a = -\delta'(1) \approx 0,5624$. *Plus précisément, pour tout* $y \geqslant 1$ *et pour toute fonction* $\xi(n) \to \infty$, *on a*

$$\sum_{\substack{n \in \mathcal{L}(x, y) \\ |\omega(n) - a \log_2 n| \leqslant \xi(n)\sqrt{\log_2 n}}} 1 \sim L(x, y) \quad (x \to \infty).$$

Le Théorème 3 permet également d'estimer, pour tout α de $]0, 1/\log 2[$ et à un facteur $(\log x)^{o(1)}$ près, le nombre des entiers de $\mathcal{L}(x, y)$ tels que $\omega(n)$ soit « proche » de $\alpha \log_2 x$.

Théorème 4. *Pour tout* α *de* $]0, 1/\log 2[$, *l'équation* $t\delta'(t) + \alpha = 0$ *possède une unique solution* $t = t_\alpha > 0$. *Soient* $T > 1$, $z > 0$, *et* (α_0, α_1) *tel que* $0 < \alpha_0 < \alpha_1 < 1/\log 2$. *On a uniformément pour* $\alpha_0 \leqslant \alpha \leqslant \alpha_1$, $1 \leqslant y \leqslant T$, $x \geqslant 16$,

$$\sum_{\substack{n \in \mathcal{L}(x, y) \\ |\omega(n) - \alpha \log_2 x| \leqslant z\sqrt{\log_2 x}}} 1 = \frac{x}{(\log x)^{\mu(\alpha)}} e^{O(\sqrt{\log_2 x})} \qquad (1.15)$$

avec $\mu(\alpha) := \delta(t_\alpha) + \alpha \log t_\alpha$.

Lorsque $\alpha = -\delta'(1)$, la formule (1.14) fournit un résultat plus précis, et en fait une formule asymptotique, pour le membre de gauche de (1.15).

La valeur $\alpha = 1$ est la plus intéressante. On a $\mu(1) \approx 0,687$. Cela fournit une réponse quantitative à la question soulevée plus haut concernant la taille de l'intersection de $\mathcal{E}$ et de l'ensemble des entiers satisfaisant au théorème de Hardy–Ramanujan : le nombre de ces exceptions atypiques vaut

$$x/(\log x)^{\mu(1)+o(1)}.$$

La Table 1 *infra* fournit une liste de valeurs pour les fonctions $\delta(t)$ at $\mu(\alpha)$.

Table 1. Valeurs approchées de $\delta(t)$ et $\mu(\alpha)$ à 5.10^{-6} près.

t	$\delta(t)$	t	$\delta(t)$	α	$\mu(\alpha)$	t_α
0	1	2	−0,22515	0	1	0+
0,1	0,90071	3	−0,53570	0,1	0,67048	0,10216
0,2	0,80495	4	−0,77444	0,2	0,48360	0,21641
0,3	0,71443	5	−0,96893	0,3	0,35638	0,35491
0,4	0,62961	6	−1,13333	0,4	0,27370	0,53488
0,5	0,55037	7	−1,27591	0,5	0,23039	0,75538
0,6	0,47633	8	−1,40189	0,56241	0,22282	1
0,7	0,40700	9	−1,51482	0,6	0,22562	1,16154
0,8	0,34192	10	−1,61722	0,7	0,26152	1,78014
0,9	0,28066	11	−1,71091	0,8	0,34325	2,92351
1,0	0,22282	12	−1,79730	0,9	0,48001	5,40823
1,1	0,16807	13	−1,87746	1,0	0,68764	12,30583
1,2	0,11610	14	−1,95226	1,1	0,99475	41,34470
1,3	0,06665	15	−2,02237	1,2	1,45971	328,17996
1,4	0,01949	16	−2,08837	1,3	2,23407	≈35185
$1/\log 2$	0	17	−2,15072	1,4	4,05733	≈2.10^{15}
1,5	−0,02558	18	−2,20981	1,44	8,13613	
1,6	−0,06874	19	−2,26597	$1/\log 2$	+∞	+∞
1,7	−0,11015	20	−2,31948			
1,8	−0,14994	+∞	−∞			
1,9	−0,18824					

2. Équations fonctionnelles

Dans tout ce qui suit, nous nous donnons un paramètre $T > 1$ et nous supposons

$$1/T \leqslant t \leqslant T, \quad 1 \leqslant y \leqslant T, \quad x \geqslant 2. \tag{2.1}$$

Toutes les constantes, implicites ou explicites, peuvent dépendre de T.

Lemme 2.1. *Sous l'hypothèse* (2.1), *on a*

$$L(x, y; t) = 1 + t \sum_{v \geqslant 1} \sum_{p^v \leqslant x} L(\min(x/p^v, p/y), y; t). \tag{2.2}$$

Démonstration. Un entier $n > 1$ compté dans $\mathcal{L}(x, y)$ s'écrit de manière unique $n = p^v m$ avec $p > ym$, $p^v m \leqslant x$, $v \geqslant 1$, et $m \in \mathcal{L}(y)$. En remarquant que $t^{\omega(n)} = t^{\omega(m)+1}$, on obtient immédiatement (2.2). □

Lemme 2.2. *Sous l'hypothèse* (2.1), *on a*

$$L(x, y; t) = t \sum_{\sqrt{xy} < p \leqslant x} L(x/p, y; t) + t \sum_{(xy)^{1/3} < p \leqslant \sqrt{xy}} L(p/y, y; t) + O(x^{2/3} (\log x)^{T-1}).$$

$$(2.3)$$

Démonstration. La contribution des nombres premiers $p \leqslant (xy)^{1/3}$ au membre de droite de (2.2) n'excède pas

$$t \sum_{p \leqslant (xy)^{1/3}} \left[\frac{\log x}{\log p} \right] L(p/y, y; t).$$

Au vu de l'estimation triviale

$$L(w, y; t) \leqslant \sum_{n \leqslant w} t^{\omega(n)} \ll_T w (\log w)^{t-1} \quad (w \geqslant 2), \tag{2.4}$$

on obtient que cette contribution est englobée par le terme d'erreur de (2.3).

Lorsque $(xy)^{1/3} < p \leqslant x$, les seules valeurs admissibles de ν dans (2.2) sont $\nu = 1$ et $\nu = 2$. Le terme principal de (2.3) prend en compte le cas $\nu = 1$. La contribution correspondant à $\nu = 2$ n'excède pas

$$\sum_{(xy)^{1/3} < p \leqslant \sqrt{xy}} L(x/p^2, y; t) \ll \sum_{p > (xy)^{1/3}} \frac{x}{p^2} (\log x)^{T-1}$$

$$\ll x^{2/3} (\log x)^{T-1},$$

où nous avons de nouveau fait appel à (2.4).   □

Lemme 2.3. *Il existe des constantes positives* K_1 *et* K_2, *ne dépendant que de* T, *telles que l'on ait sous l'hypothèse* (2.1)

$$K_1 \frac{x}{(\log x)^{\delta(t)}} \leqslant L(x, y; t) \leqslant K_2 \frac{x}{(\log x)^{\delta(t)}}. \tag{2.5}$$

Démonstration. Soit $r = r(T) = 2^{(1-\delta(1/T))/2} > 1$. Nous établissons par récurrence sur $k \geqslant 1$ qu'il existe des constantes positives A_0^- et A_0^+, ne dépendant que de T, telles que l'on ait, pour $2 \leqslant x \leqslant \exp 2^k$,

$$A_k^- \frac{x}{(\log x)^{\delta(t)}} \leqslant L(x, y; t) \leqslant A_k^+ \frac{x}{(\log x)^{\delta(t)}} \tag{2.6}$$

avec $A_k^\pm := A_0^\pm \exp\{\pm \sum_{\ell=1}^k r^{-\ell}\}$. Au vu de l'estimation triviale

$$\frac{x}{\log x} \ll t(\pi(x) - \pi(y))^+ + 1 \leqslant L(x, y; t) \leqslant \sum_{n \leqslant x} t^{\omega(n)} \ll x (\log x)^{T-1},$$

nous pouvons supposer k arbitrairement grand.

Lorsque $\exp 2^k < x \leqslant \exp 2^{k+1}$, l'hypothèse de récurrence permet d'estimer les termes $L(x/p, y; t)$ et $L(p/y, y; t)$ dans (2.3). En notant que $L(x/p, y; t) = 1$ pour $x/2 < p \leqslant x$, nous obtenons ainsi l'existence d'une constante A, dépendant au plus de T, telle que

$$L(x, y; t) \leqslant \sum_{\sqrt{xy}<p\leqslant x/2} \frac{t A_k^+ x}{p(\log(x/p))^{\delta(t)}} + \sum_{(xy)^{1/3}<p\leqslant\sqrt{xy}} \frac{t A_k^+ p}{y(\log(p/y))^{\delta(t)}} + A\frac{x}{\log x}$$

(2.7)

et

$$L(x, y; t) \geqslant t A_k^- \sum_{\sqrt{xy}<p\leqslant x/2} \frac{x}{p(\log(x/p))^{\delta(t)}} - Ax^{3/4}$$

(2.8)

Le théorème des nombres premiers permet d'estimer les sommes en p de (2.7) et (2.8) par sommation d'Abel. On obtient

$$\sum_{\sqrt{xy}<p\leqslant x/2} \frac{x}{p(\log(x/p))^{\delta(t)}} = \frac{x J(\delta(t))}{(\log x)^{\delta(t)}} + O\left(\frac{x}{\log x}\right)$$

(2.9)

et

$$\sum_{(xy)^{1/3}<p\leqslant\sqrt{xy}} \frac{p}{y(\log(p/y))^{\delta(t)}} \ll \frac{x}{(\log x)^{1+\delta(t)}}.$$

(2.10)

En reportant dans (2.7) et (2.8) et en notant que $\delta(t) \leqslant \delta(1/T) < 1$, on obtient, avec des constantes convenables $B_1(T)$, $B_2(T)$ et pour $k \geqslant k_0(T)$,

$$L(x, y; t) \leqslant \frac{A_k^+ x}{(\log x)^{\delta(t)}}\left\{t J(\delta(t)) + B_2(T)2^{-k(1-\delta(1/T))}\right\} \leqslant \frac{A_{k+1}^+ x}{(\log x)^{\delta(t)}},$$

et

$$L(x, y; t) \geqslant \frac{A_k^- x}{(\log x)^{\delta(t)}}\left\{t J(\delta(t)) - B_1(T)2^{-k(1-\delta(1/T))}\right\} \geqslant \frac{A_{k+1}^- x}{(\log x)^{\delta(t)}}$$

puisque $t J(\delta(t)) = 1$. Cela achève la démonstration du Lemme 2.3. $\qquad\square$

Lemme 2.4. *On a sous l'hypothèse* (2.1)

$$L(x, y; t) = t \int_{\sqrt{x}}^{x} L(x/u, y; t)\frac{du}{\log u} + O\left(\frac{x}{(\log x)^{1+\delta(t)}}\right).$$

(2.11)

Démonstration. Les relations (2.3) et (2.5) impliquent immédiatement

$$L(x, y; t) = t \sum_{\sqrt{xy}<p\leqslant x} L(x/p, y; t) + O\left(\frac{x}{(\log x)^{1+\delta(t)}}\right)$$

$$= t \sum_{\sqrt{x}<p\leqslant x} L(x/p, y; t) + O\left(\frac{x}{(\log x)^{1+\delta(t)}}\right).$$

La dernière somme en p vaut

$$\int_{\sqrt{x}}^{x} L(x/u, y; t) \frac{du}{\log u} + \int_{\sqrt{x}}^{x} L(x/u, y; t) \, dR(u),$$

avec $R(u) := \text{li}(u) - \pi(u) \ll u e^{-2\sqrt{\log u}}$. La seconde intégrale ci-dessus vaut

$$\int_{\sqrt{x}}^{x} \int_{1^-}^{x/u} dL(v, y; t) \, dR(u) = \int_{1^-}^{\sqrt{x}} \{R(x/v) - R(\sqrt{x})\} \, dL(v, y; t)$$

$$\ll \int_{1^-}^{\sqrt{x}} \frac{x}{v} e^{-2\sqrt{\log(x/v)}} |dL(v, y; t)|$$

$$\ll \sum_{n \leqslant \sqrt{x}} t^{\omega(n)} \frac{x}{n} e^{-2\sqrt{\log(x/n)}} \ll x e^{-\sqrt{\log x}}.$$

Cela implique bien (2.11).   □

3. Preuve du Théorème 2

Soit $T > 1$ un paramètre fixé une fois pour toutes. Étant donnés y et t satisfaisant (2.1), nous posons

$$\delta := \delta(t) \quad \text{et} \quad \alpha := \alpha(t) = \min(1, 1 - \delta(t)),$$

et nous définissons $\lambda(x)$ par la relation

$$L(x, y; t) = \frac{\lambda(x) x}{(\log x)^\delta} \quad (x > 1). \tag{3.1}$$

Comme au paragraphe précédent, nous convenons que toutes les constantes, implicites ou explicites, peuvent dépendre de T.

En opérant le changement de variables $u = x^{1-w}$ dans (2.11), il vient

$$\lambda(x) = t \int_0^{1/2} \lambda(x^w) w^{-\delta} \frac{dw}{1-w} + O((\log x)^{-1}) \tag{3.2}$$

Le Théorème 2 équivaut à la relation asymptotique

$$\lambda(x) = K_y(t) + O\left((\log x)^{-\eta(t)}\right) \quad (x \to +\infty), \tag{3.3}$$

où $K_y(t)$ est une fonction continûment dérivable de t. Il est en effet inutile de vérifier la positivité de $K_y(t)$, qui résulte trivialement de la minoration du Lemme 2.3.

Nous allons établir (3.3) en interprétant (3.2) comme une relation de convolution. Cela nécessite un nouveau changement de variable. Posons à cet effet

$$\varphi(z) := \begin{cases} \lambda(\exp\exp z) & (z > 0) \\ 0 & (z \leqslant 0) \end{cases}$$

En tenant compte de l'encadrement trivial

$$0 \leqslant \lambda(x) \leqslant (\log x)^{\delta}(1 + t) \quad (1 < x < 3)$$

pour évaluer la contribution à l'intégrale de (3.2) de l'intervalle $0 \leqslant w \leqslant 1/\log x$, on peut réécrire (3.2) sous la forme

$$\varphi(z) = \int_{-\infty}^{+\infty} \varphi(z - v)g(v)\,\mathrm{d}v + h(z) \quad (z \in \mathbb{R}) \tag{3.4}$$

où l'on a posé, notant $\varrho = \varrho(t) := 1 - \delta(t) > 0$,

$$g(v) := \begin{cases} te^{-\varrho v}(1 - e^{-v})^{-1} & (v \geqslant \log 2) \\ 0 & (v < \log 2) \end{cases} \tag{3.5}$$

et où $h(z)$ satisfait à

$$h(z) = 0 \quad (z \leqslant 0), \quad h(z) \ll e^{-\alpha z} \quad (z > 0). \tag{3.6}$$

Il est à noter que, pour chaque z, $h(z)$ est une fonction polynomiale de t, et que $\partial h(z)/\partial t$ satisfait également la majoration de (3.6).

La solution générale de l'équation de Volterra (3.4) est

$$\varphi(z) = h(z) + \int_0^{+\infty} G(v)h(z - v)\,\mathrm{d}v \tag{3.7}$$

avec

$$G(v) := \sum_{k=1}^{\infty} g^{*k}(v), \tag{3.8}$$

où g^{*k} désigne la k-ème puissance de convolution de g. La série (3.8) ne comporte, pour chaque $v > 0$, qu'un nombre fini de termes puisque

$$g^{*k}(v) = \int_{\mathbb{R}^{k-1}} \prod_{i=1}^{k-1} g(v_i)\, g\left(v - \sum_{i=1}^{k-1} v_i\right) \mathrm{d}v_1 \cdots \mathrm{d}v_{k-1} \tag{3.9}$$

est nul dès que $k \geqslant v/\log 2$. En particulier, comme $\int_{-\infty}^{+\infty} g(v)\,\mathrm{d}v = 1$ par définition de δ, on voit que

$$G(v) = 0 \quad (v < \log 2), \quad 0 \leqslant G(v) \leqslant 2tv/\log 2 \quad (v \geqslant \log 2). \tag{3.10}$$

Nous allons étudier $G(v)$ par l'intermédiaire de sa transformée de Laplace

$$\widehat{G}(s) = \int_0^\infty e^{-sv} G(v) \, dv = \frac{\widehat{g}(s)}{1 - \widehat{g}(s)}, \tag{3.11}$$

convergente pour $\Re e \, s > 0$, ainsi que l'atteste (3.10). On peut incidemment remarquer que ce dernier renseignement découle aussi de l'inégalité

$$|\widehat{g}(s)| \leqslant \widehat{g}(\sigma) < \widehat{g}(0) = 1 \quad (s = \sigma + i\tau, \ \sigma > 0)$$

et du théorème de Phragmén-Landau qui stipule que la transformée de Laplace d'une fonction positive ou nulle possède une singularité en son abscisse de convergence.

On a

$$\widehat{g}(s) = t \sum_{j=0}^\infty \frac{2^{-s-\varrho-j}}{s + \varrho + j}, \tag{3.12}$$

ce qui fournit un prolongement méromorphe de $\widehat{g}(s)$ au plan complexe tout entier. Le lemme suivant, dont nous différons la démonstration, permet de prolonger holomorphiquement $\widehat{G}(s)$ au demi-plan $\sigma \geqslant -\eta(t)$.

Lemme 3.1. *La quantité $\eta = \eta(t)$ étant définie par (1.10), on a $\widehat{g}(s) \neq 1$ pour $\sigma \geqslant -\eta$, $s \neq 0$. De plus, pour chaque $T > 1$, on a*

$$\inf_{\substack{\sigma = -\eta \\ 1/T \leqslant t \leqslant T}} |1 - \widehat{g}(s)| \gg_T 1. \tag{3.13}$$

Admettant momentanément ce résultat, nous sommes en mesure d'établir (3.3). Par la formule d'inversion de Laplace, on peut écrire pour tout $\sigma > 0$

$$G(v) = \frac{1}{2\pi i} \int_{\sigma - i\infty}^{\sigma + i\infty} \widehat{G}(s) e^{vs} \, ds. \tag{3.14}$$

Pour $\sigma \geqslant -\varrho, \ s \neq -\varrho$, on déduit de (3.12) que

$$\widehat{g}(s) = t s^{-1} 2^{1 - \varrho - s} + O(t|s|^{-2} 2^{-\varrho - \sigma}). \tag{3.15}$$

Cela implique, pour $|\tau|$ assez grand et $\sigma \geqslant -\varrho$, la formule

$$\widehat{G}(s) = \frac{t \, 2^{1 - \varrho - s}}{i \tau} + O(\tau^{-2}). \tag{3.16}$$

Reportons cette évaluation dans (3.14) et estimons la contribution à l'infini du terme principal par la seconde formule de la moyenne. Nous obtenons, pour le choix $\sigma = 1/v$, et pour toute valeur du paramètre $X > 1$,

$$G(v) = \frac{1}{2\pi i} \int_{(1/v) - iX}^{(1/v) + iX} \widehat{G}(s) e^{vs} \, ds + O\left(\frac{1}{X}\right). \tag{3.17}$$

Déplaçons alors le segment d'intégration jusqu'à $\sigma = -\eta$. Le Lemme 3.1 garantit que $s = 0$ est la seule singularité traversée par le segment d'intégration—le pôle de $\widehat{g}(s)$ en $s = -\varrho$ n'étant pas une singularité de $\widehat{G}(s) = \widehat{g}(s)/(1 - \widehat{g}(s))$. Comme

$$\widehat{g}'(0) = -\log 2 - t \sum_{j=0}^{\infty} 2^{-\varrho-j}/(\varrho + j)^2 < 0,$$

le résidu de l'intégrande en $s = 0$ vaut

$$B := -1/\widehat{g}'(0),$$

et il suit

$$G(v) = B + \frac{1}{2\pi i} \int_{\mathcal{Z}} \widehat{G}(s) e^{vs} \, \mathrm{d}s + O\left(\frac{1}{X}\right) \tag{3.18}$$

où $\mathcal{Z}$ est la ligne brisée joignant dans cet ordre les points $(1/v) - iX$, $-\eta - iX$, $-\eta + iX$, $(1/v) + iX$. La relation (3.16) permet immédiatement d'estimer la contribution à l'intégrale de (3.18) des segments horizontaux de $\mathcal{Z}$ par $O(1/X)$. Par (3.13) et (3.15), on peut majorer la contribution du segment vertical par

$$\ll \int_{-\eta-iX}^{-\eta+iX} \frac{e^{-\eta v}}{|s|} |\mathrm{d}s| \ll e^{-\eta v} \log X.$$

En choisissant $X = v_0(T) + v$ avec $v_0(T)$ assez grand, il vient

$$G(v) = B + O(v e^{-\eta v}) \quad (v \geqslant \log 2). \tag{3.19}$$

Il est maintenant facile de compléter la démonstration de (3.3). En reportant (3.19) dans (3.7), on obtient

$$\begin{aligned}
\varphi(z) &= \int_{\log 2}^{z} \{B + O(v e^{-\eta v})\} h(z - v) \, \mathrm{d}v + O(e^{-\alpha z}) \\
&= B \int_{-\infty}^{+\infty} h(w) \, \mathrm{d}w + O\left(\int_{z-\log 2}^{+\infty} |h(w)| \, \mathrm{d}w + \int_{\log 2}^{z} \frac{v \, \mathrm{d}v}{e^{\eta v} e^{\alpha(z-v)}} + e^{-\alpha z}\right) \\
&= B \int_{-\infty}^{+\infty} h(w) \, \mathrm{d}w + O(z^2 e^{-\eta z}),
\end{aligned}$$

où nous avons utilisé le fait que $\eta \leqslant \alpha$. On obtient donc (3.3) avec

$$K_y(t) = B \int_{-\infty}^{+\infty} h(w) \, \mathrm{d}w = -\widehat{h}(0)/\widehat{g}'(0). \tag{3.20}$$

La convergence uniforme de cette intégrale et de sa dérivée par rapport à t implique que $K_y(t)$ est une fonction de classe $\mathcal{C}^1$ de t, dont la dérivée est donc bornée sur $[1/T, T]$. □

Il reste à établir le Lemme 3.1. Nous consacrons à cette tâche toute la fin de ce paragraphe.

Nous considérons $\widehat{g}(s)$ avec $s = \sigma + i\tau$, $\sigma \geqslant -\eta$, $1/T \leqslant t \leqslant T$, et nous utilisons des arguments différents selon la taille de τ. Cela nous conduit à considérer les trois domaines suivants, où nous avons posé $\vartheta_0 = \pi/\log 4$, $\theta_1 = \pi/\log 2$.

$$
\begin{aligned}
(\mathcal{D}_1) & \quad |\tau| \leqslant \vartheta_0, \\
(\mathcal{D}_2) & \quad \vartheta_0 < |\tau| \leqslant \vartheta_1, \\
(\mathcal{D}_3) & \quad |\tau| > \vartheta_1.
\end{aligned}
$$

Nous allons montrer que l'on a pour $1 \leqslant j \leqslant 3$

$$
\widehat{g}(s) \neq 1 \quad (\tau \in \mathcal{D}_j,\ s \neq 0) \tag{3.21}
$$

et

$$
\inf_{\substack{\sigma = -\eta \\ \tau \in \mathcal{D}_j}} |1 - \widehat{g}(s)| \gg_T 1. \tag{3.22}
$$

Notre argument est fondé sur la minoration de $|\mathfrak{Im}\,\widehat{g}(s)|$ lorsque $j = 1$, sur la majoration de $\mathfrak{Re}\,\widehat{g}(s)$ lorsque $j = 2$, et sur celle de $|\widehat{g}(s)|$ lorsque $j = 3$. Nous pouvons supposer sans perte de généralité que $\tau \geqslant 0$ car $\widehat{g}(\bar{s}) = \overline{\widehat{g}(s)}$. Nous notons systématiquement

$$
\tau_1 := \tau \log 2.
$$

La relation (3.12) permet d'écrire

$$
\mathfrak{Im}\,\widehat{g}(s) = -t \sum_{j=0}^{\infty} \frac{2^{-\sigma-\varrho-j}}{|\varrho + s + j|^2} \{\tau \cos \tau_1 + (\varrho + \sigma + j) \sin \tau_1\}. \tag{3.23}
$$

Lorsque $0 \leqslant \tau \leqslant \vartheta_0$, on a $0 \leqslant \tau_1 \leqslant \pi/2$ et l'expression entre accolades dans (3.23) est positive ou nulle pour tout j puisque $\sigma + \varrho \geqslant \sigma + \eta \geqslant 0$. Elle est en fait strictement positive si $j \geqslant 1$ et $\tau \neq 0$. De plus, pour $-\eta \leqslant \sigma < 0$, on a

$$
\widehat{g}(\sigma) = t \int_{\log 2}^{\infty} \frac{e^{-\varrho v - \sigma v}}{1 - e^{-v}} \, dv \geqslant 2^{-\sigma} \widehat{g}(0) > 1, \tag{3.24}
$$

avec la convention que le membre de gauche doit être interprété comme $+\infty$ lorsque $\sigma = -\varrho$. Cela établit (3.21) pour $j = 1$. De plus, un argument de continuité standard permet de déduire de (3.24) que $|1 - \widehat{g}(s)| \gg_T 1$ pour $\sigma = -\eta$ et $\tau \leqslant \tau_0(T)$, où $\tau_0(T)$ est une constante suffisamment petite. Compte tenu de (3.23), on obtient donc (3.22) dans le cas $j = 1$.

Considérons ensuite le cas $\vartheta_0 < \tau \leqslant \vartheta_1$. Nous observons dans un premier temps que, par (3.12),

$$\Re e\widehat{g}(i\tau) = t \sum_{j=0}^{\infty} \frac{2^{-\varrho-j}}{|\varrho + j + i\tau|} \cos \psi_j$$

avec $\psi_j := \tau_1 + \arctan\{\tau/(\varrho + j)\} \in]\pi/2, 3\pi/2[$. On peut donc écrire

$$1 - \Re e\widehat{g}(i\tau) \geqslant 1 = \widehat{g}(0) = t \sum_{j=0}^{\infty} \frac{2^{-\varrho-j}}{\varrho + j}$$

$$\geqslant \frac{t2^{-\varrho}}{\max(1, \varrho)} \sum_{j=0}^{\infty} \frac{2^{-j}}{1 + j} = \frac{t2^{-\varrho} \log 4}{\max(1, \varrho)}. \tag{3.25}$$

Maintenant, (3.12) fournit par dérivation

$$|\widehat{g}'(s)| \leqslant t2^{-\varrho-\sigma} \sum_{j=0}^{\infty} \left(\frac{2^{-j} \log 2}{|\varrho + j + s|} + \frac{2^{-j}}{|\varrho + j + s|^2} \right). \tag{3.26}$$

Comme $\sigma + \varrho \geqslant 0$, on déduit de (3.26) que $|\widehat{g}'(s)| \leqslant c_1 t 2^{-\varrho-\sigma}$ avec

$$c_1 = \sum_{j=0}^{\infty} \left(\frac{2^{-j} \log 2}{|j + i\vartheta_0|} + \frac{2^{-j}}{|j + i\vartheta_0|^2} \right) < \frac{9}{10}.$$

Il suit, grâce à (3.25),

$$1 - \Re e\widehat{g}(s) \geqslant \frac{t2^{-\varrho}}{\max(1, \varrho)} \left\{ \log 4 - c_1 \max(1, \varrho) \int_0^{\eta} 2^u \, du \right\} > \frac{t2^{-\varrho}}{12 \max(1, \varrho)},$$

où la dernière inégalité est obtenue en observant que la quantité $\max(1, \varrho) \int_0^{\eta} 2^u du$ atteint son maximum $1/\log 2$ en $\varrho = 1$. Cela établit (3.21) et (3.22) dans le cas $j = 2$.

Il reste à examiner le cas $\tau > \vartheta_1 = \pi/\log 2$, que nous scindons en deux sous-cas, selon la place de ϱ par rapport à 1.

Lorsque $\varrho \leqslant 1$, et donc $t \leqslant 1/\log 2$, nous posons $\beta := \sigma + \varrho \geqslant 0$. Nous pouvons écrire, d'après (3.12),

$$|\widehat{g}(s)| \leqslant \frac{1}{\log 2} \left\{ \left| \frac{2^{-\beta}}{\beta + i\tau} + \frac{2^{-\beta-2}}{\beta + 2 + i\tau} \right| + \left| \frac{2^{-\beta-1}}{\beta + 1 + i\tau} \right| + \sum_{j=3}^{\infty} \frac{2^{-\beta-j}}{|\beta + j + i\tau|} \right\}$$

$$\leqslant \frac{1}{\log 2} \left\{ \sqrt{4^{-\beta} H(\beta)} + \frac{1}{2\sqrt{1 + \vartheta_1^2}} + \frac{1}{4\sqrt{9 + \vartheta_1^2}} \right\}$$

avec

$$H(\beta) := \frac{1}{16}\left|\frac{4}{\beta + i\tau} + \frac{1}{\beta + 2 + i\tau}\right|^2 = \frac{(5\beta + 8)^2 + 25\tau^2}{16(\beta^2 + \tau^2)\{(\beta + 2)^2 + \tau^2\}}.$$

La dérivée logarithmique de $4^{-\beta} H(\beta)$ n'excède pas

$$-\log 4 + \frac{10(5\beta + 8)}{(5\beta + 8)^2 + 25\tau^2} \leqslant -\log 4 + \frac{130}{64 + 25\vartheta_1^2} < 0.$$

Il suit

$$|\widehat{g}(s)| \leqslant \frac{1}{\log 2}\left\{\sqrt{H(0)} + \frac{1}{2\sqrt{1 + \vartheta_1^2}} + \frac{1}{4\sqrt{9 + \vartheta_1^2}}\right\}. \tag{3.27}$$

Or

$$H(0) = \frac{4}{\tau^2(4 + \tau^2)} + \frac{25}{16(4 + \tau^2)} \leqslant \frac{4}{\vartheta_1^2(4 + \vartheta_1^2)} + \frac{25}{16(4 + \vartheta_1^2)}.$$

En reportant cette majoration dans (3.27), on obtient

$$|\widehat{g}(s)| < \frac{2}{3} \quad (\tau > \vartheta_1, \varrho \leqslant 1). \tag{3.28}$$

Lorsque $\varrho \geqslant 1$, on a $\eta = 1/\varrho^2$ et l'on peut écrire, par (3.12),

$$|\widehat{g}(s)| \leqslant t 2^{\eta - \varrho} \sum_{j=0}^{\infty} \frac{2^{-j}}{|\varrho - \eta + j + i\vartheta_1|} \tag{3.29}$$

Maintenant, nous observons que

$$(\varrho - \eta)^2 + \vartheta_1^2 \geqslant \varrho^2 + \vartheta_1^2 - 3/2^{4/3} \geqslant \varrho^2\left(1 + \frac{7}{2}\eta\right)^2.$$

Cela permet de majorer le terme correspondant à $j = 0$ dans la somme de (3.29) par $2/\varrho(2 + 7\eta)$. Semblablement, on vérifie sans peine que

$$(\varrho - \eta + j)^2 + \vartheta_1^2 \geqslant (\varrho + j)^2 \quad (1 \leqslant j \leqslant 9).$$

Posons

$$a := t 2^{-\varrho}/\varrho, \quad b := t \sum_{j=1}^{9} 2^{-\varrho - j}/(\varrho + j), \quad c := t \sum_{j=10}^{\infty} 2^{-\varrho - j}/(\varrho + j),$$

de sorte que $a + b + c = 1$, et majorons 2^η par $1 + \eta$. Nous obtenons, en reportant les estimations précédentes dans (3.29),

$$|\widehat{g}(s)| \leqslant a\frac{2 + 2\eta}{2 + 7\eta} + b(1 + \eta) + c\frac{1 + \eta}{1 - \eta/11}.$$

Il suit $|\widehat{g}(s)| \leqslant 1 - a\eta h$ avec

$$h := \frac{5}{2 + 7\eta} - \frac{b}{a} - \frac{6c}{5a}.$$

Or on a clairement

$$b < t\sum_{j=1}^{\infty} \frac{2^{-\varrho-j}}{\varrho + 1} = \frac{t\,2^{-\varrho}}{\varrho + 1} < \frac{a}{1 + \eta}, \quad c < t\sum_{j=10}^{\infty} \frac{2^{-\varrho-j}}{\varrho} = \frac{a}{512}.$$

D'où

$$h > \frac{5}{2 + 7\eta} - \frac{1}{1 + \eta} - \frac{3}{1280} = \frac{3 - 2\eta}{(2 + 7\eta)(1 + \eta)} - \frac{3}{1280} > \frac{1}{20}.$$

Cela implique (3.21) et (3.22) pour $j = 3$, et achève ainsi la démonstration du Lemme 3.1. $\square$

4. Preuve des Théorèmes 3 et 4

Preuve du Théorème 3. Soit $F_{x,t}(z)$ le membre de gauche de (1.14). Posant

$$M = M(x; t) := a(t) \log_2 x, \quad V = V(x; t) := b(t) \log_2 x, \tag{4.1}$$

on a

$$\widehat{F}_{x,t}(u) = L(x, y; t)^{-1} \sum_{n \in \mathcal{L}(x,y)} \exp\left\{-u\frac{\omega(n) - M}{\sqrt{V}}\right\} t^{\omega(n)}. \tag{4.2}$$

Soit $R := (\log_2 x)^{\eta(t)/8}$. Nous allons établir, comme une conséquence presque immédiate de (1.11), que l'on a uniformément pour $|u| \leqslant R$, $1/T \leqslant t \leqslant T$, $1 \leqslant y \leqslant T$,

$$\widehat{F}_{x,t}(u) = \left\{1 + O\left(\frac{1}{(\log_2 x)^{1/8}}\right)\right\} e^{u^2/2}. \tag{4.3}$$

En effet, on obtient (4.3) en reportant dans (1.11) les évaluations élémentaires suivantes, où l'on a posé $s = t \exp\{-u/\sqrt{V}\}$,

$$K_y(s) = K_y(t) + O((\log_2 x)^{-1/8}),$$

$$\delta(s)\log_2 x = \left\{ \delta(t) + \left(\frac{-u}{\sqrt{V}} + \frac{u^2}{2V} \right) t\delta'(t) + \frac{u^2}{2V} t^2 \delta''(t) + O\left(\frac{u^3}{(\log_2 x)^{3/2}} \right) \right\} \log_2 x$$

$$= \delta(t)\log_2 x + \frac{uM}{\sqrt{V}} - \frac{u^2}{2} + O\left(\frac{1}{(\log_2 x)^{1/8}} \right).$$

Ainsi les conditions (i) et (ii) du Théorème A sont réalisées avec $R = (\log_2 x)^{\eta(t)/8}$, $\kappa = 1$ et $\varepsilon = (\log_2 x)^{-1/8}$. Cela fournit le résultat annoncé.   □

Preuve du Théorème 4. Nous pouvons nous borner à établir l'assertion relative à la solubilité de l'équation $t\delta'(t) + \alpha = 0$. La relation (1.15) résulte immédiatement de (1.14) et (1.11) pour le choix $t = t_\alpha$.

L'inégalité de Cauchy–Schwarz

$$J'(\delta)^2 = \left(\int_0^{1/2} \log(1/v) v^{-\delta} \frac{dv}{1-v} \right)^2$$

$$< \int_0^{1/2} v^{-\delta} \frac{dv}{1-v} \int_0^{1/2} \log^2(1/v) v^{-\delta} \frac{dv}{1-v} = J(\delta) J''(\delta),$$

valable pour tout δ de $]-\infty, 1[$, implique que J'/J est une fonction strictement croissante sur cet intervalle. Comme il résulte d'un calcul standard que

$$\frac{J'}{J}(-\infty) = \log 2, \quad \frac{J'}{J}(1-) = +\infty,$$

on voit que J/J' est une bijection de $]-\infty, 1[$ sur $]0, 1/\log 2[$. L'assertion requise concernant t_α en résulte puisque l'on a par dérivation de (1.8)

$$-t\delta'(t) = J(\delta(t))/J'(\delta(t)) \quad (t > 0).$$ □

Bibliographie

1. K. Alladi, "An Erdős-Kac theorem for integers free of large prime factors," *Acta Arithmetica* **49** (1987), 81–105.
2. J.D. Bovey, "On the size of prime factors of integers," *Acta Arithmetica* **33** (1977), 65–80.
3. P. Erdős, "On some properties of prime factors of integers," *Nagoya Math. J.* **27** (1966), 617–623.
4. P. Erdős et G. Tenenbaum, "Sur les fonctions arithmétiques liées aux diviseurs consécutifs," *J. Number Theory* **31** (1989), 285–311.
5. R.R. Hall et G. Tenenbaum, *Divisors*, Cambridge Tracts in Mathematics no. 90, Cambridge, 1988.
6. H. Maier et G. Tenenbaum, "On the set of divisors of an integer," *Invent. Math.* **76** (1984), 121–128.
7. A. Stef, "L'ensemble exceptionnel dans la conjecture d'Erdős concernant la proximité des diviseurs," Doctorat d'université, Département de mathématiques de l'Université Henri-Poincaré Nancy 1 (1992).
8. A. Stef et G. Tenenbaum, "Inversion de Laplace quantitative," prépublication.
9. G. Tenenbaum, "Sur un problème de crible et ses applications," *Ann. Sci. Éc. Norm. Sup.* (4) **19** (1986), 1–30.
10. G. Tenenbaum, "Une inégalité de Hilbert pour les diviseurs," *Indag. Mathem.*, N.S., 2(1) (1991), 105–114.
11. G. Tenenbaum, "Sur un problème de crible et ses applications, 2. Corrigendum et étude du graphe divisoriel," *Ann. Sci. Éc. Norm. Sup.* (4) **28** (1995), 115–127.

The Ramanujan Journal, 2, 185–199 (1998)

The Berry–Esseen Bound in the Theory of Random Permutations

E. MANSTAVIČIUS *   eugenijus.manstavicius@maf.vu.lt
Department of Mathematics, Vilnius University, Naugarduko str. 24, LT-2600 Vilnius, Lithuania

Professori Paulo Erdös in Memoriam

Received April 11, 1997; Accepted October 31, 1997

Abstract. The convergence rate in the central limit theorem for linear combinations of the cycle lengths of a random permutation is examined. It is shown that, in contrast to the Berry-Esseen theorem, the optimal estimate in terms of the sum of the third absolute moments has the exponent 2/3.

Keywords: random permutation, Berry-Esseen theorem, Central limit, remainder term, random mappings, symmetric group

AMS Classification Numbers: Primary – 60C05; Secondary – 60F05, 05A05

1. Introduction and results

Statistical group theory and probabilistic number theory are the fields to which P.Erdös has contributed many pioneering and enduring works. Studying his papers written mainly with P.Turán as well as more recent articles which deal with the value distribution problems of maps defined on the symmetric group $\mathbb{S}_n$ we could not shake off an impression that this direction has much in common with probabilistic number theory, nevertheless, the interaction between them is rather poor. We came to an opinion that in developing of the analytic tools, number theory is a bit ahead than similar branches of discrete mathematics. For instance, the survey [4] considered as the most comprehensive paper on analytic approaches of discrete mathematics can be compared to the Selberg– Delange method used to investigate mean values of multiplicative functions but we hardly could find an analogous influence of the method taking its background in the G.Halász' papers [6] or [7]. The articles [9], [14], [15] comprise a rare exception. Now solving the problem of the remainder term estimation in the central limit theorem we demonstrate other possibilities of this approach.

Let $\sigma \in \mathbb{S}_n$ be an arbitrary permutation and

$$\sigma = \kappa_1 \cdots \kappa_\omega \tag{1}$$

be its (unique up to the order) expression by the product of the independent cycles κ and $\omega = \omega(\sigma)$ be the number of the cycles comprising σ. Denote

$$\nu_n(\ldots) = (n!)^{-1} \#\{\sigma \in \mathbb{S}_n : \ldots\}.$$

*   Partially supported by Grant from Lithuanian Foundation of Studies and Science.

In 1942 V.L.Goncharov [5] proved that

$$\nu_n\big(\omega(\sigma) - \log n < x\sqrt{\log n}\big) \to \Phi(x) := \frac{1}{\sqrt{2\pi}}\int_{-\infty}^{x} e^{-u^2/2}\, du\,.$$

Here and in what follows, the limit is taken with respect to $n \to \infty$. Starting a very fascinating series of papers on the asymptotic distribution of the group-theoretic order of the random permutation σ, P.Erdös and P.Turán [3] considered the sum $s(\sigma)$ of natural logarithms of different lengths $l(\kappa)$ of the cycles comprising σ. It was shown that

$$\nu_n\big(s(\sigma) - (1/2)\log^2 n < (1/\sqrt{3})\, x\, \log^{3/2} n\big) \to \Phi(x)\,.$$

The convergence rate in this relation was estimated by J.-L.Nicolas [16]. An improvement of the convergence rate estimates in the central limit theorem for the group-theoretic order function was given by A.D.Barbour and S.Tavaré [2]. Other relevant references can be found in the book [10] and in the recent lecture [18]. In addition, we note that random permutations, not necessarily taken with equal probabilities, comprise a rather significant object in applied mathematics (see [1] and the references therein). So we hope that our remark, though written in purely theoretical style, will be useful for those interested in analytic problems of the applied probability theory.

In what follows we adopt a few definitions from probabilistic number theory. Having in mind the examples of functions $\omega(\sigma)$ and $s(\sigma)$, we call the map $h : \mathbb{S}_n :\to \mathbb{R}$ *additive* if it satisfies the relation $h(\sigma) = h(\kappa_1) + \cdots + h(\kappa_\omega)$ for each σ having the expression (1). Similarly, the map $f : \mathbb{S}_n :\to \mathbb{C}$ satisfying the equality $f(\sigma) = f(\kappa_1)\cdots f(\kappa_\omega)$ is called *multiplicative*. Further, the function $g : \mathbb{S}_n \to \mathbb{C}$ will be called *class dependent*, or shortly, CD function if its values on cycles depend only on their lenghts, e.g., there exists a function $\widehat{g} : \mathbb{N} \to \mathbb{C}$ such that $g(\kappa) = \widehat{g}(l(\kappa))$. To argue the definition, we remind that each $\sigma \in \mathbb{S}_n$ belongs to a class of conjugate elements, which we denote by $\bar{m} := (m_1, \ldots, m_n)$ with $0 \le m_k \le n/k$ and $1m_1 + \cdots + nm_n = n$. The relation $\sigma \in \bar{m}$ means that σ consists of m_k cycles of the length k, $1 \le k \le n$. The CD additive and multiplicative functions have the representations

$$h(\sigma) = \sum_{k=1}^{n} \widehat{h}(k) m_k, \quad f(\sigma) = \prod_{k=1}^{n} \widehat{f}(k)^{m_k} \qquad (2)$$

with $m_k = m_k(\sigma)$. The general task is to describe these functions when the values $\widehat{h}(k)$ or $\widehat{f}(k)$, $k \ge 1$ are given. The problem of weak convergence to a limit law of

$$\nu_n\big(h(\sigma) - \alpha(n) < x\beta(n)\big),$$

where $\alpha(n)$ and $\beta(n) > 0$ are suitably chosen normalizing sequences, was considered in the paper [14]. According to the result of V.L.Goncharov [5], the distribution with respect to ν_n of the random variable $m_k(\sigma)$ tends to the Poissonian law having the parameter $1/k$ for each fixed k. Hence the choice of

$$\alpha(n) = \sum_{k=1}^{n} \frac{\widehat{h}(k)}{k}, \quad \beta(n) = \left(\sum_{k=1}^{n} \frac{\widehat{h}^2(k)}{k}\right)^{1/2}$$

called *standard normalization* should be considered at the first place. That motivates the normalization used afterwards. We quote the following partial result.

THEOREM A ([14]). *Let* $h_n(\sigma)$ *be a sequence of real CD additive functions satisfying the condition*

$$\sum_{k=1}^{n} \frac{\widehat{h}_n^2(k)}{k} = 1 \tag{3}$$

and

$$A(n) := \sum_{k=1}^{n} \frac{\widehat{h}_n(k)}{k}.$$

If the Lindeberg type condition

$$\sum_{\substack{k \le n \\ |\widehat{h}_n(k)| \ge \epsilon}} \frac{\widehat{h}_n^2(k)}{k} = o(1) \tag{4}$$

holds for each $\epsilon > 0$, *then*

$$\nu_n(x) := \nu_n\big(h_n(\sigma) - A(n) < x\big) = \Phi(x) + o(1) \tag{5}$$

uniformly in $x \in \mathbb{R}$ *and also*

$$\frac{1}{n!} \sum_{\sigma \in \mathbb{S}_n} \big(h_n(\sigma) - A(n)\big)^2 = 1 + o(1). \tag{6}$$

Having in mind the Berry–Esseen estimate in the central limit theorem for sums of independent random variables (see V.V.Petrov [17], Chapter 5), we expect that the remainder in (5) can be estimated in terms of

$$L_n := \sum_{k=1}^{n} \frac{|\widehat{h}_n(k)|^3}{k}.$$

Observe that the relation $L_n = o(1)$ implies also the condition (4). It appears that dependence of the random variables $m_k(\sigma)$, $1 \le k \le n$ involved by (2) in the function $h_n(\sigma)$ makes a substantial influence.

In what follows, let the symbols O or $\ll$ contain absolute constants when there is no other indication, and

$$D_n = \sum_{\substack{1 \le k, l \le n \\ k+l > n}} \frac{\widehat{h}_n(k)\widehat{h}_n(l)}{kl}.$$

In contrast to the above mentioned Berry–Esseen bound, we have the following results.

THEOREM 1. *Let* $h_n(\sigma)$, $n \geq 1$ *be a sequence of real CD additive functions satisfying the condition (3). Then*

$$R'_n := \sup_{x \in \mathbb{R}} \left| \nu_n(x) - \Phi(x) - \frac{D_n x}{2\sqrt{2\pi}} e^{-x^2/2} \right| \ll L_n.$$

COROLLARY. *We have*

$$R_n := \sup_{x \in \mathbb{R}} |\nu_n(x) - \Phi(x)| \ll L_n^{2/3}.$$

There exists a sequence of CD additive functions satisfying the condition (3) and $L_n = o(1)$ *but such that*

$$R_n \gg L_n^{2/3}.$$

THEOREM 2. *Let* $h_n(\sigma)$, $n \geq 1$ *be a sequence of real CD additive functions normalized so that*

$$\sum_{k=1}^{n} \frac{\widehat{h}_n^2(k)}{k} - D_n = 1. \tag{7}$$

Then with the same centralizing sequence $A(n)$, *we have*

$$R_n := \sup_{x \in \mathbb{R}} |\nu_n(x) - \Phi(x)| \ll L_n.$$

Theorems 1 and 2 are analogous to the results obtained by A.Mačiulis [13] for additive functions defined on $\mathbb{N}$. The proofs are based upon the Esseen inequality connecting the convergence rate of distribution functions to their characteristic functions and analysis of the last. If $g(\sigma) := \exp\{ith(\sigma)\}$, $t \in \mathbb{R}$ and, as above, the function $\widehat{g}$ is defined by $\widehat{g}(l(\kappa)) = g(\kappa)$, then the main difficulty is to find asymptotic formulae for

$$M_n(g) := \frac{1}{n!} \sum_{\sigma \in \mathbb{S}_n} g(\sigma) = \sum_{\bar{m}} \prod_{k=1}^{n} \left(\frac{\widehat{g}(k)}{k} \right)^{m_k} \frac{1}{m_k!}$$

uniform in parameters of g. Moreover, we have

$$\exp \left\{ \sum_{k=1}^{\infty} \frac{\widehat{g}(k) z^k}{k} \right\} = \sum_{n=0}^{\infty} M_n(g) z^n, \quad |z| < 1. \tag{8}$$

So, our task reduces to a problem in function theory. We hope that the analysis of the relations between the coefficients of the series in (8) done in the next two sections has independent interest.

2.   The first analytic formula

Let $f(k)$, $k \geq 1$ be complex numbers, depending, maybe, on n or other parameters. Denote

$$F(z) = \exp\left\{ \sum_{k=1}^{\infty} \frac{f(k)z^k}{k} \right\} =: \sum_{n=0}^{\infty} M_n z^n, \quad |z| < 1. \tag{9}$$

We will obtain asymptotic expressions of M_n in terms of $f(k)$. Since the values of $f(k)$, when $k > n$, make no influence onto M_n, we assume them equal to one. The remainder in the formula obtained in this section will involve the quantity

$$\rho(n,p) = \left(\sum_{k \leq n} \frac{|f(k) - 1|^p}{k} \right)^{1/p}$$

where $p > 1$. Put

$$L(z) = \sum_{k \leq n} \frac{f(k) - 1}{k} z^k, \quad z = re^{i\tau} := e^{-1/n + i\tau}, \quad \tau \in \mathbb{R}.$$

Let

$$I_j(n) = \frac{1}{2\pi i} \int_{|z|=r} \frac{(L(z) - L(1))^j}{(1 - z)z^{n+1}} \, dz, \quad j = 0, 1, \ldots.$$

Calculating the coefficients of the integrand, we have $I_0(n) = 1$, $I_1(n) = 0$, and

$$I_2(n) = - \sum_{\substack{1 \leq k, l \leq n \\ k+l > n}} \frac{(f(k) - 1)(f(l) - 1)}{kl}.$$

We have the following result.

THEOREM 3. *Let $p > 1$. There exists sufficiently small $\delta = \delta(p)$ such that, if*

$$\rho := \rho(n,p) \leq \delta, \tag{10}$$

then

$$M_n = \exp\{L(1)\}\left(1 + \sum_{j=2}^{N-1} \frac{I_j(n)}{j!} + O(\rho^N + n^{-c})\right)$$

for each $N \geq 2$ with some constant $c = c(p) > 0$. The constant in the symbol O also depends on p only.

The proof of Theorem 3 goes along the lines drawn up by A.Mačiulis in the paper [13]. At first we prove few auxilliary results. The following estimate of the norm of a polynomial is perhaps known, but we have failed to find it in the literature.

LEMMA 1. *Let* $r = e^{-1/n}$, $s > \max\{2, p/(p-1)\}$, *and* $p > 1$. *For each polynomial*

$$P(z) = \sum_{k=1}^{n} a_k z^k, \quad a_k \in \mathbb{C},$$

we have

$$\|P(z)\|_s := \left(\int_0^{2\pi} |P(re^{i\tau})|^s \, d\tau \right)^{1/s} \leq C(s,p) n^{1-1/s} \left(\sum_{k=1}^{n} \frac{|a_k|^p}{k} \right)^{1/p}$$

$$=: C(s,p) n^{1-1/s} Q.$$

The constant $C(s,p)$ *depends only on* s *and* p.

Proof: The main idea takes its backgroud in the G.Halász' paper [6]. Let $\alpha = \min\{2, p\}$, $\beta = \alpha/(\alpha - 1)$, and $z = re^{i\tau}$. Observe that

$$\sum_{k=1}^{n} |a_k|^\alpha \leq n Q^\alpha, \quad |P(z)| \leq nQ.$$

Define

$$\Omega_j = \{\tau \in [0, 2\pi] : |P(ze^{i\tau})| > 2^{-j} nQ\}, \quad j = 0, 1, \ldots, j_0.$$

Hence using the partition

$$[0, 2\pi] = ([0, 2\pi] \setminus \Omega_{j_0}) \bigcup_{j=1}^{j_0} (\Omega_j \setminus \Omega_{j-1}),$$

we have

$$\int_0^{2\pi} |P(re^{i\tau})|^s \, d\tau \leq n^s Q^s \left(2\pi 2^{-j_0 s} + 2^s \sum_{j=1}^{j_0} 2^{-js} \mu \Omega_j \right),$$

where $\mu\Omega$ stands for the Lebesgue measure of the set Ω. We see that the assertion of Lemma 1 will follow from the estimate

$$\mu \Omega_j \ll n^{-1} j 2^{\beta j} \tag{11}$$

with $\beta < s$. Here and in what follows the constant in the symbol $\ll$ depends at most on s and p.

In order to prove (11) when $j \geq 1$, we choose the points τ_l, $l = 1, \ldots, n_j$ by induction. Let

$$\tau_1 = \inf \Omega_j, \quad \tau_{l+1} = \inf\{\tau \in \Omega_j : \tau \geq \tau_l + 1/n\}.$$

Then $\mu\Omega_j \leq n_j/n$, and it remains to prove the estimate

$$n_j \ll j2^{\beta j}.\tag{12}$$

We can suppose that $n_j \geq 3$. If $z_l = \exp\{-i \arg P(re^{i\tau_l})\}$, then

$$2^{-j}nQn_j \leq \sum_{l=1}^{n_j} z_l P(re^{i\tau_l}) = \sum_{k=1}^{n} a_k r^k \sum_{l=1}^{n_j} z_l e^{ik\tau_l} \leq$$

$$\leq \left(\sum_{k=1}^{n} |a_k|^\alpha\right)^{1/\alpha} \left(\sum_{k=1}^{\infty} r^k \left|\sum_{l=1}^{n_j} z_l e^{ik\tau_l}\right|^\beta\right)^{1/\beta} \leq$$

$$\leq n^{1/\alpha} Q\left(n_j^{\beta-2} \sum_{k=1}^{\infty} r^k \left|\sum_{l=1}^{n_j} z_l e^{ik\tau_l}\right|^2\right)^{1/\beta}.\tag{13}$$

The double sum on the right hand side equals

$$\Sigma := \sum_{l=1}^{n_j} \sum_{m=1}^{n_j} z_l \bar{z}_m \sum_{k=1}^{\infty} r^k e^{ik(\tau_l - \tau_m)} \leq$$

$$\leq 3 \sum_{l=1}^{n_j} \sum_{k=1}^{\infty} r^k + 4 \sum_{\substack{l,m=1 \\ 1/n \leq \tau_l - \tau_m \leq \pi}}^{n_j} \left|\sum_{k=1}^{\infty} r^k e^{ik(\tau_l - \tau_m)}\right|.$$

Observing that $\tau_{m+k} - \tau_m \geq k/n$, we proceed

$$\Sigma \ll nn_j + n_j \sum_{k=1}^{n_j-1} \max_{k/n \leq \tau \leq \pi} |1 - re^{i\tau}|^{-1} \ll nn_j \log n_j.$$

The last estimate and (13) imply $n_l \log n_j \ll 2^{j\beta}$. Hence we obtain (12).
    Lemma 1 is proved.   ∎

Let $\tau_0 = \min\{e^{1/\rho}, \sqrt{n}\}/n$ and

$$l = \{z : |z| = r := e^{-1/n}\}, \quad l_0 = \{z \in l : |\tau| := |\arg z| \leq \tau_0\},$$

$$l_1 = \{z \in l : \tau_0 < |\tau| \leq \pi\}.$$

The constants in the symbols O or $\ll$ will depend at most on p provided that δ is chosen smaller than some constant depending on p.

LEMMA 2. *Let* $p > 1$, $1/p + 1/q = 1$. *Then* $L(z) - L(1) \ll \rho \log^{1/q}(2 + |\tau|n)$ *and, for* $|\tau| \leq \tau_0$,

$$\exp\{L(z) - L(1)\} = \sum_{j=0}^{N-1} \frac{(L(z) - L(1))^j}{j!} + O\left(\frac{|L(z) - L(1)|^N}{N!}\right).$$

Proof: We start with the inequalities

$$|L(z) - L(1)| \leq \sum_{k \leq n} \frac{|f(k) - 1|}{k} r^k |e^{ik\tau} - 1| + \frac{1}{n} \sum_{k \leq n} |f(k) - 1|$$

$$\leq \rho(n) \left(\Psi_n(\tau)^{1/q} + 1 \right), \tag{14}$$

where

$$\Psi_n(\tau) = \sum_{k=1}^{\infty} \frac{r^{kq} |e^{ik\tau} - 1|^q}{k}.$$

Expanding into the Fourier series (see [11] or [8], Exercise 34), we have

$$|1 - e^{ix}|^q = a(q) + \sum_{\substack{m=-\infty \\ m \neq 0}}^{\infty} a_m(q) e^{imx},$$

$$a(q) = -\sum_{\substack{m=-\infty \\ m \neq 0}}^{\infty} a_m(q) = \frac{2^q \Gamma((1+q)/2)}{\sqrt{\pi} \Gamma((2+q)/2)},$$

for $x \in \mathbb{R}$, $q \geq 1$ with $a_m(q) \in \mathbb{R}$, $a_m(q) \ll |m|^{-2}$. Hence as in [12], we obtain

$$\Psi_n(\tau) = a(q) \log \frac{|1 - z^q|}{1 - r^q} + \sum_{\substack{m=-\infty \\ m \neq 0}}^{\infty} a_m(q) \log \frac{|1 - z^q|}{|1 - r^q e^{im\tau}|}$$

$$\leq a(q) \log \frac{|1 - z^q|}{1 - r^q} + O(1).$$

Inserting this estimate into (14) and analyzing the logarithmic function, we obtain the first assertion of Lemma 2.

By virtue of $\log^{1/q}(2 + |\tau|n) \ll \rho^{-1/q}$ when $|\tau| \leq \tau_0$, the second estimate follows from the first one.

Lemma 2 is proved.   ∎

Proof: [Proof of Theorem 3] According to Cauchy's formula,

$$M_n = \frac{1}{2\pi in} \int_{|z|=r} \frac{F'(z)}{z^n} \, dz.$$

We recall that $f(k) = 1$ when $k > n$. Using the notations, we obtain

$$M_n = \frac{\exp\{L(1)\}}{2\pi in} \int_{|z|=r} \frac{\exp\{L(z) - L(1)\}}{(1 - z)z^n} \sum_{k=1}^{\infty} f(k) z^{k-1} \, dz$$

$$= \frac{\exp\{L(1)\}}{2\pi in} \int_{|z|=r} \frac{\exp\{L(z) - L(1)\}}{(1 - z)z^n} \left(\frac{1}{1 - z} + L'(z) \right) dz. \tag{15}$$

When $z \in l_1$, we have $|1 - z| \gg |\tau|$, and by Lemma 1, $\exp\{L(z) - L(1)\} \ll (n|\tau|)^\rho$. Thus,

$$J_1 := \frac{1}{n} \int_{z \in l_1} \frac{\exp\{L(z) - L(1)\}}{(1 - z)^2 z^n} \, dz \ll (n\tau_0)^{\rho-1} \ll n^{(\delta-1)/2} + e^{-1/\rho}. \qquad (16)$$

Similarly using Cauchy's inequality, Lemma 1 with s satisfying its condition and $t > 1$ such that $1/s + 1/t = 1$, we obtain

$$\begin{aligned} J_2 &:= \frac{1}{n} \int_{z \in l_1} \frac{\exp\{L(z) - L(1)\}}{(1 - z)z^n} L'(z) \, dz \\ &\ll \frac{1}{n} \left(\int_{z \in l_1} \left| \frac{\exp\{L(z) - L(1)\}}{1 - z} \right|^t |dz| \right)^{1/t} \|L'(z)\|_s \\ &\ll n^{-1+\rho} \tau_0^{\rho-1/s} \cdot \rho n^{1-1/s} \ll n^{-1/2s} + e^{-1/s\rho} \end{aligned} \qquad (17)$$

provided that $\delta \le 1/2s$.

We now consider the integral in (15) when $z \in l_0$. Applying Lemma 1, we have

$$\begin{aligned} J_0 &:= \frac{1}{2\pi i n} \int_{l_0} \frac{1}{(1 - z)^2 z^n} \sum_{j=0}^{N-1} \frac{(L(z) - L(1))^j}{j!} \, dz \\ &+ \frac{1}{2\pi i n} \int_{l_0} \frac{L'(z)}{(1 - z)z^n} \sum_{j=0}^{N-2} \frac{(L(z) - L(1))^j}{j!} \, dz + R \\ &=: J_{01} + J_{02} + R, \end{aligned} \qquad (18)$$

where

$$\begin{aligned} R &\ll \frac{1}{nN!} \int_{l_0} \frac{|L(z) - L(1)|^N}{|1 - z|^2} |dz| \\ &+ \frac{1}{n(N-1)!} \int_{l_0} \frac{|L(z) - L(1)|^{N-1}}{|1 - z|} |L'(z)| \, |dz| =: R' + R''. \end{aligned}$$

It follows from Lemma 2 that

$$R' \ll \frac{n\rho^N}{N!} \int_0^{\tau_0} \frac{\log^{N/q}(2 + n\tau)}{(1 + n\tau)^2} \, d\tau \ll \rho^N. \qquad (19)$$

As estimating J_2, we obtain

$$R'' \ll \frac{\rho^{N-1}}{(N-1)!} \left(\int_0^{\tau_0} \frac{\log^{t(N-1)/q}(2 + n\tau)}{(1 + n\tau)^t} \, d\tau \right)^{1/t} \|L'(z)\|_s \ll \rho^N. \qquad (20)$$

We extend the integrals J_{01} and J_{02} over the region l_1. While

$$\frac{1}{nj!} \int_{l_1} \frac{|L(z) - L(1)|^j}{|1 - z|^2} |dz| \ll \rho^j (n\tau_0)^{-1/2} \ll \rho^j (n^{-1/4} + e^{-1/2\rho})$$

and

$$\frac{1}{nj!} \int_{l_1} \frac{|L(z) - L(1)|^j}{|1 - z|} |L'(z)| \, |dz|$$

$$\ll \frac{1}{nj!} \left(\int_{l_1} \frac{|L(z) - L(1)|^{jt}}{|1 - z|^t} \, |dz| \right)^{1/t} \|L'(z)\|_s$$

$$\ll \rho^{j+1} (n\tau_0)^{-(t-1)/2} \ll \rho^{j+1} (n^{-c_1} + e^{-c_1/\rho})$$

with $c_1 = c_1(p) > 0$, we obtain from (18), (19), and (20)

$$J_0 := \frac{1}{2\pi i n} \int_l \frac{1}{(1 - z)^2 z^n} \sum_{j=0}^{N-1} \frac{(L(z) - L(1))^j}{j!} \, dz$$

$$+ \frac{1}{2\pi i n} \int_l \frac{L'(z)}{(1 - z) z^n} \sum_{j=0}^{N-2} \frac{(L(z) - L(1))^j}{j!} \, dz + O(n^{-c_2} + \rho^N)$$

$$= \sum_{j=1}^{N-1} \frac{I_j}{j!} + O(n^{-c_2} + \rho^N),$$

where $c_2 = c_2(p) > 0$. Inserting the estimates (16), (17), and the last one into (15), we end the proof of Theorem 3. ∎

3. The second analytic formula

Now we will compensate the shortage of Theorem 3 appearing in the case when the quantity $\rho(p)$ is large. We will derive another asymptotic formula with the remainder estimate in terms of

$$\mu_n^2 := \frac{1}{n} \sum_{k=1}^{n} |f(k) - 1|^2$$

and

$$E(u) := \exp\left\{ 2 \sum_{\substack{k=1 \\ |f(k)-1|>u}}^{n} \frac{|f(k) - 1|}{k} \right\}$$

with $u \geq 0$. All other previous notation remain the same.

THEOREM 4. *We have*

$$M_n = \exp\{L(1)\}\left(1 + O\left((\mu_n + n^{-1})^{1/2} E(3/8)\right)\right).$$

The constant in the symbol O is absolute.

The proof goes along the similar lines as that of Theorem 3, though we need auxilliary results. Some of the ideas of the proof have been previously used in the papers [7] and [11].

LEMMA 3. *We have*

$$\|L'(z)\|_2 \leq \sqrt{2\pi n}\, \mu_n .$$

Proof: Apply the Parseval equality.   ∎

LEMMA 4. *We have*

$$\exp\{|L(z) - L(1)|\} \ll_u E(u)\left|\frac{1-z}{1-r}\right|^{4u/\pi}$$

for $z = re^{i\tau}$ and each $u \geq 0$.

Proof: Observe that

$$\sum_{k=1}^{n} \frac{|r^k e^{i\tau k} - 1|}{k} \ll 1 + \sum_{k=1}^{n} \frac{|e^{i\tau k} - 1|}{k}$$

and apply the Fourier expansions used in the proof of Lemma 2 with $q = 1$. So we deduce

$$|L(z) - L(1)| \ll \frac{4u}{\pi} \log \frac{|1-z|}{1-r} + \log E(u) + u$$

for $u \geq 0$. Hence follows the desired estimate.
  Lemma 4 is proved.   ∎

Proof of Theorem 4: We start with the formula (15). Now it is easier, than in the proof of Theorem 3, to estimate the integral

$$J_3 := \frac{1}{n} \int_{|z|=r} \frac{|\exp\{L(z) - L(1)\}|}{|1-z|} |L'(z)| \, |dz|$$

$$\ll \frac{1}{n} \|L'(z)\|_2 \left(\int_{|z|=r} \frac{|\exp\{L(z) - L(1)\}|^2}{|1-z|^2} |dz| \right)^{1/2}$$

We obtain from Lemma 3 and Lemma 4 with $u = 3/8$

$$J_3 \ll E(3/8)\mu_n \left(n^{-1+3/\pi} \int_{|z|=r} |1 - z|^{3/\pi - 2} |dz| \right)^{1/2} \ll E(3/8)\mu_n .$$

Let now $l_3 = \{z \in l : |\tau| \le K/n\}$, $K = \min\{\mu_n^{-1}, n\}$, and $l_4 = l \setminus l_3$. By Lemma 4 with $u = \pi/8$, we have

$$
\begin{aligned}
J_4 &:= \frac{1}{n} \int_{l_4} \frac{1 + |\exp\{L(z) - L(1)\}|}{|1 - z|^2} |dz| \\
&\ll \frac{1}{K} + E(\pi/8) n^{-1/2} \int_{l_4} |1 - z|^{-3/2} |dz| \ll E(3/8) K^{-1/2} .
\end{aligned}
$$

Similar by applying $|L(z) - L(1)| \le n\mu_n |1 - z|$, we obtain

$$
\begin{aligned}
J_5 &:= \frac{1}{n} \int_{l_3} \frac{|L(z) - L(1)| \exp\{|L(z) - L(1)|\}}{|1 - z|^2} |dz| \\
&\ll E(3/8) \mu_n \int_{l_3} |1 - z|^{-1/2} |dz| \ll E(3/8) \mu_n \sqrt{K} .
\end{aligned}
$$

Inserting these obtained estimates into (15) and recalling the choice of K, we have

$$
\begin{aligned}
M_n &= \exp\{L(1)\}\big(1 + O(J_3 + J_4 + J_5)\big) \\
&\quad \exp\{L(1)\}\big(1 + O((\mu_n + n^{-1})^{1/2} E(3/8))\big) .
\end{aligned}
$$

Theorem 4 is proved.   ∎

4.   Estimation of the convergence rate

Proof of Theorem 1:   We use a generalization of the Esseen inequality (see [17], Theorem 2, Chapter 5.2). Let

$$
\varphi_n(t) := \frac{\exp\{-itA(n)\}}{n!} \sum_{\sigma \in \mathcal{S}_n} \exp\{ith_n(\sigma)\}, \quad t \in \mathbb{R} .
$$

We have

$$
R_n' \ll \frac{1}{T} + \int_{|t| \le T} |\varphi_n(t) - e^{-t^2/2}(1 + \frac{t^2 D_n}{2})| \frac{dt}{|t|} \tag{21}
$$

where $T > 0$.

In order to obtain asymptotic formulas for $\varphi_n(t)$, we take $f(k) = \exp\{it\widehat{h}_n(k)\}$, $1 \le k \le n$ and apply Theorem 3 with $p = 3$, $N = 3$ and Theorem 4. Now $\rho \le |t| L_n^{1/3}$. Put $T_1 = \delta L_n^{-1/3}$, where $\delta > 0$ is sufficiently small to guarantee the validity of the formula in Theorem 3 in the region $|t| \le T_1$. Observe that the condition (3) implies

$$
1 = \sum_{k=1}^{n} \frac{\widehat{h}_n^2(k)}{k} \le L_n^{2/3} \left(\sum_{k=1}^{n} \frac{1}{k} \right)^{1/3}
$$

and hence $L_n \gg (\log n)^{-1/2}$. We can also suppose that $L_n = o(1)$. We obtain from Theorem 3

$$\varphi_n(t) = \exp\{-\frac{t^2}{2} + \frac{\Theta}{6}|t|^3 L_n\}\Big(1 -$$

$$- \frac{1}{2}\sum_{\substack{k,l=1 \\ k+l>n}}^{n} \frac{(\exp\{it\widehat{h}_n(k)\} - 1)(\exp\{it\widehat{h}_n(l)\} - 1)}{kl} + O(|t|^3 L_n)\Big) \quad (22)$$

in the region $L_n \le |t| \le T_1$ with $|\Theta| \le 1$.

Analysis of the double sum, say $S_n(t)$, in (22) requires more calculations. Let

$$a_k = \sum_{n-k<l\le n} \frac{1}{l}, \quad \Sigma(s) = \sum_{k=1}^{n} \frac{a_k^s}{k}.$$

Using the relation

$$\sum_{k\le n/2} \frac{a_k^s}{k} \to \int_0^{1/2} \frac{(-\log(1-x))^s}{x}\,dx < \infty, \quad s > 0,$$

we obtain $\Sigma(s) \ll_s 1$ for each fixed $s > 0$. Now as in [13], we have

$$S_n(t) = it\sum_{k=1}^{n} \frac{\exp\{it\widehat{h}_n(k)\} - 1}{k} \sum_{n-k<l\le n} \frac{\widehat{h}_n(l)}{l}$$

$$+ O\Big(|t|^3 L_n^{2/3} \sum_{k=1}^{n} \frac{|\widehat{h}_n(k)|}{k} a_k^{1/3}\Big) =$$

$$= -t^2 D_n + O\Big(|t|^3 \sum_{k=1}^{n} \frac{\widehat{h}_n^2(k)}{k} \sum_{n-k<l\le n} \frac{|\widehat{h}_n(l)|}{l}\Big)$$

$$+ O(|t|^3 L_n \Sigma(1/2)^{2/3}) = -t^2 D_n + O(|t|^3 L_n).$$

Inserting the last formula into (22), we obtain

$$\varphi_n(t) = e^{-t^2/2}(1 + t^2 D_n/2) + O(|t|^3 e^{-t^2/4} L_n) \quad (23)$$

in the region $L_n \le |t| \le T_1$.

Let $|t| \le (64L_n)^{-1} =: T$, then rough estimation of the terms in the formula obtained in Theorem 4 yields

$$\varphi_n(t) \ll \exp\Big\{-\frac{t^2}{2} + \frac{|t|^3 L_n}{6} + 2(\frac{8}{3})^2|t|^3 L_n\Big\} \le e^{-t^2/4}. \quad (24)$$

It follows from (6) that

$$\varphi_n(t) - 1 \ll |t| \left(\frac{1}{n!} \sum_{\sigma \in \mathcal{S}_n} (h_n(\sigma) - A(n))^2 \right)^{1/2} \ll |t|. \tag{25}$$

Splitting the region $|t| \leq T$ of integration in (21) into intervals $|t| \leq L_n$, $L_n \leq |t| \leq T_1$, $T_1 \leq |t| \leq T$ and using (25), (23), and (24) respectively, we obtain the desired estimate. Theorem 1 is proved. ∎

Proof of Corollary: The first estimate follows from Theorem 1 and the inequality

$$D_n \leq L_n^{1/3} \sum_{k=1}^{n} \frac{|\widehat{h}_n(k)|}{k} a_k^{2/3} \leq L_n^{2/3} \Sigma(2/3) \leq L_n^{2/3}$$

by the estimate above.

To prove the second assertion of Corollary, we construct the following example. Let $d(1) = 1$,

$$d(k) = \begin{cases} \log^{-2/5} k, & 2 \leq k \leq n/2, \\ 1, & n/2 < k \leq n, \end{cases}$$

then

$$\sum_{k=1}^{n} \frac{d^3(k)}{k} = c + o(1), \quad \beta_n^2 := \sum_{k=1}^{n} \frac{d^2(k)}{k} = 5 \log^{4/5} n + O(1).$$

If $\widehat{h}(k) := d(k)/\beta_n$, then $L_n \sim 5^{-3/2} c \log^{-6/5} n$ and

$$D_n \geq \frac{1}{\beta_n^2} \left(\sum_{n/2 < k \leq n} \frac{1}{k} \right)^2 \geq \frac{(\log 2)^2}{2\beta_n^2} \geq c_1 L_n^{2/3}$$

with $c_1 > 0$ provided n is sufficiently large. Thus, for the sequence of CD additive functions defined by $h(\kappa) = \widehat{h}(l(\kappa))$, Theorem 1 yields $R_n \gg L_n^{2/3}$. ∎

Proof of Theorem 2: As earlier, we may assume that $L_n = o(1)$. Then also $D_n = o(1)$. The estimates (24) and (25) but (23) remain valid.

Using the condition (7) instead of (3), we derive from Theorem 3

$$\varphi_n(t) = e^{-t^2/2} + O\left(|t|^3 e^{-t^2/4} L_n\right) \tag{26}$$

in the region $L_n \leq |t| \leq T_1$.

Now the traditional form of the Esseen inequality yields

$$R_n \ll \frac{1}{T} + \int_{|t| \leq T} |\varphi_n(t) - e^{-t^2/2}| \frac{dt}{|t|}$$

where $T = (64L_n)^{-1}$. Using the formulae (25), (24), and (26) in the regions $|t| \leq L_n$, $T_1 \leq |t| \leq T$, and $L_n \leq |t| \leq T_1$, respectively, we obtain the desired estimate. Theorem 2 is proved. ∎

Acknowledgments

The results of the paper were contributed at the Kyoto conference on analytic number theory in 1996. We thank the organizers and the Lithuanian Open Society Foundation for the financial support to attend it. We remain also deeply indebted to Professor Y.Motohashi for the warm hospitality shown during the meeting.

References

1. Aratia, R., Barbour, A.D., Tavaré, S., "Poisson process approximations for the Ewens sampling formula," *Ann. Appl. Probab.* **2**(1992), 3, pp. 519–535.
2. Barbour, A.D., Tavaré, S., "A rate for the Erdös-Turán law," *Combinatorics, Prob. Comput.* **3**(1994), pp. 167–176.
3. Erdös, P., Turán, P., "On some problems of a statistical grouptheory I," *Zeitschr. für Wahrscheinlichkeitstheorie und verw. Gebiete* **4**(1965), pp. 175–186.
4. Flajolet, P., Odlyzko, "A., Singularity analysis of generating functions," *SIAM J. Discrete Math.* **3**(1990), 2, pp. 216–240.
5. Goncharov, V.L., "On the distribution of cycles in permutations," *Dokl. Acad. Nauk SSSR* **35**(1942), 9, pp. 299–301 (Russian).
6. Halász, G., "Über die Mittelwerte multiplikativer zahlentheoretisher Funktionen," *Acta Math. Acad. Sci. Hung.* **19**(1968), pp. 365–403.
7. Halász, G., "On the distribution of additive and mean values of multiplicative arithmetic functions," *Studia Sci. Math. Hung.* **6**(1971), pp. 211–233.
8. Hall, R.R., Tenenbaum, G., *Divisors*, Cambridge Tracts in Mathematics, 90, 1988.
9. Indlekofer, K.-H., Manstavičius, E., "Additive and multiplicative functions on arithmetical semigroups," *Publicationes Mathematicae Debrecen* **45**(1994), 1–2, pp. 1–17.
10. Kolchin, V.F., *Random Mappings*, Optimization Software, Inc. New York, 1986.
11. Mačiulis, A., "Mean value of multiplicative functions," *Lith. Math. J.* **28**(1988), 2, pp. 221–229.
12. Mačiulis, A., "The mean values of multiplicative functions defined on a semigroup," In: *New Trends in Probab. and Statistics, vol. 2. Analytic and Probabilistic Methods in Number Theory*, F.Schweiger and E.Manstavičius (Eds), VSP/TEV, Utrecht/Vilnius, 1992, pp. 121–133.
13. Mačiulis, A., "The exact order of the convergence rate in the central limit theorem for additive functions," *Lith. Math. J.* **33**(1993), 3, pp. 243–254.
14. Manstavičius, E., "Additive and multiplicative functions on random permutations," *Lith. Math. J.* **36**(1996), 4, pp. 400–408.
15. Manstavičius, E., Skrabutėnas, R., "Summation of values of multiplicative functions on semigroups," *Lith. Math. J.* **33**(1993), 3, pp. 255-264.
16. Nicolas, J.L., "Distribution statistique de l'ordre d'un element du groupe symetrique," *Acta Math. Hung.* **45**(1985), 1-2, pp. 96–84.
17. Petrov, V.V., *Sums of Independent Random variables*, Moscow, "Nauka", 1972 (Russian).
18. Vershik, A.M., "Asymptotic combinatorics and algebraic analysis," in: *Proceedings of the International Congress of Mathematicians, Zürich, 1994*, Birkhäuser, Basel, 1995, pp. 1384–1394.

THE RAMANUJAN JOURNAL 2, 201–217 (1998)

Products of Shifted Primes. Multiplicative Analogues of Goldbach's Problems, II

P.D.T.A. ELLIOTT   pdtae@euclid.colorado.edu
Department of Mathematics, University of Colorado, Boulder, CO 80309-0395

In memory of my friend and colleague, Paul Erdős

Received April 29, 1997; Accepted January 16, 1998

Abstract. A fixed power of each positive integer has a product representation using shifted primes $N - p$.

Key words: shifted primes, products, Goldbach

1991 Mathematics Subject Classification: 11N99, 11N05

1. Introduction

In 1917, Hardy and Ramanujan proved that most integers n have about $\log \log n$ distinct prime factors [11]. Seventeen years later a new proof by Turán vitalized their result, showing it to exemplify a general phenomenon [15]. From these papers grew the Probabilistic Theory of Numbers, with signal achievements by Erdős and Wintner, Erdős and Kac, Kubilius and others. Accounts of this extensive discipline may be found in Kac [12], Kubilius [13], and Elliott [4].

In a 1936 study of their particular functions, Erdős adapted the approach of Hardy and Ramanujan to show that shifted primes $p - 1$ usually have about $\log \log p$ distinct prime divisors [8].

The understanding of general arithmetic functions developed in the sixty years since allows us a dual procedure. We view arithmetic functions as characters on the multiplicative group of positive rationals and develop an harmonic analysis. Looking towards a celebrated conjecture of Goldbach we set about factorizing a given integer, not in terms of the primes but in terms of the shifted primes $N - p$, $p < N$.

The harmonic analysis delivers a result of a general nature. In this paper I show how to particularize it, using sieves in a manner pioneered by Erdős.

I thank the referee for a careful reading of the text.

2. Statement of results

In a previous paper I formulated three conjectures.

Conjecture 1. *If N is a sufficiently large positive integer, then every rational r/s with $1 \le r \le s \le \log N$, $(rs, N) = 1$, has a representation of the form*

$$\frac{r}{s} = \frac{N - p}{N - q}, \quad p, q \ prime, \ p < N, q < N.$$

Conjecture 2. *There is a positive integer k so that in the above notation and terms there are representations*

$$\frac{r}{s} = \prod_{i=1}^{k} (N - p_i)^{\varepsilon_i}, \quad \varepsilon_i = +1 \ or \ -1.$$

Conjecture 3. *There are representations of this type, but with the number k, of factors needed, possibly varying with r and s.*

Towards the third conjecture I here establish two results.

Theorem 1. *There is an integer k so that if $c > 0$, $N > N_0(c)$, then every integer in the range $1 \le m \le (\log N)^c$, $(m, N) = 1$, has a representation*

$$m^k = \prod_{p \le N/2} (N - p)^{d_p} \tag{1}$$

with p prime and d_p integral.

A value can be computed for k and I return to consider it later. I note that every shifted prime $N - p$ in such a product representation is at least $N/2$ and so large compared to m. All large prime factors of these shifted primes, in fact all prime factors exceeding m, cancel away. Granted the existence of a representation (1), an algorithm to effect it is indicated in [5, Ch. 15].

Theorem 2. *There is a positive γ so that the product representation of Theorem 1 holds for each prime value of m in the range $(1, N^\gamma)$, $(m, N) = 1$, with at most one exception.*

Let M be the product of the primes not exceeding N^γ and not dividing N, with the exceptional prime of Theorem 2 removed if it exists. The prime number theorem shows that $M = \exp((1 + o(1))N^\gamma (\gamma \log N)^{-1})$ as $N \to \infty$. Since each $N - p$ has $O(\log N (\log \log N)^{-1})$ distinct prime factors, in the product representation of M^k guaranteed by Theorem 2 we have

$$\sum_{p \le N/2} |d_p| > N^\gamma (\log N)^{-2}, \quad N \ge N_1.$$

The argument to establish Theorems 1 and 2 does not take the size of the represented integer m much into account. In order to reduce the number of terms in the representing products (1) it would seem necessary to do so.

3. Background

Theorems 1 and 2 rest upon the following result, obtained by harmonic analysis.

Let $0 < \delta < 1$, N be a positive integer, P a set of primes not exceeding N and coprime to N,

$$|P| = \sum_{p \in P} 1 \geq \delta \pi(N) > 0.$$

Let Q_1 be the multiplicative group generated by the positive integers n not exceeding N and satisfying $(n, N) = 1$, Γ the subgroup of Q_1 generated by the $N - p$ with p in P, G_1 the quotient group Q_1/Γ.

Lemma 1.   *If N is sufficiently large in terms of δ, then we may remove a set of primes q, not exceeding N, with $\sum q^{-1} \leq c_1(\delta)$, such that G, the subgroup of G_1 generated by the rationals in Q_1 with no q factor, satisfies $|G| \leq c_2(\delta)$.*

Proof: Lemma 1 is Theorem 1 of [7]. In fact that theorem asserts the existence of a positive integer D, a subgroup L of G with $|L| \leq 4/\delta$, and a group homomorphism $(\mathbb{Z}/D\mathbb{Z})^* \to G/L$ which makes the following diagram commute.

$$(\mathbb{Z}/D\mathbb{Z})^*$$

$$\nearrow \qquad\qquad \searrow$$

$$Q_3 \qquad \longrightarrow \qquad G/L$$

Here Q_3 is the subgroup of Q_1 when the q-factors are removed, D is (in an obvious sense) coprime to Q_3, $(\mathbb{Z}/D\mathbb{Z})^*$ is the multiplicative group of reduced residue classes (mod D), the maps $Q_3 \to (\mathbb{Z}/D\mathbb{Z})^*$, $Q_3 \to G \to G/L$ are canonical. Values for D and the $c_j(\delta)$ may be determined, but not for the individual q. In particular, the representability of an integer m as a product of the $N - p$ with p in P, depends essentially upon the residue class (mod D) to which m belongs.   □

On the face of it Lemma 1 already delivers a strong version of Theorems 1 and 2. For example, let P contain all primes in the interval $[2, N/2]$, and let m be comprised of primes not exceeding N, not dividing N, and not among the corresponding exceptional q. Then there is a representation

$$m^{|G|} = \prod_{p \leq N/2} (N - p)^{d_p}$$

with d_p integral. However, the exceptional q are not precisely located, and whether a particular integer is divisible by any of them is not immediately apparent. Moreover, the exceptional primes q may vary with N.

Given a particular positive integer d, $(d, N) = 1$ a natural procedure is to look for a prime p, not exceeding N, which satisfies $p \equiv N (\text{Mod } d)$, $(N - p)d^{-1}$ coprime to N and not divisible by any q. Since the q may cover all primes in an interval $(N^{\varepsilon}, N]$, $0 < \varepsilon < 1$, the procedure amounts to representing N in the form $p + s$, where every divisor of the integer s is at most N^ε in size. This is a problem of independent difficulty. See, for example, Friedlander [10], Baker and Harman [1].

Adaption of this line to my present circumstances would require much calculation and offer limited hope of success. Rather, I consider the problem of removing exceptional primes of the type q from a result such as Lemma 1, in more general terms.

4. A basic inequality

In this section I begin a proof of Theorem 1 that does not seek to minimize the value of the exponent k.

Let $\pi(x, D, r)$ denote the number of primes not exceeding x which lie in the residue class $r \pmod D$.

Lemma 2. *For any $A > 0$ there is a B such that*

$$\sum_{D \leq x^{1/2}(\log x)^{-B}} \max_{(r,D)=1} \max_{y \leq x} \left| \pi(y, D, r) - \frac{Li(y)}{\phi(D)} \right| \ll x(\log x)^{-A}.$$

Proof: See Bombieri [2].   □

Lemma 3.

$$\pi(x, D, r) \leq \frac{2x}{\phi(D)\log(x/D)}$$

uniformly for $(r, D) = 1$, $1 < D < x$.

Proof: For this version of the Brun-Titchmarsh theorem see Montgomery and Vaughan [14].   □

Lemma 4. *Let $0 < u < v$. Then*

$$\sum_{\substack{y_1 < n \leq y_2 \\ (n,N)=1}} \frac{1}{n} = \frac{\phi(N)}{N}\left(\log \frac{y_2}{y_1} + O(1)\right)$$

uniformly for $N^u \leq y_1 \leq y_2 \leq N^v$, and positive integers N.

Proof: The Möbius function μ satisfies

$$\sum_{d|n} \mu(d) = \begin{cases} 1 & \text{if } n = 1, \\ 0 & \text{if } n > 1. \end{cases}$$

In terms of this function the interval $(y_1, z]$ contains

$$\sum_{y_1 < n \le z} \sum_{d|(N,n)} \mu(d) = \sum_{d|N} \left(\left[\frac{z}{d} \right] - \left[\frac{y_1}{d} \right] \right)$$

integers prime to N. Removing the square brackets introduces an error not exceeding the number of squarefree divisors of N, and therefore $O(N^\varepsilon)$ for each fixed $\varepsilon > 0$. Provided ε is fixed at a sufficiently small value, an integration by parts gives

$$\sum_{\substack{y_1 < n \le y_2 \\ (n,N)=1}} n^{-1} = O\left(y_1^{-1/2}\right) + \int_{y_1}^{y_2} z^{-2} \sum_{\substack{y_1 < n \le z \\ (n,N)=1}} 1 \, dz$$

$$= \frac{\phi(N)}{N} \left(\log \frac{y_2}{y_1} - 1 + \frac{y_1}{y_2} \right) + O\left(y_1^{-1/2}\right).$$

We complete the proof by appealing to the bound $\phi(N) \gg N (\log \log N)^{-1}$, valid for all sufficiently large positive integers N. $\qquad\square$

Lemma 5. *Let $0 < \alpha < \beta < 1/2$, $c > 0$, $c_1 > 0$, Let A be a set of integers prime to N, and for which*

$$\sum_{\substack{N^\alpha < a \le N^\beta \\ a \in A}} \frac{1}{a} \ge \frac{c_1 \phi(N) \log N}{N}.$$

Then there is a positive c_2, depending at most upon the four initial parameters, so that for all sufficiently large N, the integers d, $(d, N) = 1$ which can be expressed in the form $(N - p)(am)^{-1}$, $p \le N/2$, $N^\alpha < a \le N^\beta$, $a \in A$, satisfy

$$\sum_{N^{1-\beta} < d < N^{1-\alpha}} \frac{1}{d} \ge \frac{c_2 \phi(N) \log N}{N}$$

uniformly in integers m, $(m, N) = 1$, $1 \le m \le (\log N)^c$.

Proof: Let $r(d)$ denote the number of representations of d in the form $(N - p)(am)^{-1}$, $p \le N/2$, $N^\alpha < a \le N^\beta$, $a \in A$. Then

$$\sum_{}^{*} r(d) \ge \sum_{\substack{N^\alpha < a \le N^\beta \\ a \in A}} \sum_{\substack{p \le N/2 \\ p \equiv N \pmod{am}}} 1,$$

where $*$ denotes that d belongs to the interval $(N^{1-\beta}(2(\log N)^c)^{-1}, N^{1-\alpha})$. We estimate the double sum by Lemma 2. For any fixed c_0 it is at least

$$\sum_{\substack{N^\alpha < a \le N^\beta \\ a \in A}} \frac{Li(N/2)}{\phi(am)} + O\left(\frac{N(\log N)^{-c_0}}{\phi(m)} \right). \qquad (2)$$

Note that $mM^\beta \le N^\beta (\log N)^c$; it is here that we need the constraint $\beta < 1/2$. Moreover, if m is appreciably larger than a power of $\log N$, then Lemma 1 will not guarantee an error small enough to allow the factor $\phi(m)$ to be extracted.

To estimate the sum in (2) from below we apply the Cauchy-Schwarz inequality:

$$\left(\sum_{\substack{N^\alpha < a \le N^\beta \\ a \in A}} \frac{1}{am} \right)^2 \le \sum_{\substack{N^\alpha < a \le N^\beta \\ a \in A}} \frac{1}{\phi(am)} \sum_{\substack{N^\alpha < a \le N^\beta \\ a \in A}} \frac{\phi(am)}{(am)^2}. \tag{3}$$

For each fixed m, the function $n \mapsto \phi(nm)(n\phi(m))^{-1}$ is multiplicative in n. We represent it as a Dirichlet convolution $1 * h$, where the function 1 is 1 on every integer. Note that $h(p^k) = 0$ if $k \ge 2$, or $k = 1$ and $p \mid m$. On the primes not dividing m, $|h(p)| \le p^{-1}$. Allowing the second of the bounding sums at (3) to run over all the integers prime to n in the interval $[N^\alpha, N^\beta]$ we see that the sum does not exceed

$$\frac{\phi(m)}{m^2} \sum_{\substack{s \le N^\beta \\ (s,N)=1}} \frac{h(s)}{s} \sum_{\substack{s^{-1}N^\alpha < t \le s^{-1}N^\beta \\ (t,N)=1}} \frac{1}{t}. \tag{4}$$

Those terms with $s \le N^{\alpha/2}$ we estimate by Lemma 4 to contribute

$$\frac{\phi(m)}{m^2} \frac{\phi(N)}{N} ((\beta - \alpha) \log N + O(1)) \sum_{\substack{s \le N^{\alpha/2} \\ (s,N)=1}} \frac{h(s)}{s}.$$

Since this is only to be an upper bound we extend the range of s to infinity, noting that the resulting innersum has an Euler product representation:

$$\sum_{\substack{s=1 \\ (s,N)=1}}^{\infty} \frac{h(s)}{s} = \prod_{(p,mN)=1} \left(1 - \frac{1}{p^2} \right).$$

Those terms with $s > N^{\alpha/2}$ contribute to (4) an amount

$$\ll \frac{\phi(m)}{m^2} \sum_{\substack{s > N^{\alpha/2} \\ (s,N)=1}} \frac{|h(s)|}{s} \log N \ll \frac{\phi(m)}{m^2} \frac{\log N}{N^{\alpha/4}} \sum_{s=1}^{\infty} \frac{|h(s)|}{s^{1/2}}.$$

The sum of the $s^{-1/2}|h(s)|$ has an absolutely convergent Euler product. Altogether the expression (4) is at most

$$(1 + o(1)) \frac{\phi(m)}{m^2} \prod_{(p,mN)=1} \left(1 - \frac{1}{p^2} \right) \frac{\phi(N)}{N} (\beta - \alpha) \log N, \quad N \to \infty,$$

uniformly for $1 \le m \le (\log N)^c$, $(m, N) = 1$.

Taking into account the hypothesis on the set of integers A we obtain the uniform lower bound

$$\sum_{}^{*} r(d) \geq (1 + o(1)) \frac{1}{2} c_1^2 (\beta - \alpha)^{-1} \phi(m)^{-1} \prod_{(p,mN)=1} \left(1 - \frac{1}{p^2}\right)^{-1} \phi(N),$$

as $N \to \infty$.

Again by the Cauchy-Schwarz inequality

$$\left(\sum_{}^{*} r(d)\right)^2 \leq \sum_{r(d)>0}^{*} \frac{1}{d} \sum_{}^{*} dr(d)^2.$$

To continue the proof we estimate the final sum from above:

$$r(d) = \sum_{\substack{N-p=dma \\ p \leq N/2, a \in A \\ N^\alpha < a \leq N^\beta}} 1 \leq \sum_{\substack{N-p=dmw \\ p \leq N/2, w \in Z}} 1 \leq \pi(N, dm, N)$$

$$\leq 2N(\phi(dm) \log N/dm)^{-1},$$

the third step by appeal to Lemma 3. Since every $dm \leq N^{1-\alpha}(\log N)^c$, for all sufficiently large N,

$$\sum_{}^{*} dr(d)^2 \leq \frac{16N^2}{(\alpha \log N)^2} \sum_{}^{*} \frac{d}{\phi(dm)^2}.$$

The innersum over d here may be treated like the innersum over a in the upper bound at (3). In this case $d \mapsto (d\phi(m)\phi(dm)^{-1})^2$ is multiplicative in d. We express it as $1 * z$. Then $z(p^k) = 0$ if $k \geq 2$, $z(p) = (2p-1)(p-1)^{-2}$ if $p \nmid m$, $z(p) = 0$ otherwise. In terms of the function z

$$\sum_{}^{*} \frac{d}{\phi(dm)^2} = \frac{1}{\phi(m)^2} \sum_{\substack{s \leq N^{1-\alpha} \\ (s,N)=1}} \frac{z(s)}{s} \sum_{\substack{\Delta_1 < t \leq \Delta_2 \\ (t,N)=1}} \frac{1}{t} \qquad (5)$$

with $\Delta_1 = N^{1-\beta}(2s(\log N)^c)^{-1}$, $\Delta_2 = s^{-1}N^{1-\alpha}$. Those terms with $s \leq N^{1/4}$ contribute not more than

$$\frac{1}{\phi(m)^2} \sum_{\substack{s \leq N^{1/4} \\ (s,N)=1}} \frac{z(s)}{s} \frac{\phi(N)}{N} \left(\log \frac{\Delta_2}{\Delta_1} + O(1)\right)$$

$$\leq (1 + o(1)) \frac{(\beta - \alpha)\phi(N) \log N}{\phi(m)^2 N} \sum_{(s,N)=1} \frac{z(s)}{s}, \qquad N \to \infty.$$

The final sum has an Euler product representation $\Pi(1 + (2p-1)(p(p-1)^2)^{-1})$ taken over the primes not dividing mN. The terms of (5) with $s > N^{1/4}$ contribute a negligible amount to the total.

Altogether

$$\sum_{d}^{*} \frac{1}{d} \geq (1 + o(1))\theta \frac{\phi(N)}{N} \log N, \quad N \to \infty,$$

with

$$\theta = \left(\frac{1}{2}c_1^2(\beta - \alpha)^{-1}\phi(m)^{-1} \prod_{(p,mN)=1}\left(1 - \frac{1}{p^2}\right)^{-1}\right)^2 \frac{(\alpha\phi(m))^2}{16(\beta - \alpha)}$$

$$\times \prod_{(p,mN)=1}\left(1 + \frac{2p-1}{p(p-1)^2}\right)^{-1}$$

$$= \frac{c_1^4\alpha^2}{64(\beta - \alpha)^3} \prod_{(p,mN)=1}\left(1 + \frac{2p-1}{p(p-1)^2}\right)^{-1}$$

$$\geq \frac{c_1^4\alpha^2}{64(\beta - \alpha)^3} \prod_{p \geq 2}\left(1 + \frac{2p-1}{p(p-1)^2}\right)^{-1} > 0,$$

uniformly in m, $1 \leq m \leq (\log N)^c$, $(m, N) = 1$.

To tidy up the range of d, note that for each $\varepsilon > 0$ and all sufficiently large values of N,

$$\sum_{\substack{N^{1-\beta}(2(\log N)^c)^{-1} < d \leq N^{1-\beta} \\ (d,N)=1}} \frac{1}{d} \leq \sum_{\substack{N^{1-\beta-\varepsilon} < D \leq N^{1-\beta} \\ (d,N)=1}} \frac{1}{d} \leq \frac{2\varepsilon\phi(N)\log N}{N},$$

the last step by Lemma 4.

Lemma 5 is established.   □

We can recast Lemma 5 in terms of local (logarithmic) densities. Let $0 < u < v$. Define a local density of a set of rationals E by

$$d(u, v, E) = \left(\sum_{\substack{N^u < n \leq N^v \\ (n,N)=1}} \frac{1}{n}\right)^{-1} \left(\sum_{\substack{N^u < n \leq N^v \\ (n,N)=1, n \in E}} \frac{1}{n}\right).$$

This notion of density depends upon the integer N, viewed as 'large compared to u, v'. After Lemma 4 we see that

$$d(u, v, E) = \frac{(1 + o(1))N}{(v - u)\phi(N)\log N} \sum_{\substack{N^u < n \leq N^v, (n,N)=1 \\ n \in E}} \frac{1}{n},$$

as $N \to \infty$.

Let B denote the set of rationals of the form $(N - p)a^{-1}$, $p \leq N/2$, $a \in A$; $m^{-1}B$ those of the form $(N - p)(am)^{-1}$, $p \leq N/2$, $a \in A$; in all cases with $N^\alpha < a \leq N^\beta$. Lemma 5 asserts that if $d(\alpha, \beta, A) \geq \gamma$, then $d(1 - \beta, 1 - \alpha, m^{-1}B) \geq (1 + o(1))(4e)^{-3}\alpha^2\gamma^4$ uniformly in m, $1 \leq m \leq (\log N)^c$, $(m, N) = 1$. In this formulation β plays only the role of restricting α to the interval $(0, 1/2)$.

5. Application of a sieve

I manufacture a suitable sequence A using Lemma 1.

Lemma 6. *Let a_n, $n = 1, \ldots, N$, be a sequence of rational integers, repetition allowed. Let r be a positive real number, and let $p_1 < p_2 < \cdots < p_s \leq r$ be rational primes. Set $Q = p_1 \cdots p_s$. If $d \mid Q$ then let*

$$\sum_{\substack{n=1 \\ a_n \equiv 0 \pmod d}}^{N} 1 = \eta(d)X + R(N, d),$$

where X, $R(N, d)$ are real numbers, $X \geq 0$, and $\eta(d_1 d_2) = \eta(d_1)\eta(d_2)$ whenever d_1 and d_2 are mutually prime divisors of Q.

Assume that for each prime p, $0 \leq \eta(p) < 1$.

Let $I(N, Q)$ denote the number of a_n, $n = 1, \ldots, N$, which are coprime to Q. Then

$$I(N, Q) = (1 + 2\theta_1 H)X \prod_{p \mid Q}(1 - \eta(p)) + 2\theta_2 \sum_{\substack{d \mid Q \\ d \leq z^3}} 3^{\omega(d)}|R(N, d)|$$

holds uniformly for $r \geq 2$, $\max(\log r, S) \leq \frac{1}{8} \log z$, where $|\theta_1| \leq 1$, $|\theta_2| \leq 1$ and

$$H = \exp\left(-\frac{\log z}{\log r}\left(\log\left(\frac{\log z}{S}\right) - \log\log\left(\frac{\log z}{S}\right) - \frac{2S}{\log z}\right)\right)$$

$$S = \sum_{p \mid Q} \frac{\eta(p)}{1 - \eta(p)} \log p.$$

When these conditions are satisfied $2H \leq \exp(-0.0006)$.

Proof: This is a particular case of Lemma 2.1 from [4].   □

In what follows $T(r)$ will denote a product of primes, each prime not exceeding r.

Lemma 7. *If $0 < \varepsilon \leq 10^{-5}$ and y is sufficiently large in terms of ε, then*

$$\sum_{\substack{n \leq y \\ (n, T(y^\varepsilon))=1}} 1 \geq (1 - \varepsilon^2)y \prod_{p \mid T(y^\varepsilon)}\left(1 - \frac{1}{p}\right)$$

uniformly in $T(y^\varepsilon)$.

The bound on ε is not of particular significance.

Proof: We apply Lemma 6 to the integers in the interval $[1, y]$, with $X = y$, $\eta(d) = d^{-1}$, and each $R(N, d)$ not exceeding 1 in absolute value. If $r = y^\varepsilon$, $z = \exp(\varepsilon^{1/2} \log y)$, then

$$S = \sum_{p|Q} \frac{\log p}{p - 1} \leq \sum_{p \leq y^\varepsilon} \frac{\log p}{p} + O(1) \leq \varepsilon \log y + O(1).$$

For all y sufficiently large (in terms of ε) the condition $\max(\log r, S) \leq \frac{1}{8} \log z$ is amply satisfied. Moreover,

$$\log\left(\frac{\log z}{S}\right) - \log\log\left(\frac{\log z}{S}\right) - \frac{2S}{\log z} \geq \log 8 - \log\log 8 - \frac{1}{4} > 1,$$

so that (still in the notation of Lemma 6)

$$2H \leq 2\exp(-\varepsilon^{-1/2}) < 2(5!\varepsilon^{5/2}).$$

The sum involving $3^{\omega(d)}$ is

$$\ll z^3 \sum_{\substack{d|Q \\ d \leq z^3}} \frac{3^{\omega(d)}}{d} \ll z^3 \prod_{p \leq r}\left(1 + \frac{3}{p}\right) \ll y^{3\sqrt{\varepsilon}}(\log y)^3$$

and may be absorbed into the main term by replacing $5!\varepsilon^{5/2}$ with $\varepsilon^2/2$ and (further) assuming y sufficiently large in terms of ε.

This completes the proof of Lemma 7. Note that with $-\varepsilon^2$ replaced by ε^2, the inequality of Lemma 7 goes in the other direction.   □

Lemma 8.   *If* $10^{10}\varepsilon \leq 1$, *then*

$$\sum_{\substack{y^\varepsilon < n \leq y \\ (n, T(y^\varepsilon))=1}} \frac{1}{n} \geq (1 - 2\sqrt{\varepsilon}) \prod_{p|T(y^\varepsilon)}\left(1 - \frac{1}{p}\right)\log y$$

uniformly for $T(y^\varepsilon)$ *and all* y *sufficiently large in terms of* ε.

Proof: An integration by parts gives a representation

$$\int_{y^\varepsilon}^y z^{-2} \sum_{\substack{n \leq z \\ (n, T(y^\varepsilon))=1}} 1\,dz + O\left(\prod_{p|T(y^\varepsilon)}\left(1 - \frac{1}{p}\right)\right) \tag{6}$$

for the sum to be estimated. Note that a $T(y^\varepsilon)$ is a $T(y)$ also.

Over the range $y^{\sqrt{\varepsilon}} < z \leq y$ we appeal to Lemma 7, replacing ε, y in that lemma by $\sqrt{\varepsilon}, z$. The corresponding contribution to the integral in (6) is at least

$$(1 - \varepsilon) \prod_{p|T(y^\varepsilon)} \left(1 - \frac{1}{p}\right) \int_{y^{\sqrt{\varepsilon}}}^{y} \frac{dz}{z}.$$

We abandon the remaining range of z-values. □

Again we note that there is an upper bound corresponding to the inequality of Lemma 8. Indeed, with a little extra argument we obtain a bound

$$\sum_{\substack{y^{1/2} < n \leq y \\ (n, T(y)) = 1}} \frac{1}{n} \ll \prod_{p|T(y)} \left(1 - \frac{1}{p}\right) \log y$$

uniform in $T(y)$ for $y \geq 2$. It is desirable to remove the restraint $y^{1/2} < n$.

Lemma 9.

$$\sum_{\substack{n \leq y \\ (n, T(y)) = 1}} \frac{1}{n} \ll \prod_{p|T(y)} \left(1 - \frac{1}{p}\right) \log y$$

uniformly in $T(y)$ for $y \geq 2$.

Proof: Arguing via Selberg's sieve, as in Lemma 8, apparently will not work without serious modification. I use a different approach.

Let $\sigma = 1 + (\log y)^{-1}$. Then the sum to be estimated does not exceed

$$e \sum_{(n, T(y)) = 1} \frac{1}{n^\sigma} \leq e \prod_{p|T(y)} \left(1 - \frac{1}{p^\sigma}\right) \zeta(\sigma).$$

In turn, the product exceeds $\prod(1 - p^{-1})$, taken over the p dividing $T(y)$, by a factor at most

$$\prod_{p|T(y)} \left(1 - \frac{1}{p^\sigma}\right)\left(1 - \frac{1}{p}\right)^{-1} \leq \exp\left(\sum_{p \leq y} \left\{\frac{1}{p}\left(1 - \frac{1}{p^{\sigma-1}}\right) + O\left(\frac{1}{p^2}\right)\right\}\right)$$

$$\ll \exp\left(\sum_{p \leq y} \frac{(\sigma - 1)\log p}{p} \exp\left((\sigma - 1)\log p\right)\right) \ll \exp\left(\sum_{p \leq y} \frac{\log p}{p \log y}\right) \ll 1,$$

since $1 - e^{-t} \leq |t|e^{|t|}$ for all real t.

In view of the asymptotic relation $(\delta - 1)\zeta(\delta) \to 1$ as $\delta \to 1+$, the assertion of Lemma 9 is justified.   □

It follows from Lemma 9 that

$$\sum_{\substack{y^\varepsilon < q \le y \\ q|T(y)}} \sum_{\substack{y^\varepsilon < n \le y,(n,T(y^\varepsilon))=1 \\ n \equiv 0 (\text{mod } q)}} \frac{1}{n} \le \sum_{\substack{y^\varepsilon < q \le y \\ q|T(y)}} \frac{1}{q} \sum_{\substack{m \le yq^{-1} \\ (m,T(y^\varepsilon))=1}} \frac{1}{m}$$

which, extending the innersum to y, is

$$\ll \sum_{\substack{y^\varepsilon < q \le y \\ q|T(y)}} \frac{1}{q} \prod_{p|T(y^\varepsilon)} \left(1 - \frac{1}{p}\right) \log y, \quad y \ge 2.$$

Together with Lemma 8 this yields a bound

$$\sum_{\substack{y^\varepsilon < n \le y \\ (n,T(y))=1}} \frac{1}{n} \ge \prod_{p|T(y^\varepsilon)} \left(1 - \frac{1}{p}\right) \log y \left(1 - 2\sqrt{\varepsilon} - c_3 \sum_{\substack{y^\varepsilon < q \le y \\ q|T(y)}} \frac{1}{q}\right) \tag{7}$$

for some positive absolute constant c_3, each $\varepsilon \le 10^{-10}$, all $T(y^\varepsilon)$ restricted from $T(y)$, and all y sufficiently large in terms of ε.

To apply this result usefully we need

$$2c_3 \sum_{\substack{y^\varepsilon < q \le y \\ q|T(y)}} \frac{1}{q} \le 1, \tag{8}$$

say. The final bracket in the lower bound (7) will then be at least $1/4$.

6. Proof of Theorem 1

We apply Lemma 1 with P the set of primes not exceeding $N/2$, and not dividing N. For N sufficiently large we may take $\delta = 1/4$ in the hypotheses of the Lemma. Let A run through the integers made up of primes not exceeding N, not dividing N, and not amongst the q described in Lemma 1. Let $\varepsilon = 10^{-10}$. For every positive integer t

$$\sum_{j=1}^{t} \sum_{N^{\varepsilon^{j+1}} < q \le N^{\varepsilon^j}} \frac{1}{q} \le c_1 \left(\frac{1}{4}\right).$$

There is a j, $1 \le j \le t$, for which

$$\sum_{N^{\varepsilon^{j+1}} < q \le N^{\varepsilon^j}} \frac{1}{q} \le \frac{1}{t} c_1 \left(\frac{1}{4}\right).$$

Proof: We choose $t > c_1(1/4)$, to satisfy (8) with some $y = N^{\varepsilon^j}$, and $T(y)$ the product of those q and the prime divisors of N that do not exceed y. This is possible since so long as y exceeds a fixed power of N

$$\sum_{p|N, p>y^\varepsilon} \frac{1}{p} \leq \frac{\log N}{\varepsilon y \log y} = o(1), \quad N \to \infty.$$

For the same value of j,

$$\sum_{\substack{y^\varepsilon < n \leq y, (n,N)=1 \\ \forall q:(n,q)=1}} \frac{1}{n} \geq \frac{\phi(N)}{4N} \prod \left(1 - \frac{1}{q}\right) \log y. \tag{9}$$

The hypothesis on A, in Lemma 5, is satisfied with $\beta = \varepsilon^j$, $\alpha = \varepsilon^{j+1}$, c_1 (of Lemma 5) $= \frac{1}{4}\prod(1 - \frac{1}{q})\varepsilon^{j+1}$. The value of j may vary with N, but as there are only finitely many j values, we may confine ourselves to those N with the same (temporarily fixed) value of j. In the notation following Lemma 5,

$$d(1 - \beta, 1 - \alpha, m^{-1}B) \geq \frac{1}{2}(4e)^{-3}\alpha^2 \left(\frac{1}{4}\prod\left(1 - \frac{1}{q}\right)\right)^4$$

uniformly for $1 \leq m \leq (\log N)^c$, $(m, N) = 1$ and all N sufficiently large.

Let the integer s exceed the reciprocal of this lower bound.

The sets $m^{-r}B$, $r = 0, 1, \ldots, s - 1$, cannot all be disjoint, otherwise the local density of their union in the interval $(N^{1-\beta}, N^{1-\alpha}]$ will for all large N exceed 1. Therefore, $m^{-i}(N - p_1)a_1^{-1} = m^{-j}(N - p_2)a_2^{-1}$ for some $0 \leq i < j \leq s - 1$.

With k_0 the least common multiple of the integers not exceeding $s - 1$ there is a representation

$$m^{k_0} = ((N - p_1)^{-1}(N - p_2)a_1a_2^{-1})^{k_0(j-i)^{-1}}.$$

Lemma 1 guarantees to each $a_i^{|G|}$ a product representation by the $N - p$, where $P \leq N/2$, $(p, N) = 1$. With $k = k_0|G|$ Theorem 1 is proved. □

7. Lowering the bound for k

Let q run through a (possibly infinite) set of primes for which $\sum q^{-1} \leq K$. Erdős and Ruzsa [9], showed that for all x sufficiently large in terms of K, the interval $[1, x]$ contains at least $d(K)x$ integers none of which is divisible by a q. Their proof employs a Buchstab identity and delivers for $d(K)$ a value $\exp(-\exp(c_4 K))$, $c_4 > 0$. They show that the double exponential of K is then best possible.

In the proof of Theorem 1 I do not need a uniform lower bound, only an interval $(N^\alpha, N^\beta]$, $0 < \alpha < \beta < 1$ to contain many integers n free of factors q. This allows me to increase (what corresponds to) $d(K)$ to a value of the type $c_5 \exp(-K)$, a single exponential of K.

Moreover, the requirement $(n, N) = 1$ cannot be directly encompassed by the result of Erdős and Ruzsa since $\sum p^{-1}$, taken over the prime divisors p of N, may well become unbounded with N.

The value of k in Theorem 1 is partly controlled by the factor $\Pi (1 - q^{-1})$ appearing in the lower bound (9). The method of proof may be (elaborately) modified so that the particular value of $d(K)$ implicit in a bound such a (9) becomes irrelevant; it matters only that it not be too small as a function of N. In this way a choice $k \leq 16 (= 4^2)$ and with further effort $k \leq 9 (= 3^2)$ can be achieved in the product representation (1), uniformly for $1 \leqq m \leq \log \log N$, $(m, N) = 1$. The details warrant a separate account.

8. Primes in arithmetic progression

The proof of Theorem 2 requires a lower bound on the sum

$$\sum_{\substack{N^\alpha < a \leq N^\beta \\ a \in \bar{A}}} \pi(N/2, am, N) \tag{10}$$

that appears at the beginning of the proof of Lemma 5, uniform in integers m up to a power of N rather than to the power of $\log N$ appropriate for Theorem 1. Bombieri's theorem in Lemma 2 no longer suffices. We replace it with the next two results.

Lemma 10. *There are c_6, c_7, positive reals such that $\pi(x, m, r) \geq c_7 x (\phi(m) \log x)^{-1}$ uniformly for $(r, m) = 1$, $2 \leq m \leq x^{c_6}$, $m \not\equiv 0 \pmod{m_0}$, where the exceptional modulus m_0 satisfies: Given $F > 0$, $m_0 > (\log x)^F$ for x sufficiently large in terms of F.*

Lemma 11.

$$\sum_{\substack{D \leq Qq^{-1} \\ (D,q)=1, p|D \Rightarrow p>\omega}} \max_{(a,D)=1} \max_{y \leq x} \left| \sum_{\substack{n \leq y \\ n \equiv a \pmod{Dq}}} \Lambda(n) - \frac{1}{\phi(D)} \sum_{\substack{n \leq y \\ n \equiv a \pmod{q}}} \Lambda(n) \right|$$

$$\ll \frac{\tau(q) x (\log x)^6}{q} \left(\frac{1}{\omega} + \frac{Q}{x^{1/2}} + \frac{q^{1/2}}{x^{1/6}} \right)$$

holds uniformly for $2 \leq \omega \leq Q \leq x$ and positive integers q.

Proof of Lemmas 10 and 11:   The estimate of Lemma 11 is indicated in [6], p. 188. The detailed proof of an earlier version may be found as Lemma 3 of [3]. This same reference contains an appropriate version of Lemma 10.

The condition $(D, q) = 1$ in Lemma 11 may be omitted. Since $\Lambda(n) \ll \log n$ and $\phi(D) \gg D(\log \log D)^{-1}$, those terms for which $(D, q) > 1$ contribute to the sum of the maxima

$$\ll x(\log x)^2 \sum_{\substack{D \le Q \\ (D,q)>1}} \frac{1}{Dq} \ll \frac{x(\log x)^2}{q} \sum_{\substack{t|q \\ \ell|t \Rightarrow \ell > \omega}} \sum_{\substack{D \le Q \\ (D,q)=t}} \frac{1}{D}$$

$$\ll \frac{x(\log x)^2}{q} \sum_{\substack{t|q \\ \ell|t \Rightarrow \ell > \omega}} \frac{1}{t} \sum_{s \le Qt^{-1}} \frac{1}{s}$$

$$\ll \frac{x(\log x)^3}{q} \sum_{\substack{t|q \\ \ell|t \Rightarrow \ell > \omega}} \frac{1}{t} \ll \frac{\tau(q)x(\log x)^3}{q\omega},$$

an amount which may be absorbed into the existing upper bound.   □

For $\omega \ge 2$ let n_1 denote that part of the integer n comprised of powers of the primes not exceeding ω; $n = n_1 n_2$.

Lemmas 10 and 11 are coupled by the following simple result.

Lemma 12.

$$\sum_{\substack{n \le N, (n,N)=1 \\ n_1 > \omega^c}} \frac{1}{n} \ll \frac{\phi(N) \log N}{Nc}$$

uniformly for $2 \le \omega \le N$.

Proof: The sum to be estimated does not exceed

$$\frac{1}{c \log \omega} \sum_{\substack{n \le N \\ (n,N)=1}} \frac{\log n_1}{n} = \frac{1}{c \log \omega} \sum_{\ell \le \omega} \frac{\log \ell^s}{\ell^s} \sum_{\substack{m \le N\ell^{-s} \\ (m,N\ell)=1}} \frac{1}{m}.$$

The innersum is by Lemma 9 $\ll \phi(N)N^{-1} \log N$. The sum over the prime-powers ℓ^s is

$$\sum_{\ell \le \omega} \left(\frac{\log \ell}{\ell} + O\left(\frac{1}{\ell^{3/2}}\right) \right) \ll \log \omega.$$

The asserted bound is clear.   □

Remark. In this proof logarithmic density shows to advantage. A more elaborate argument employing Lemma 6 yields the bound

$$\sum_{\substack{n \le y, (n,N)=1 \\ n_1 > \omega^c}} 1 \ll \frac{\phi(N)y}{Nc}$$

uniformly for $(N\phi(N)^{-1})^4 \le y \le N$, $\omega \ge 2$, $c \ge 4$.

9. Proof of Theorem 2

The proof follows that for Theorem 1 save for a large change in the first step, the estimation from below of the sum at (10). I sketch the details of this change.

Let $\omega = (\log N)^8$. For $c > 0$ the contribution to the sum (10) arising from terms with $a_1 > \omega^c$ is by Lemma 3

$$\ll \frac{N}{\log N} \sum_{\substack{N^\alpha < a \leq N^\beta \\ a_1 > \omega^c}} \frac{1}{\phi(am)}.$$

A simple modification of Lemma 9 gives

$$\sum_{\substack{n \leq N \\ (n,N)=1}} \frac{n}{\phi(n)^2} \ll \frac{\phi(N) \log N}{N}.$$

Noting that $\phi(am) \geq \phi(a)\phi(m)$, application of the Cauchy-Schwarz inequality followed by that of Lemma 12 shows the above contribution to be $\ll \phi(N)c^{-1/2}$.

For the remaining terms in (10) there is the estimate

$$\frac{1}{\log N} \sum_{\substack{N^\alpha < a \leq N^\beta \\ a \in A, a_1 \leq \omega^c}} \sum_{\substack{n \leq N/2 \\ n \equiv N(\text{mod } am)}} \Lambda(n) + o\left(\frac{\phi(N)}{\phi(m)}\right), \quad N \to \infty,$$

uniformly in A. Considering together the terms with a given value of a_1, Lemma 11 allows us to replace the double sum by

$$\frac{1}{\log N} \sum_{\substack{N^\alpha < a \leq N^\beta \\ a \in A, a_1 \leq \omega^c}} \frac{1}{\phi(a_2)} \sum_{\substack{n \leq N/2 \\ n \equiv N(\text{mod } a_1 m)}} \Lambda(n) + O\left(\frac{\tau(m)N(\log \log N)^2}{m(\log N)^2}\right) \tag{11}$$

certainly if m is a prime power not exceeding $N^{1/4}$, $\beta < 1/2$, $\gamma < 1/3$. Note that for a prime power m, $\tau(m) \ll \log N$.

In the generalization of Theorem 1 we may confine ourselves to primes exceeding ω^c. We call any integer made up of such primes standard.

We choose for the parameter F in Lemma 10 a value greater than $8c$. This ensures that if a typical $a_1 m$, with m standard, is divisible by m_0, then $(m, m_0) > 1$. In particular, m_0 will be divisible by some prime p_0 exceeding ω^c. Any further standard m' such that $a_1' m'$ is divisible by m_0 will also be divisible by p_0. Let $\varepsilon > 0$. Lemma 10 shows that for all standard integers m not exceeding $N^{c_6 - \varepsilon}$, save possibly for the multiples of p_0, the double sum at (11) is at least

$$(1 + o(1))c_7 \frac{N}{2 \log N} \sum_{\substack{N^\alpha < a \leq N^\beta \\ a \in A, a_1 \leq \omega^c}} \frac{1}{\phi(am)}.$$

At the expense of a further error of $\ll \phi(N)c^{-1/2}$ the condition $a_1 \leq \omega^c$ can be removed.

It will be arranged that the sum of the $\phi(am)^{-1}$ taken over a suitable sequence A restricted to $(N^\alpha, N^\beta]$ exceeds a multiple of $(N\phi(m))^{-1}\phi(N) \log N$. We may fix c large enough that

$$\sum_{\substack{N^\alpha < a \leq N^\beta \\ a \in A}} \pi\left(\frac{N}{2}, am, N\right) \geq \frac{c_7}{3} \frac{N}{\log N} \sum_{\substack{N^\alpha < a \leq N^\beta \\ a \in A}} \frac{1}{\phi(am)}$$

holds uniformly for all N sufficiently large, and all standard prime powers not exceeding $N^{c_6-\varepsilon}$ and not divisible by the prime p_0. The exceptional prime p_0 may vary with N.

The proof continues as for Theorem 1.

To effect a reasonably small value of k in the representation (1) taken over the wider range of Theorem 2, not only must the refinements mentioned in Section 7 be adopted, a study must be made of the constants c_6 and c_7 appearing in Lemma 10.

Acknowledgment

Partially supported by NSF contract DMS 9530690.

References

1. R. Baker and G. Harman, "The Brun-Titchmarsh Theorem on average," in *Analytic Number Theory Proc. Conf. in Honor of H. Halberstam*, B.C. Berndt, H.G. Diamond, and A.J. Hildebrand, eds., vol. 1; *Progress in Math.*, Birkhäuser, Boston-Basel-Berlin, vol. 138, pp. 39–103, 1996.
2. E. Bombieri, "On the large sieve," *Mathematika* 12 (1965), 201–225.
3. P.D.T.A. Elliott, "On two conjectures of Kátai," *Acta Arith.* 30 (1976), 35–39.
4. P.D.T.A. Elliott, "Probabilistic number theory, I. Mean-value theorems, II Central limit theorems," Grundlehren der Math. Wiss. 239, 240, Springer-Verlag, New York, Heidelberg, Berlin, 1979, 1980, respectively.
5. P.D.T.A. Elliott, "Arithmetic functions and integer products," *Grund. der Math. Wiss.* 272, Springer Verlag, New York, Berlin, Heidelberg, Tokyo, 1985.
6. P.D.T.A. Elliott, "The multiplicative group of rationals generated by the shifted primes, I.," *J. reine angew. Math.* 463 (1995), 169–216.
7. P.D.T.A. Elliott, "Second plenary address," *International Conference on Analytic Number Theory*, Kyoto, Japan, 1996.
8. P. Erdős, "On the normal number of prime factors of $p-1$ and some related problems concerning Euler's ϕ-function," *Quart. Journ. Math.*, Oxford 6 (1935), 205–213.
9. P. Erdős and I.Z. Ruzsa, "On the small sieve I; Shifting by primes," *J. Number Theory* 12 (1980), 385–394.
10. J.B. Friedlander, "Shifted primes without large prime factors," *Number Theory and Its Applications*, Kluwer, pp. 393–401, 1989.
11. G.H. Hardy and S. Ramanujan, "The normal number of prime factors of a number n," *Quart. Journ. Math.* (Oxford) 48 (1917), 76–92.
12. M. Kac, "Statistical independence in probability, analysis and number theory," Carus. Math. Monograph No. 12, MAA, Wiley and Sons, 1959.
13. J. Kubilius, "Probabilistic methods in the theory of numbers," Amer. Math. Soc. Translations of Math. Monographs, no. 11, Providence, 1964.
14. H. Montgomery and R. Vaughan, "On the large sieve," *Mathematika* 20 (1973), 119–134.
15. P. Turán, "On a theorem of Hardy and Ramanujan," *Journ. London Math. Soc.* 9 (1934), 274–276.

At least infinitely often for all N sufficiently large ...

$$\sum \phi(a/m)$$

... holds uniformly for all N sufficiently large ...

Acknowledgment

Partially supported by NSF contract DMS 9530690.

References

The Ramanuian Journal. 2. 219–223 (1998)

On Products of Shifted Primes

P. BERRIZBEITIA

Departamento de Matematicas, Universidad Simon Bolivar, Apartado 8900, Caracas 1080-A, Venezuela

P.D.T.A. ELLIOTT   pdtae@eudid.colorado.edu
Department of Mathematics, University of Colorado, Boulder, Colorado 80309–0395

In memory of Paul Erdős
Received May 21, 1997; Accepted Juen 25, 1997

Abstract. Any rational representable as a product of shifted primes $p + 1$ or their reciprocals, has a representation with exactly 19 terms.

Keywords: shifted primes, products, rationals

AMS Classification Numbers: Primary – 11N99; Secondary – 11N05

1. Introduction and statement of results

Let Q^* be the multiplicative group of positive rationals, Γ the subgroup generated by the shifted primes $p + 1$, G the quotient group Q^*/Γ.

In a recent paper the second author proved that G has order $|G| \leq 3$, and that there is an integer k so that every positive rational r has a representation

$$r^{|G|} = (p_1 + 1)^{\varepsilon_1} \cdots (p_k + 1)^{\varepsilon_k}, \tag{1}$$

where the p_i are primes, not necessarily distinct, and each ε_i has the value 1 or -1, [3]. The argument did not furnish a value for k. Doubtless, G is trivial and k should be 2, forming a multiplicative analogue of the conjectured infinitude of prime pairs.

Answering a question of Erdös, we show by means of a largely combinatorial argument that one may always take $k = 19$ and if $|G| > 1$, then a smaller value. In fact we establish more.

A product of shifted primes and their reciprocals will be called a *group product*.

Theorem 1 *Every group product has a representation with 19 terms.*

Theorem 2 *If $|G| = 2$, then 9 terms suffice.*

Theorem 3 *If $|G| = 3$, then 5 terms suffice.*

2. Towards Theorem 1

For each positive integer k, S_k will denote the set of group products utilising at most k shifted primes, counted multiply. As a set, Γ is the union of the S_k.

We employ the following criterion.

Lemma 1 *If $S_i = S_{i+1}$, then Γ is S_i.*

It will be convenient to denote by $E - F$ the set of all elements in E but not F.

Proof of Lemma 1: If γ is in $S_{t+1} - S_t$ for some $t > i$, and γ has the form $(p_1 + 1)^{\varepsilon_1} \cdots$ $(p_{t+1} + 1)^{\varepsilon_{t+1}}$, with each $\varepsilon_i = \pm 1$, then $(p_1 + 1)^{\varepsilon_1} \cdots (p_{i+1} + 1)^{\varepsilon_{i+1}}$ is in $S_{i+1} - S_i$, contrary to assumption. Thus the S_j with $j \geq i$ coincide, and with Γ itself. ∎

To each set of rationals, S, we attach the lower asymtotic density

$$d(S) = \liminf_{x \to \infty} x^{-1} \sum_{\substack{n \leq x \\ n \in S}} 1$$

counting over the positive integers in S. If two sets do not contain a common integer, then the density of their union is at least as large as the sum of their individual densities.

rS will denote the set obtained by multiplying each element of S by r.

Lemma 2 *For each positive integer m, $d(m^{-1}S_2) \geq 1/4$.*

Proof of Lemma 2: Meyer and Tenenbaum, [4]. ∎

A modification of our argument will show that any explicit uniform lower bound in Lemma 2 leads to a version of Theorem 1 and an explicit value for k in (1).

Lemma 3 *If $i \geq 4$ and the integer m belongs to $S_i - S_{i-1}$, then $d(S_{i+2}) \geq d(S_{i-3}) + d(m^{-1}S_2)$.*

Proof of Lemma 3: The sets S_{i-3} and $m^{-1}S_2$ are disjoint, and contained in S_{i+2}. ∎

Lemma 4 *If $t \geq 3$ and $S_t \neq S_{t-1}$, then each of $S_i - S_{i-1}$, $2 \leq i \leq [\frac{1}{2}(t+1)]$ contains an integer.*

Proof of Lemma 4: Let $(p_1 + 1) \cdots (p_u + 1)\{(q_1 + 1) \cdots (q_v + 1)\}^{-1}$ belong to $S_t - S_{t-1}$. Without loss of generality $u \geq v$. Since $u + v = t$ and t is an integer, $u \geq [\frac{1}{2}(t+1)]$. Then the product of the $(p_j + 1)$, $1 \leq j \leq i$, belongs to $S_i - S_{i-1}$, $2 \leq i \leq [\frac{1}{2}(t+1)]$. ∎

Lemma 5 *For any positive rational* γ, $d(\gamma S_2) > 0$.

Proof of Lemma 5: Let $\gamma = ab^{-1}$ with a, b mutually prime integers. The number of integers in γS_2 that do not exceed a given x is at least as large as the number of integers in $b^{-1} S_2$ that do not exceed x/a, and we may appeal to Lemma 2. ■

Lemma 6 *If* $k \geq 4$ *and* $S_k \neq S_{k-1}$, *then* $d(S_{k+2}) > d(S_{k-3})$.

Proof of Lemma 6: Following the proof of Lemma 3, $d(S_{k+2}) \geq d(S_{k-3}) + d(\gamma S_2)$, and we apply Lemma 5. ■

Proof of Theorem 1: Suppose that $S_{19} \neq S_{20}$. Then Lemma 4 shows that each of $S_i - S_{i-1}$, $2 \leq i \leq 10$, contains an integer. Let y be an integer in $S_5 - S_4$, z an integer in $S_{10} - S_9$. The sets S_2, $y^{-1} S_2$, $z^{-1} S_2$, $(yz)^{-1} S_2$ are mutually disjoint. By Lemma 2, $d(S_{17}) \geq 1$. Bearing in mind our hypothesis, Lemma 6 now gives $d(S_{22}) > d(S_{17}) \geq 1$, which is impossible. Hence $S_{19} = S_{20}$ and, from Lemma 1, Γ is S_{19}.

This argument shows each positive rational r to have a representation (1) with k at most 19. If we require the shifted primes $p_i + 1$ in the representation to exceed a given bound, then both the assertion and its proof remain valid. To gain a representation (1) with k exactly 19 we appeal to the properties of particular primes.

If s belongs to Γ, then $s/4$ also belongs to Γ, hence to S_{19}. Since $s = 4(s/4)$, we may fill to a product representation of s with exactly 20 terms by using the representations $1 = (2+1)(2+1)^{-1}$, $4 = 3 + 1 = (11+1)(2+1)^{-1}$.

For each $i \geq 1$ let $\bar{S}_i$ denote the set of group products in Γ employing exactly i shifted primes, again counted multiply. With the exception of Lemma 1 our various propositions remain valid if every S_i is replaced by $\bar{S}_i$. Note that simple representations of 1 ensure $\bar{S}_i$ to be a subset of $\bar{S}_j$ for every $j \geq i + 2$. We thus arrive at the assertion $\bar{S}_{19} = \bar{S}_{20}$ and Theorem 1 is established. ■

3. Towards Theorems 2 and 3

For the proofs of Theorems 2 and 3 we use a little more.

We say that a set of rationals, S, has asymptotic density α if

$$\alpha = \lim_{x \to \infty} x^{-1} \sum_{\substack{n \leq x, \\ n \in S}} 1$$

exists, again counting integers. In particular, $d(S)$ will then have the value α.

Lemma 7 *Each coset of* Γ *has asymptotic density* $|G|^{-1}$.

Proof: Consider the homomorphism $g : Q^* \to Q^*/\Gamma = G$. By Lemma 9.5 of Ruzsa, [5], those integers mapped onto any particular coset of Γ have an asymptotic density. Moreover, the various densities will all be $|G|^{-1}$ if G has a set of generators γ for which the series $\sum q^{-1}$, $g(q) = \gamma$, taken over primes q, diverge.

In the present situation G is cyclic. The condition can fail to be satisfied only if the sum $\sum q^{-1}$ taken over all primes not in Γ, converges. Then each positive integer D has infinitely many representations of the form $p + 1 = 2Ds$, with p prime and s not divisible by any q. Indeed, a stronger result, employing a basic version of Dirichlet's theorem on primes in arithmetic progression together with the sieve method of Selberg, may be found on pp. 17–19 of the proof of Theorem 2 in [1]. The rôle of the q_i there is here played by the q.

In particular, every even integer $2D$ belongs to Γ. Then every integer belongs to Γ, $|G| = 1$ and Lemma 7 is clearly valid.

For his proof Ruzsa employs a theorem of Wirsing, [6]. Note the remarks made in Elliott, [2], p. 393.   ∎

Proof of Theorem 2: If $S_9 \neq S_{10}$, then Lemma 4 shows each of $S_i - S_{i-1}$, $2 \leq i \leq 5$, to contain an integer. Let m be an integer in $S_5 - S_4$. By Lemma 3 with Lemma 2, $d(S_7) \geq d(S_2) + d(m^{-1}S_2) \geq 1/2$. However, Lemma 6 then gives $d(S_{12}) > d(S_7) \geq 1/2$, which contradicts Lemma 7. Therefore $S_9 = S_{10}$ and, by appeal to Lemma 1, Γ is S_9. We continue as for the proof of Theorem 1.   ∎

Proof of Theorem 3: Let γ belong to $S_6 - S_5$. Dirichlet's theorem guarantees a prime p so that $(p + 1)\gamma$ is an integer m, then in $S_7 - S_4$. One of $S_t - S_{t-1}$, $5 \leq t \leq 7$, contains m. Again applying Lemma 3 with Lemma 2: $d(S_{t+2}) \geq d(S_{t-3}) + d(m^{-1}S_2) \geq 1/2$, which contradicts Lemma 7. Therefore $S_5 = S_6$, and by appeal to Lemma 1, $\Gamma = S_5$. We continue as for the proof of Theorem 1.   ∎

4.   Concluding remarks

(i) Suppose only that the density $d(m^{-1}S_2)$ in Lemma 2 uniformly exceeds $c > 0$. With no regard for efficiency we may argue that if $S_{10k-1} \neq S_{10k}$, then $d(S_{5k+2}) \geq (k+1)c$. For $h = [c^{-1}]$ this is impossible. Hence Γ is S_{10k-1}, with $10k - 1 \leq 10c^{-1} - 1$.

(ii) Lemma 7 is of interest for itself. However, it was pointed out to us by Juan Carlos Peral that variants of the argument for Theorem 1 enable appeal to Lemma 7 to be here avoided. Indeed, in the notation of the proof of Theorem 2, once $d(S_{12}) > 1/2$ is assured, the union of disjoint sets S_{12}, $a^{-1}S_2$, $(ma)^{-1}S_2$, where a is any integer not in Γ, has density greater than 1.

Likewise in the proof of Theorem 3, arrived at an integer m in $S_7 - S_4$, we may consider the disjoint sets S_2, $m^{-1}S_2$, $(am)^{-1}S_2$, $a^{-1}S_2$, $a^{-2}S_2$ for some integer a not in Γ. Here we (again) use the fact that a group of order 3 is cyclic.

Acknowledgments

Second author partially supported by NSF Contract DMS–9530690.

References

1. P.D.T.A. Elliott. "A conjecture of Kátai," *Acta Arith.* **26** (1974), 11–20.
2. P.D.T.A. Elliott. *Arithmetic functions and integer products*, Grund. der math. Wiss **272**, Springer–Verlag, New York, Berlin, Heidelberg, Tokyo, 1985.
3. P.D.T.A. Elliott. "The multiplicative group of rationals generated by the shifted primes," *J. reine angew. Math.* **463** (1995), 169–216.
4. J. Meyer, G. Tenenbaum. "Une remarque sur la conjecture de Schinzel," *Bull. Sc. Math.* 2e série **108** (1984), 437–444.
5. I.Z. Ruzsa. "General multiplicative functions," *Acta Arith.* **32** (1977), 313–247.
6. E. Wirsing. "Das asymptotiche Verhalten von Summen über multiplikativer Funktionen, II." *Acta Math. Acad. Sci. Hung.* **18** (1967), 411–467.

THE RAMANUJAN JOURNAL 2, 225–245 (1998)
© 1998 Kluwer Academic Publishers.

On Large Values of the Divisor Function

P. ERDŐS

J.-L. NICOLAS   jlnicola@in2p3.fr
Institut Girard Desargues, UPRES-A-5028, Mathématiques, Bat. 101, Université Claude Bernard (LYON 1), F-69622 Villeurbanne cédex, France

A. SÁRKÖZY*   sarkozy@cs.elte.hu
Eötvös Loránd University, Dept. of Algebra and Number Theory, H-1088 Budapest, Múzeum krt. 6-8, Hungary

Jean-Louis Nicolas and Andràs Sárközy dedicate this paper to the memory of Paul Erdős

Received June 10, 1997; Accepted January 7, 1998

Abstract. Let $d(n)$ denote the divisor function, and let $D(X)$ denote the maximal value of $d(n)$ for $n \leq X$. For $0 < z \leq 1$, both lower and upper bounds are given for the number of integers n with $n \leq X$, $zD(X) \leq d(n)$.

Key words: division function, highly composite numbers, maximal order

1991 Mathematics Subject Classification: Primary 11N56

1. Introduction

Throughout this paper, we shall use the following notations: $\mathbf{N}$ denotes the set of the positive integers, $\pi(x)$ denotes the number of the prime numbers not exceeding x, and p_i denotes the ith prime number. The number of the positive divisors of $n \in \mathbf{N}$ is denoted by $d(n)$, and we write

$$D(X) = \max_{n \leq X} d(n).$$

Following Ramanujan we say that a number $n \in \mathbf{N}$ is highly composite, briefly h.c., if $d(m) < d(n)$ for all $m \in \mathbf{N}$, $m < n$. For information about h.c. numbers, see [13, 15] and the survey paper [11].

The sequence of h.c. numbers will be denoted by $n_1, n_2, \ldots : n_1 = 1, n_2 = 2, n_3 = 4$, $n_4 = 6, n_5 = 12, \ldots$ (for a table of h.c. numbers, see [13, Section 7, or 17]. For $X > 1$, let $n_k = n_{k(X)}$ denote the greatest h.c. number not exceeding X, so that

$$D(X) = d\big(n_{k(X)}\big).$$

*Research partially supported by Hungarian National Foundation for Scientific Research, Grant No. T017433 and by C.N.R.S, Institut Girard Desargues, UPRES-A-5028.

It is known (cf. [13, 8]) that n_k is of the form $n_k = p_1^{r_1} p_2^{r_2} \cdots p_\ell^{r_\ell}$, where $r_1 \geq r_2 \geq \cdots \geq r_\ell$,

$$\ell = (1 + o(1)) \frac{\log X}{\log \log X}, \tag{1}$$

$$r_i = (1 + o(1)) \frac{\log p_\ell}{\log 2 \log p_i} \quad \left(\text{for } X \to \infty \text{ and } \frac{\log p_i}{\log p_\ell} \to 0 \right) \tag{2}$$

and, if m is the greatest integer such that $r_m \geq 2$,

$$p_m = p_\ell^\theta + O\left(p_\ell^{\tau_0 \theta}\right) \tag{3}$$

where

$$\theta = \frac{\log(3/2)}{\log 2} = 0.585\ldots \tag{4}$$

and τ_0 is a constant <1 which will be given later in (8).

For $0 < z \leq 1$, $X > 1$, let $S(X, z)$ denote the set of the integers n with $n \leq X$, $d(n) \geq zD(X)$. In this paper, our goal is to study the function $F(X, z) = \mathrm{Card}(S(X, z))$.

In Section 4, we will study $F(X, 1)$, further we will prove (Corollary 1) that for some $c > 0$ and infinitely many X's with $X \to +\infty$, we have $F(X, z) = 1$ for all z and X satisfying

$$1 - \frac{1}{(\log X)^c} < z \leq 1.$$

Thus, to have a non trivial lower bound for $F(X, z)$ for *all* X, one needs an assumption of the type $z < 1 - f(X)$, cf. (6).

In Section 2, we shall give lower bounds for $F(X, z)$. Under a strong, but classical, assumption on the distribution of primes, the lower bound given in Theorem 1 is similar to the upper bound given in Section 3. The proofs of the lower bounds will be given in Section 5: in the first step we construct an integer $\hat{n} \in S(X, z)$ such that $d(\hat{n})$ is as close to $zD(X)$ as possible. This will be done by using diophantine approximation of θ (defined by (4)), following the ideas of [2, 8]. Further, we observe that slightly changing large prime factors of $\hat{n}$ will yield many numbers n not much greater than $\hat{n}$, and so belonging to $S(X, z)$. The proof of the upper bound will be given in Section 7. It will use the superior h.c. numbers, introduced by Ramanujan (cf. [13]). Such a number N_ε maximizes $d(n)/n^\varepsilon$. The problem of finding h.c. numbers is in fact an optimization problem

$$\max_{n \leq x} d(n)$$

and, in this optimization problem, the parameter ε plays the role of a Lagrange multiplier. The properties of the superior h.c. numbers that we shall need will be given in Section 6.

In [10, p. 411], it was asked whether there exists a positive constant c such that, for n_j large enough,

$$\frac{d(n_{j+1})}{d(n_j)} \leq 1 + \frac{1}{(\log n_j)^c}.$$

In Section 8, we shall answer this question positively, while in Section 4 we shall prove that for infinitely many n_j, one has $d(n_{j+1})/d(n_j) \geq 1 + (\log n_j)^{-0.71}$.

We are pleased to thank J. Rivat for communicating us reference [1].

2. Lower bounds

We will show that

Theorem 1. *Assume that τ is a positive number less than 1 and such that*

$$\pi(x) - \pi(x - y) > A\frac{y}{\log x} \quad for \; x^\tau < y < x \tag{5}$$

for some $A > 0$ and x large enough. Then for all $\varepsilon > 0$, there is a number $X_0 = X_0(\varepsilon)$ such that, if $X > X_0(\varepsilon)$ and

$$\exp(-(\log X)^\lambda) < z < 1 - \log X)^{-\lambda_1} \tag{6}$$

where λ is any fixed positive real number <1 and λ_1 a positive real number ≤ 0.03, then we have:

$$F(X, z) > \exp\left((1 - \varepsilon)\min\{2(A\log 2\log X\log(1/z))^{1/2}, 2(\log X)^{1-\tau}\log\log X\log(1/z)\}\right). \tag{7}$$

Note that (5) is known to be true with

$$\tau = \tau_0 = 0.535 \quad and \quad A = 1/20 \tag{8}$$

(cf. [1]) so that we have

$$F(X, z) > \exp((1 - \varepsilon)2(\log X)^{0.465}\log\log X\log(1/z))$$

for all z satisfying (6), and assuming the Riemann hypothesis, (5) holds for all $\tau > 1/2$ so that

$$F(X, z) > \exp((\log X)^{1/2-\varepsilon}\log(1/z))$$

for all $\varepsilon > 0$, X large enough and z satisfying (6). Moreover, if (5) holds with some $\tau < 1/2$ and $A > 1 - \varepsilon/2$ (as it is very probable), then for a fixed z we have

$$F(X, z) > \exp((2 - \varepsilon)((\log 2)(\log X)\log(1/z))^{1/2}). \tag{9}$$

In particular,

$$F(X, 1/2) > \exp((1 - \varepsilon)(\log 2)(\log X)^{1/2}). \tag{10}$$

While we need a very strong hypothesis to prove (9) for all X, we will show without any unproved hypothesis that, for fixed z and with another constant in the exponent, it holds for infinitely many $X \in \mathbf{N}$:

Theorem 2. *If z is a fixed real number with $0 < z < 1$, and $\varepsilon > 0$, then for infinitely many $X \in \mathbf{N}$ we have*

$$F(X, z) > \exp((1 - \varepsilon)(\log 4 \log X \log(1/z)^{1/2}) \tag{11}$$

so that, in particular

$$F(X, 1/2) > \exp((1 - \varepsilon)\sqrt{2} \, \log 2(\log X)^{1/2}). \tag{12}$$

We remark that the constant factor $\sqrt{2} \log 2$ on the right hand side could be improved by the method used in [12] but here we will not work out the details of this. It would also be possible to extend Theorem 2 to all z depending on X and satisfying (6).

3. Upper bounds

We will show that:

Theorem 3. *There exists a positive real number γ such that, for $z \geq 1 - (\log X)^{-\gamma}$, as $X \to +\infty$ we have*

$$\log F(X, z) = O\big((\log X)^{(1-\gamma)/2}\big), \tag{13}$$

and if λ, η are two real numbers, $0 < \lambda < 1, 0 < \eta < \gamma$, we have for

$$1 - (\log X)^{-\gamma+\eta} \geq z \geq \exp(-(\log X)^{\lambda}), \tag{14}$$

and X large enough:

$$F(X, z) \leq \exp\left(\frac{24}{\sqrt{1 - \gamma}}(\log(1/z) \log X)^{1/2}\right). \tag{15}$$

The constant γ will be defined in Lemma 5 below. One may take $\gamma = 0.03$. Then for $z = 1/2$, (15) yields

$$\log F(X, 1/2) \leq 21(\sqrt{\log X})$$

which, together with the results of Section 2, shows that the right order of magnitude of $\log F(X, 1/2)$ is, probably, $\sqrt{\log X}$.

4. The cases $z = 1$ and z close to 1

Let us first define an integer n to be largely composite (l.c.) if $m \leq n \Rightarrow d(m) \leq d(n)$. S. Ramanujan has built a table of l.c. numbers (see [14, p. 280 and 15, p. 150]). The distribution of l.c. numbers has been studied in [9], where one can find the following results:

Proposition 1. *Let $Q_\ell(X)$ be the number of l.c. numbers up to X. There exist two real numbers $0.2 < b_1 < b_2 < 0.5$ such that for X large enough the following inequality holds:*

$$\exp((\log X)^{b_1}) \leq Q_\ell(X) \leq \exp((\log X)^{b_2}).$$

We may take any number $< (1 - \frac{\log 3/2}{\log 2})/2 = 0.20752$ for b_1, and any number $> (1 - \gamma)/2$ with $\gamma > 0.03$ defined in Lemma 5, for b_2.

From Proposition 1, it is easy to deduce:

Theorem 4. *There exists a constant $b_2 < 0.485$ such that for all X large enough we have*

$$F(X, 1) \leq \exp((\log X)^{b_2}). \tag{16}$$

There exists a constant $b_1 > 0.2$ such that, for a sequence of X tending to infinity, we have

$$F(X, 1) \geq \exp((\log X)^{b_1}). \tag{17}$$

Proof: $F(X, 1)$ is exactly the number of l.c. numbers n such that $n_k \leq n \leq X$. Thus $F(X, 1) \leq Q_\ell(X)$ and (16) follows from Proposition 1.

The proof of Proposition 1 in [9, Section 3] shows that for any $b_1 < 0.207$, there exists an infinite number of h.c. numbers n_j such that the number of l.c. numbers between n_{j-1} and n_j (which is exactly $F(n_j - 1, 1)$) satisfies $F(n_j - 1, 1) \geq \exp((\log n_j)^{b_1})$ for n_j large enough, which proves (17). $\square$

We shall now prove:

Theorem 5. *Let (n_j) be the sequence of h.c. numbers. There exists a positive real number a, such that for infinitely many n_j's, the following inequality holds:*

$$\frac{d(n_j)}{d(n_{j-1})} \geq 1 + \frac{1}{(\log n_j)^a}. \tag{18}$$

One may take any $a > 0.71$ in (18).

Proof: Let X tend to infinity, and define $k = k(X)$ by $n_k \leq X < n_{k+1}$. By [8], the number $k(X)$ of h.c. numbers up to X satisfies

$$k(X) \leq (\log X)^\mu \tag{19}$$

for X large enough, and one may choose for μ the value $\mu = 1.71$, cf. [10, p. 411 or 11, p. 224]. From (19), the proof of Theorem 5 follows by an averaging process: one has

$$\prod_{\sqrt{X} < n_j \leq X} \frac{d(n_j)}{d(n_{j-1})} = \frac{D(X)}{D(\sqrt{X})}.$$

The number of factors in the above product is $k(X) - k(\sqrt{X}) \leq k(X)$ so that there exists $j, k(\sqrt{X}) + 1 \leq j \leq k(X)$, with

$$\frac{d(n_j)}{d(n_{j-1})} \geq \left(\frac{D(X)}{D(\sqrt{X})} \right)^{1/k(X)}. \tag{20}$$

But it is well known that $\log D(X) \sim \frac{(\log 2)(\log X)}{\log \log X}$, and thus

$$\log(D(X)/D(\sqrt{X})) \sim \frac{\log 2}{2} \frac{\log X}{\log \log X}.$$

Observing that $X < n_j^2$, it follows from (19) and (20) for X large enough:

$$\frac{d(n_j)}{d(n_{j-1})} \geq \exp\left(\frac{1}{3} \frac{1}{(\log X)^{\mu-1} \log \log X} \right)$$

$$\geq \exp\left(\frac{1}{3} \frac{1}{(2 \log n_j)^{\mu-1} \log(2 \log n_j)} \right)$$

$$\geq \exp\left(\frac{1}{(\log n_j)^a} \right) \geq 1 + \frac{1}{(\log n_j)^a}$$

for any $a > \mu - 1$, which completes the proof of Theorem 5.   □

A completely different proof can be obtained by choosing a superior h.c. number for n_j and following the proof of Theorem 8 in [7, p. 174], which yields $a = \frac{\log(3/2)}{\log 2} = 0.585\ldots$ See also [10, Proposition 4].

Corollary 1. *For $c > 0.71$, there exists a sequence of values of X tending to infinity such that $F(X, z) = 1$ for all z, $1 - 1/(\log X)^c < z \leq 1$.*

Proof: Let us choose $X = n_j$, with n_j satisfying (18), and $c > a$. For all $n < X$, we have

$$d(n) \leq d(n_{j-1}) \leq \frac{d(n_j)}{1 + (\log n_j)^{-a}} = \frac{D(X)}{1 + (\log X)^{-a}} < zD(X).$$

Thus $S(X, z) = \{n_j\}$, and $F(X, z) = 1$.   □

5. Proofs of the lower estimates

Proof of Theorem 1: Let us denote by α_i/β_i the convergents of θ, defined by (4). It is known that θ cannot be too well approximated by rational numbers and, more precisely, there exists a constant κ such that

$$|q\theta - p| \gg q^{-\kappa} \tag{21}$$

for all integers $p, q \neq 0$ (cf. [4]). The best value of κ

$$\kappa = 7.616 \tag{22}$$

is due to G. Rhin (cf. [16]). It follows from (21) that

$$\beta_{i+1} = O\left(\beta_i^{\kappa}\right). \tag{23}$$

Let us introduce a positive real number δ which will be fixed later, and define $j = j(X, \delta)$ so that

$$\beta_j \leq (\log X)^{\delta} < \beta_{j+1}. \tag{24}$$

By Kronecker's theorem (cf. [6], Theorem 440), there exist two integers α and β such that

$$\left| \beta\theta - \alpha - \frac{\log z}{\log 2} - \frac{2}{\beta_j} \right| < \frac{2}{\beta_j} \tag{25}$$

and

$$\frac{\beta_j}{2} \leq \beta \leq \frac{3\beta_j}{2}. \tag{26}$$

Indeed, as α_j and β_j are coprime, one can write B, the nearest integer to $(\beta_j \frac{\log z}{\log 2} + 2)$, as $B = u_1\alpha_j - u_2\beta_j$ with $|u_1| \leq \beta_j/2$, and then $\alpha = \alpha_j + u_2$ and $\beta = \beta_j + u_1$ satisfy (25). With the notation of Section 1, we write

$$\hat{n} = n_k \frac{p_{m+1} p_{m+2} \cdots p_{m+\beta}}{p_\ell p_{\ell-1} \cdots p_{\ell-\alpha+1}} \tag{27}$$

for X large enough. By (26), (24), and (6), (25) yields

$$\alpha \leq \beta\theta + \frac{\log(1/z)}{\log 2} \ll \max((\log X)^{\delta}, (\log X)^{\lambda}) \tag{28}$$

and

$$\alpha \geq \beta\theta - \frac{\log z}{\log 2} - \frac{4}{\beta_j} > \beta\theta - \frac{6}{\beta} + \frac{\log(1/z)}{\log 2} > 0$$

for X large enough. Thus, if we choose $\delta < 1$, from (3) and (1) we have $r_\ell = r_{\ell-1} = \cdots = r_{\ell-\alpha+1} = 1$. By (1) and the prime number theorem, we also have

$$p_\ell \sim \log X \tag{29}$$

and by (3), we have $r_{m+1} = r_{m+2} = \cdots = r_{m+\beta} = 1$ so that, by (25),

$$d(\hat{n}) = d(n_k)\frac{(3/2)^\beta}{2^\alpha} = d(n_k)\exp(\log 2(\beta\theta - \alpha)) \geq zd(n_k) = zD(X). \tag{30}$$

Now we need an upper bound for $\hat{n}/n_k$. First, it follows from (5) that for $i = o(m)$ we have

$$p_{m+i} - p_m \leq \max\left(p_{m+i}^\tau, \frac{i}{A}\log p_{m+i}\right) \tag{31}$$

and consequently,

$$\prod_{i=1}^\beta \frac{p_{m+i}}{p_m} = \exp\left(\sum_{i=1}^\beta \log\frac{p_{m+i}}{p_m}\right) \leq \exp\left(\sum_{i=1}^\beta \frac{p_{m+i} - p_m}{p_m}\right)$$

$$\leq \exp\left(\frac{\beta}{p_m}\max\left(p_{m+\beta}^\tau, \frac{\beta}{A}\log p_{m+\beta}\right)\right)$$

$$\leq \exp(O(\max((\log X)^{\delta+\theta(\tau-1)}, (\log X)^{2\delta-\theta}\log\log X))) \tag{32}$$

by (26), (24), (3) and (1). Similarly, we get

$$\prod_{i=0}^{\alpha-1} \frac{p_\ell}{p_{\ell-i}} \leq \exp\left(\frac{\alpha}{p_{\ell-\alpha+1}}\max\left(p_\ell^\tau, \frac{\alpha}{A}\log p_\ell\right)\right)$$

$$\leq \exp\left(O\left(\max\left(\frac{(\log X)^\delta - \log z}{(\log X)^{1-\tau}}, \frac{((\log X)^\delta - \log z)^2}{\log X}\log\log X\right)\right)\right) \tag{33}$$

by (28). Further, it follows from (3) and (25) that

$$\frac{p_m^\beta}{p_\ell^\alpha} = p_\ell^{\beta\theta-\alpha}\left(1 + O(p_\ell^{(\tau-1)\theta})\right)^\beta \leq p_\ell^{\frac{\log z}{\log 2} + \frac{4}{\beta_j}}\exp(O(\beta p_\ell^{(\tau-1)\theta}))$$

$$\leq \exp\left\{\left(\frac{\log z}{\log 2}\log p_\ell\right) + \frac{4\log p_\ell}{\beta_j} + \frac{\beta}{p_\ell^{(1-\tau)\theta}}\right\}. \tag{34}$$

It follows from (23) and (24) that

$$\beta_j \gg (\log X)^{\delta/\kappa}. \tag{35}$$

Multiplying (32), (33) and (34), we get from (27) and (29):

$$\hat{n}/n_k \leq \exp\left\{(1 + o(1))\frac{\log z\log\log X}{\log 2}\right\} \tag{36}$$

if we choose δ in such a way that the error terms in (32), (33) and (34) can be neglected. More precisely, from (6) and (36), δ should satisfy:

$$\delta + \theta(\tau - 1) < -\lambda_1$$
$$2\delta - \theta < -\lambda_1$$
$$\kappa\lambda_1 < \delta < 1.$$

It is possible to find such a δ if λ_1 satisfies

$$\lambda_1 < \min\left(\frac{(1-\tau)\theta}{1+\kappa}, \frac{\theta}{1+2\kappa}\right).$$

(4), (8) and (22) yield $\lambda_1 < 0.03157$.
    For convenience, let us write

$$\hat{n} = p_1^{\hat{r}_1} p_2^{\hat{r}_2} \cdots p_t^{\hat{r}_t} \tag{37}$$

with, by (27), $t = \ell - \alpha$. It follows from (1) and (28) that

$$t = (1 + o(1))\frac{\log X}{\log \log X}; \quad p_t \sim \log X \tag{38}$$

and from (24) and (26) that

$$\hat{r}_i = 1 \quad \text{for } i \geq t - t^{9/10}. \tag{39}$$

Now, consider the integers v satisfying

$$P(t, v) \overset{\text{def}}{=} \frac{p_{t+1}p_{t+2}\cdots p_{t+v}}{p_{t-v+1}p_{t-v+2}\cdots p_t} \leq \exp\left((1-\varepsilon)\frac{\log(1/z)\log X}{\log 2}\right) \tag{40}$$

and

$$v \leq t^{9/10}. \tag{41}$$

By a calculation similar to that of (32) and (33), by (5) and the prime number theorem, for all v satisfying (41) and for all $1 \leq i \leq v$ we have:

$$\frac{p_{t+i}}{p_{t-v+i}} = 1 + \frac{p_{t+i} - p_{t-v+i}}{p_{t-v+i}} \leq 1 + (1+o(1))\frac{1}{p_t}\max\left(p_{t+v}^\tau, \frac{v}{A}\log p_{t+v}\right)$$
$$= 1 + (1+o(1))\frac{1}{t}\max\left(t^\tau(\log t)^{\tau-1}, \frac{v}{A}\right)$$

so that, by (38), the left hand side of (40) is

$$P(t, v) = \prod_{i=1}^{v} \frac{p_{t+i}}{p_{t-v+i}}$$

$$\leq \exp\left(v(1 + o(1)) \frac{1}{t} \max\left(t^{\tau} (\log t)^{\tau-1}, \frac{v}{A} \right) \right)$$

$$= \exp\left((1 + o(1)) v \frac{\log \log X}{\log X} \max\left(\frac{(\log X)^{\tau}}{\log \log X}, \frac{v}{A} \right) \right)$$

$$= \exp\left((1 + o(1)) v \max\left((\log X)^{\tau-1}, \frac{v}{A} \frac{\log \log X}{\log X} \right) \right). \tag{42}$$

By (42), (40) follows from

$$\exp\left((1 + o(1)) v \max\left((\log X)^{\tau-1}, \frac{v}{A} \frac{\log \log X}{\log X} \right) \right) < \exp\left(\left(1 - \frac{\varepsilon}{2} \right) \frac{\log(1/z) \log X}{\log 2} \right). \tag{43}$$

An easy computation shows that with

$$\left(1 - \frac{5\varepsilon}{6} \right) \min\left(\left(\frac{A \log X}{\log 2} \log(1/z)^{1/2} \right), (\log X)^{1-\tau} \frac{\log \log X}{\log 2} \log(1/z) \right)$$

in place of v both (41) and (43) hold. Thus fixing v now as the greatest integer v satisfying (41) and (43), we have

$$v > \left(1 - \frac{3\varepsilon}{4} \right) \min\left(\left(\frac{A \log X}{\log 2} \log(1/z)^{1/2} \right), (\log X)^{1-\tau} \frac{\log \log X}{\log 2} \log(1/z) \right). \tag{44}$$

Then it follows from (39) and (41) that

$$\hat{r}_{t-v+i} = 1 \quad \text{for } i = 1, 2, \ldots, v. \tag{45}$$

Let now $\mathcal{A}$ denote the set of the integers a of the form

$$a = 2^{\hat{r}_1} p_2^{\hat{r}_2} \cdots p_{t-v}^{\hat{r}_{t-v}} p_{i_1} \cdots p_{i_v} \quad \text{where } t - v + 1 \leq i_1 < i_2 < \cdots < i_v \leq t + v. \tag{46}$$

Then, by (37), (46) and (30) we have

$$d(a) = d(\hat{n}) \geq zD(X). \tag{47}$$

Moreover, by (40) and (36) such an a satisfies

$$a = \frac{p_{i_1} p_{i_2} \cdots p_{i_v}}{p_{t-v+1} p_{t-v+2} \cdots p_t} \hat{n} \leq P(t, v)\hat{n} \leq n_k. \tag{48}$$

It follows from (47) and (48) that $a \in S(X, z)$ and

$$F(X, z) \geq |\mathcal{A}|. \tag{49}$$

The numbers $i_1, i_2, \ldots, i_v$ in (46) can be chosen in $\binom{2v}{v}$ ways so that

$$|\mathcal{A}| = \binom{2v}{v} > \exp\left(\left(1 - \frac{\varepsilon}{8} \right) (\log 4) v \right). \tag{50}$$

Now (7) follows from (44), (49) and (50), and this completes the proof of Theorem 1. □

Proof of Theorem 2: By a theorem of Selberg [19, 9], if the real function $f(x)$ is increasing, $f(x) > x^{1/6}$ and $\frac{f(x)}{x} \searrow 0$, then there are infinitely many integers y such that

$$\pi(y + f(y)) - \pi(y) \sim \frac{f(y)}{\log y} \quad \text{and} \quad \pi(y) - \pi(y - f(y)) \sim \frac{f(y)}{\log y}. \tag{51}$$

We use this result with $f(y) = (1 - \frac{\varepsilon}{3}) \log y (\frac{y \log(1/z)}{\log 4})^{1/2}$ and for a y value satisfying (51), define t by

$$p_t \le y < p_{t+1}. \tag{52}$$

Further, we define β_j (instead of (24)) so that $\beta_j \ge \frac{4 \log 2}{\varepsilon \log(1/z)}$ and α, β by (25) and (26); we set $\ell = t + \alpha$ and choose $X = n_k$ a h.c. number whose greatest prime factor is p_ℓ (such a number exists, see [13] or (59), (60) below). We define $\hat{n}$ by (27), and (30) and (38) still hold, while (36) becomes

$$\frac{\hat{n}}{n_k} \le \exp\left((1 + o(1)) \log \log X \left(\frac{\log z}{\log 2} + \frac{4}{\beta_j} \right) \right)$$

$$\le \exp\left((1 + o(1)) \frac{\log \log X}{\log 2} \log z (1 - \varepsilon) \right)$$

$$\le \exp\left(\frac{\log \log X}{\log 2} \log z \left(1 - \frac{\varepsilon}{2} \right) \right) \tag{53}$$

for X large enough. Let v denote the greatest integer with

$$p_{t+v} \le y + f(y) \quad \text{and} \quad p_{t-v} \ge y - f(y), \tag{54}$$

so that by the definition of y we have

$$v \sim \frac{f(y)}{\log y}. \tag{55}$$

By (38) and (52), we have

$$y \sim \log x. \tag{56}$$

Moreover, by (38), (54) and (55), we have

$$P(t, v) \stackrel{\text{def}}{=} \prod_{i=1}^{v} \frac{p_{t+i}}{p_{t-v+i}} \le \left(\frac{y + f(y)}{y - f(y)} \right)^v$$

$$\le \exp\left((1 + o(1)) \frac{f(y)}{\log \log X} \log\left(1 + 2 \frac{f(y)}{y} \right) \right)$$

$$= \exp\left((2 + o(1)) \frac{f^2(y)}{y \log \log X} \right) = \left(\frac{1}{\log 2} + o(1) \right) \left(1 - \frac{\varepsilon}{3} \right)^2 \log \log X \log(1/z). \tag{57}$$

It follows from (53) and (57) that $P(t, v) < n_k/\hat{n}$ for X large enough and ε small enough.

Again, as in the proof of Theorem 1, we consider the set $\mathcal{A}$ of the integers a of the form (48). Then as in the proof of Theorem 1, by using (38) and (55) finally we obtain

$$F(X, z) \geq |\mathcal{A}| = \binom{2v}{v} > \exp\left(\left(1 - \frac{\varepsilon}{3}\right)(\log 4)v\right)$$

$$> \exp((1 - \varepsilon)(\log 4)^{1/2}(\log X)^{1/2}(\log(1/z))^{1/2})$$

which completes the proof of Theorem 2.   □

6. Superior highly composite numbers and benefits

Following Ramanujan (cf. [13]) we shall say that an integer N is superior highly composite (s.h.c.) if there exists $\varepsilon > 0$ such that for all positive integer M the following inequality holds:

$$d(M)/M^\varepsilon \leq d(N)/N^\varepsilon. \tag{58}$$

Let us recall the properties of s.h.c. numbers (cf. [13], [7, p. 174], [8–11]). To any $\varepsilon, 0 < \varepsilon < 1$, one can associate the s.h.c. number:

$$N_\varepsilon = \prod_{p \leq x} p^{\alpha_p} \tag{59}$$

where

$$x = 2^{1/\varepsilon}, \quad \varepsilon = (\log 2)/\log x \tag{60}$$

and

$$\alpha_p = \left\lfloor \frac{1}{p^\varepsilon - 1} \right\rfloor. \tag{61}$$

For $i \geq 1$, we write

$$x_i = x^{\log(1+1/i)/\log 2} \tag{62}$$

and then (61) yields:

$$\alpha_p = i \iff x_{i+1} < p \leq x_i. \tag{63}$$

A s.h.c. number is h.c. thus from (1) we deduce:

$$x \sim \log N_\varepsilon. \tag{64}$$

Let $P > x$ be the smallest prime greater than x. There is a s.h.c. number N' such that $N' \leq NP$ and $d(N') \leq 2d(N)$.

Definition. Let $\varepsilon, 0 < \varepsilon < 1$, and N_ε satisfy (58). For a positive integer M, let us define the benefit of M by

$$\text{ben } M = \varepsilon \log \frac{M}{N_\varepsilon} - \log \frac{d(M)}{d(N_\varepsilon)}. \tag{65}$$

From (58), we have ben $M \geq 0$. Note that ben N depends on ε, but not on N_ε: If $N^{(1)}$ and $N^{(2)}$ satisfy (58), (65) will give the same value for ben M if we set $N_\varepsilon = N^{(1)}$ or $N_\varepsilon = N^{(2)}$.

Now, let us write a generic integer:

$$M = \prod_p p^{\beta_p},$$

for $p > x$, let us set $\alpha_p = 0$, and define:

$$\text{ben}_p(M) = \varepsilon(\beta_p - \alpha_p) \log p - \log\left(\frac{\beta_p + 1}{\alpha_p + 1}\right). \tag{66}$$

From the definition (61) of α_p, we have $\text{ben}_p(M) \geq 0$, and (65) can be written as

$$\text{ben } M = \sum_p \text{ben}_p(M). \tag{67}$$

If $\beta_p = \alpha_p$, we have $\text{ben}_p(M) = 0$. If $\beta_p > \alpha_p$, let us set

$$\varphi_1 = \varphi_1(\varepsilon, p, \alpha_p, \beta_p) = (\beta_p - \alpha_p)\left(\varepsilon \log p - \log \frac{\alpha_p + 2}{\alpha_p + 1}\right) = (\beta_p - \alpha_p)\varepsilon \log\left(\frac{p}{x_{\alpha_p + 1}}\right)$$

$$\psi_1 = \psi_1(\alpha_p, \beta_p) = (\beta_p - \alpha_p)\log\left(1 + \frac{1}{\alpha_p + 1}\right) - \log\left(1 + \frac{\beta_p - \alpha_p}{\alpha_p + 1}\right).$$

We have

$$\text{ben}_p(M) = \varphi_1 + \psi_1,$$

$\varphi_1 \geq 0, \psi_1 \geq 0$ and $\psi_1(\alpha_p, \alpha_p + 1) = 0$. Similarly, for $\beta_p < \alpha_p$, let us introduce:

$$\varphi_2 = \varphi_2(\varepsilon, p, \alpha_p, \beta_p) = (\alpha_p - \beta_p)\left(\log \frac{\alpha_p + 1}{\alpha_p} - \varepsilon \log p\right) = (\alpha_p - \beta_p)\varepsilon \log\left(\frac{x_{\alpha_p}}{p}\right)$$

$$\psi_2 = \psi_2(\alpha_p, \beta_p) = (\alpha_p - \beta_p)\log\left(1 - \frac{1}{\alpha_p + 1}\right) - \log\left(1 - \frac{\alpha_p - \beta_p}{\alpha_p + 1}\right).$$

We have $\varphi_2 \geq 0, \psi_2 \geq 0, \psi_2(\alpha_2, \alpha_p - 1) = 0$. Moreover, observe that ψ_1 is an increasing function of $\beta_p - \alpha_p$, and ψ_2 is an increasing function of $\alpha_p - \beta_p$, for α_p fixed.

We will prove:

Theorem 6. *Let $x \to +\infty$, ε be defined by (60) and N_ε by (59). Let $\lambda < 1$ be a positive real number, μ a positive real number not too large ($\mu < 0.16$) and $B = B(x)$ such that*

$x^{-\mu} \leq B(x) \leq x^{\lambda}$. *Then the number of integers M such that the benefit of M (defined by (65)) is smaller than B, satisfies*

$$v \leq \exp\left(\frac{23}{\sqrt{1-\mu}}\sqrt{Bx}\right) \tag{68}$$

for x large enough.

In [9], an upper bound for v was given, with $B = x^{-\gamma}$. In order to prove Theorem 6, we shall need the following lemmas:

Lemma 1. *Let $p_1 = 2$, $p_2 = 3, \ldots, p_k$ be the kth prime. For $k \geq 2$ we have $k \log k \geq 0.46 p_k$.*

Proof: By [18] for $k \geq 6$ we have

$$p_k \leq k(\log k + \log\log k) \leq 2k \log k$$

and the lemma follows after checking the cases $k = 2, 3, 4, 5$.   □

Lemma 2. *Let $p_1 = 2$, $p_2 = 3, \ldots, p_k$ be the kth prime. The number of solutions of the inequality*

$$p_1 x_1 + p_2 x_2 + \cdots + p_k x_k + \cdots \leq x \tag{69}$$

in integers $x_1, x_2, \ldots$, is $\exp((1+o(1))\frac{2\pi}{\sqrt{3}}\sqrt{\frac{x}{\log x}})$.

Proof: The number $T(n)$ of partitions of n into primes satisfies (cf. [5]) $\log T(n) \sim \frac{2\pi}{\sqrt{3}}\sqrt{\frac{n}{\log n}}$, and the number of solutions of (69) is $\sum_{n \leq x} T(n)$.   □

Lemma 3. *The number of solutions of the inequality*

$$x_1 + x_2 + \cdots + x_r \leq A \tag{70}$$

in integers $x_1, \ldots, x_r$ is $\leq (2r)^A$.

Proof: Let $a = \lfloor A \rfloor$. It is well known that the number of solutions of (70) is

$$\binom{r+a}{a} = \frac{r+a}{a}\frac{r+a-1}{a-1}\cdots\frac{r+2}{2}\frac{r+1}{1} \leq (r+1)^a \leq (2r)^a.$$

                                    □

Proof of Theorem 6: Any integer M can be written as

$$M = \frac{A}{D}N_\varepsilon, \quad (A, D) = 1 \text{ and } D \text{ divides } N_\varepsilon.$$

First, we observe that, if p^y divides A and ben $M \leq B$, we have for x large enough:

$$y \leq x. \tag{71}$$

Indeed, by (61), we have

$$\alpha_p \leq \frac{1}{p^\varepsilon - 1} \leq \frac{1}{\varepsilon \log p} = \frac{\log x}{\log 2 \log p} \leq \frac{\log x}{(\log 2)^2} \leq 3 \log x.$$

It follows that

$$B \geq \text{ben } M \geq \text{ben}_p(AN_\varepsilon) \geq \psi_1(\alpha_p, \alpha_p + y)$$
$$= y \log\left(1 + \frac{1}{\alpha_p + 1}\right) - \log\left(1 + \frac{y}{\alpha_p + 1}\right)$$
$$\geq \frac{y}{\alpha_p} - \log(1 + y) \geq \frac{y}{3 \log x} - \log(1 + y),$$

and since $B \leq x^\lambda$, this inequality does not hold for $y > x$ and x large enough.

Further we write $A = A_1 A_2 \cdots A_6$ with $(A_i, A_j) = 1$ and

$$p \mid A_1 \Longrightarrow p > 2x$$
$$p \mid A_2 \Longrightarrow x < p \leq 2x$$
$$p \mid A_3 \Longrightarrow 2x_2 < p \leq x$$
$$p \mid A_4 \Longrightarrow x_2 < p \leq 2x_2$$
$$p \mid A_5 \Longrightarrow 2x_3 < p \leq x_2$$
$$p \mid A_6 \Longrightarrow p \leq 2x_3,$$

where x_2 and x_3 are defined by (62). Similarly, we write $D = D_1 D_2 \ldots D_5$, with $(D_i, D_j) = 1$ and

$$p \mid D_1 \Longrightarrow x/2 < p \leq x$$
$$p \mid D_2 \Longrightarrow x_2 < p \leq x/2$$
$$p \mid D_3 \Longrightarrow x_2/2 < p \leq x_2$$
$$p \mid D_4 \Longrightarrow 2x_3 < p \leq x_2/2$$
$$p \mid D_5 \Longrightarrow p \leq 2x_3.$$

We have

$$\text{ben } M = \sum_{i=1}^{6} \text{ben}(A_i N_\varepsilon) + \sum_{i=1}^{5} \text{ben}(N_\varepsilon/D_i),$$

and denoting by v_i (resp. v_i') the number of solutions of

$$\text{ben}(A_i N_\varepsilon) \leq B \quad (\text{resp. ben}(N_\varepsilon/D_i) \leq B),$$

we have

$$\nu \leq \prod_{i=1}^{6} \nu_i \prod_{i=1}^{5} \nu_i'. \tag{72}$$

In (72), we shall see that the main factors are ν_2 and ν_1' and the other ones are negligible.

Estimation of ν_2. Let us denote the primes between x and $2x$ by $x < P_1 < P_2 < \cdots < P_r \leq 2x$, and let

$$A_2 = P_1^{y_1} P_2^{y_2} \cdots P_r^{y_r}, \quad y_i \geq 0.$$

From the Brun-Titchmarsh inequality, it follows for $i \geq 2$ that

$$i = \pi(P_i) - \pi(x) \leq 2\frac{P_i - x}{\log(P_i - x)} \leq 2\frac{P_i - x}{\log 2(i-1)}$$

and it follows from Lemma 1:

$$P_i - x \geq \frac{i}{2}\log 2(i-1) \geq \frac{i \log i}{2} \geq 0.23\, p_i.$$

By (60) and (61) we have $\alpha_{P_i} = 0$ and

$$\text{ben}(A_2 N_\varepsilon) \geq \sum_{i=2}^{r} \varphi_1(\varepsilon, P_i, 0, y_i) = \sum_{i=2}^{r} \varepsilon y_i \log(P_i/x)$$

$$\geq \sum_{i=2}^{r} \varepsilon y_i \frac{P_i - x}{P_i} \geq \sum_{i=2}^{r} \frac{\varepsilon y_i}{2x}(P_i - x) \geq \sum_{i=2}^{r} 0.115\frac{\varepsilon y_i}{x}\, p_i.$$

By (71), the number of possible choices for y_1 is less than $(x+1)$, so that ν_2 is certainly less than $(x+1)$ times the number of solutions of:

$$\sum_{i=2}^{\infty} p_i y_i \leq \frac{Bx}{\varepsilon(0.115)} \leq 12.6Bx \log x,$$

and, by Lemma 2,

$$\nu_2 \leq (x+1) \exp\left\{ (1+o(1))\frac{2\pi}{\sqrt{3}}\sqrt{\frac{12.6Bx \log x}{\log(Bx)}} \right\} \leq \exp\left(\frac{13\sqrt{Bx}}{\sqrt{1-\mu}}\right).$$

Estimation of ν_1. First we observe that, if a large prime P divides M and ben $M \leq B$ then we have:

$$B \geq \text{ben } M \geq \text{ben}_p(M) \geq \varphi_1(\varepsilon, P, 0, \beta_p) \geq \varepsilon \log(P/x),$$

so that

$$P \le x \exp(B/\varepsilon) = x \exp\left(\frac{B \log x}{\log 2}\right).$$

If λ is large, we divide the interval $[0, \lambda]$ into equal subintervals: $[\lambda_i, \lambda_{i+1}], 0 \le i \le s-1$, such that $\lambda_{i+1} - \lambda_i < \frac{1-\lambda}{2}$. We set $T_0 = 2x, T_i = x \exp(x^{\lambda_i})$ for $1 \le i \le s - 1$, and $T_s = x \exp(\frac{B \log x}{\log 2})$. If $\lambda < \frac{1}{3}$, there is just one interval in the subdivision. Further, we write $A_1 = a_1 a_2 \ldots a_s$ with $p \mid a_i \implies T_{i-1} < p \le T_i$, and if we denote the number of solutions of ben $(a_i N_\varepsilon) \le B$ by $v_1^{(i)}$ clearly we have

$$v_1 \le \prod_{i=1}^{s} v_1^{(i)}.$$

To estimate $v_1^{(i)}$ let us denote the primes between T_{i-1} and T_i by $T_{i-1} < P_1 < \cdots < P_r \le T_i$, and let $a_i = P_1^{y_1} \cdots P_r^{y_r}$. We have

$$B \ge \text{ben}(a_i N_\varepsilon) \ge \sum_{i=1}^{r} \varphi_1(\varepsilon, P_i, 0, y_i) = \sum_{i=1}^{r} \varepsilon y_i \log \frac{P_i}{x}$$

$$\ge \sum_{i=1}^{r} \varepsilon y_i \log \frac{T_{i-1}}{x}.$$

If $i = 1$, $T_0 = 2x$, this implies $\sum_{i=1}^{r} y_i \le \frac{B(\log x)}{(\log 2)^2} \le 3B \log x$, and by Lemma 3,

$$v_1^{(1)} \le \exp(3B \log x \log(2r)) \le \exp(3B \log x \log T_1) \le \exp((1 + o(1))Bx^{\lambda_1}).$$

If $i > 1$, we have $\sum_{i=1}^{r} y_i \le \frac{B}{\varepsilon x^{\lambda_{i-1}}}$, and by Lemma 3,

$$v_1^{(i)} \le \exp\left(\frac{B}{\varepsilon x^{\lambda_{i-1}}} \log T_i\right) \le \exp\{(1 + o(1))Bx^{\lambda_i - \lambda_{i-1}}\},$$

and from the choice of the λ_i's, one can easily see that, for $B \le x^\lambda$, $v_1 = \prod_{i=1}^{s} v_1^{(i)}$ is negligible compared with v_2.

The other factors of (72) are easier to estimate:

Estimation of v_3. Let us denote the primes between $2x_2$ and x by $2x_2 < P_r < P_{r-1} < \cdots < P_1 \le x$. By (62) and (4), $x_2 = x^\theta$, and by (63), $\alpha P_i = 1$. Let us write $A_3 = P_1^{y_1} \cdots P_r^{y_r}$. We have

$$B \ge \text{ben}(A_3 M) \ge \sum_{i=1}^{r} \varphi_1(\varepsilon, P_i, 1, 1 + y_i) = \sum_{i=1}^{r} \varepsilon y_i \log \frac{P_i}{x_2} \ge \sum_{i=1}^{r} \frac{(\log 2)^2}{\log x} y_i.$$

So, $\sum_{i=1}^{r} y_i \le B \log x/(\log 2)^2 \le 3B \log x$, and by Lemma 3,

$$v_3 \le \exp(3B \log x \log(2r)) \le \exp(3B(\log x)^2).$$

Estimation of v_4. Replacing x by x_2 the upper bound obtained for v_2 becomes:

$$v_2 = \exp(O(\sqrt{Bx_2})) = \exp(O(\sqrt{Bx^\theta})).$$

Estimation of v_5. Replacing x by x_2, the upper bound obtained for v_3 becomes:

$$v_5 \le \exp(3B \log x \log x_2) = \exp(3\theta B (\log x)^2).$$

Estimation of v_6. Let $p_1, p_2, \ldots, p_r \le 2x_3$ be the first primes and write $A_6 = p_1^{y_1} p_2^{y_2} \cdots p_r^{y_r}$. By (71), $y_i \le x$, and thus by (62),

$$v_6 \le (x + 1)^r \le (x + 1)^{x_3} = \exp(x^{1-\theta} \log(x + 1))$$

and for $B \ge x^{-\mu}$ and $\mu < 0.16$, this is negligible compared with v_2.

Estimation of v_1'. Let us denote the primes between $\frac{x}{2}$ and x by $\frac{x}{2} < P_r < P_{r-1} < \cdots < P_1 \le x$, and let $D_1 = P_1^{y_1} \cdots P_r^{y_r}$. We have $\alpha_{P_i} = 1$ and since D_1 divides N_ε, $y_i = 0$ or 1. By a computation similar to that of v_2, we obtain

$$B \ge \text{ben} \, \frac{N_\varepsilon}{D_1} \ge \sum_{i=2}^{r} \varphi_2(\varepsilon, P_i, 1, y_i) = \sum_{i=2}^{r} \varepsilon y_i \log \frac{x}{P_i} \ge \sum_{i=2}^{r} \varepsilon y_i \frac{x - P_i}{x},$$

and by using the Brun-Titchmarsch inequality and Lemma 1, it follows that

$$\sum_{i=2}^{r} p_i y_i \le \frac{Bx}{0.23\varepsilon} \le 6.3 \, Bx \log x.$$

Thus, as y_1 can only take 2 values, by Lemma 2 we have

$$v_1' \le 2 \exp((1 + o(1)) \frac{2\pi}{\sqrt{3}} \sqrt{\frac{6.3 Bx \log x}{\log(Bx)}} \le \exp(9.2\sqrt{Bx}).$$

Estimation of v_2'. By an estimation similar to that of v_3, replacing φ_1 by φ_2 and using Lemma 3, we get

$$v_2' \le \exp(3B \log^2 x).$$

Estimation of v_3'. Replacing x by x_2, it is similar to that of v_1' and we get

$$v_3' = \exp(O(\sqrt{Bx_2})).$$

Estimation of v_4'. Replacing x by x_2, we get, as for v_2',

$$v_4' \le \exp(3B \log x \log x_2) = \exp(3\theta B \log^2 x).$$

Estimation of v_5'.   As we have seen for v_6, we have

$$D_5 = p_1^{y_1} \cdots p_r^{y_r}$$

with $y_i \le \alpha_{p_i} \le 3 \log x$ and $r \le \pi(2x_3) \le x_3$. Thus

$$v_5' \le (1 + 3 \log x)^r \le \exp(x^{1-\theta} \log(1 + 3 \log x)).$$

By formula (68) and the estimates of v_i and v_i', the proof of Theorem 6 is completed. $\square$

By a more careful estimate, it would have been possible to improve the constant in (68). However, using the Brun-Titchmarsch inequality we loose a factor $\sqrt{2}$, and we do not see how to avoid this loss. A similar method was used in [3]. Also, the condition $\mu < 0.16$ can be replaced easily by $\mu < 1$.

7.   Proof of Theorem 3

We shall need the following lemmas:

Lemma 4.   *Let n_j the sequence of h.c. numbers. There exists a positive real number c such that for j large enough, the following inequality holds:*

$$\frac{n_{j+1}}{n_j} \le 1 + \frac{1}{(\log n_j)^c}.$$

Proof: This result was first proved by Erdős in [2]. The best constant c is given in [8]:

$$c = \frac{\log(15/8)}{\log 8}(1 - \tau_0) = 0.1405\ldots$$

with the value of τ_0 given by (8). $\square$

Lemma 5.   *Let n_j be a h.c. number, and N_ε the superior h.c. number preceding n_j. Then the benefit of n_j (defined by (65)) satisfies:*

$$\text{ben } n_j = O((\log n_j)^{-\gamma}).$$

Proof: This is Theorem 1 of [8]. The value of γ is given by

$$\gamma = \theta(1 - \tau_0)/(1 + \kappa) = 0.03157\ldots$$

where θ, τ_0 and κ are defined by (4), (8) and (22). $\square$

To prove Theorem 3, first recall that n_k is defined so that

$$n_k \le X < n_{k+1}. \tag{73}$$

We define N_ε as the largest s.h.c. number $\leq n_k$. Now let $n \in S(X, z)$. We get from (65):

$$\text{ben } n = \varepsilon \log \frac{n}{N_\varepsilon} - \log \frac{d(n)}{d(N_\varepsilon)},$$

$$\text{ben } n_k = \varepsilon \log \frac{n_k}{N_\varepsilon} - \log \frac{d(n_k)}{d(N_\varepsilon)}$$

and, subtracting,

$$\text{ben } n = \text{ben } n_k + \varepsilon \log \frac{n}{n_k} - \log \frac{d(n)}{d(n_k)}.$$

But $n \in S(X, z)$ so that $n \leq X$ and $d(n) \geq z d(n_k)$. Thus

$$\text{ben } n \leq \text{ben } n_k + \varepsilon \log \frac{X}{n_k} + \log(1/z).$$

By (73) and Lemma 4, we have $n_k \sim X$, and by (60), (64), (73) and Lemma 4, we have

$$\varepsilon \log \frac{X}{n_k} \leq \varepsilon \log \frac{n_{k+1}}{n_k} \leq \frac{1}{(\log X)^{c+o(1)}}.$$

By Lemma 5,

$$\text{ben } n \leq B = \log \frac{1}{z} + O(\log X)^{-\gamma}.$$

Applying Theorem 6 completes the proof of Theorem 3.   □

8. An upper bound for $d(n_{j+1})/d(n_j)$

We will prove:

Theorem 7.   *There exists a constant $c > 0$ such that for n_j large enough, the inequality*

$$\frac{d(n_{j+1})}{d(n_j)} \leq 1 + \frac{1}{(\log n_j)^c}$$

holds. Here c can be chosen as any number less than γ defined in Lemma 5.

Proof: Let N_ε the s.h.c. number preceding n_j. We have by Lemma 5 ben $(n_j) = O((\log n_j)^{-\gamma})$ and $\text{ben}(n_{j+1}) = O((\log n_j)^{-\gamma})$. Further, it follows from (65) that

$$\log \frac{d(n_{j+1})}{d(n_j)} = \varepsilon \log \frac{n_{j+1}}{n_j} + \text{ben}(n_{j+1}) - \text{ben}(n_j) \leq \log \frac{n_{j+1}}{n_j} + \text{ben}(n_{j+1})$$

which, by using Lemma 4 and Lemma 5, completes the proof of Theorem 7.   □

References

1. R.C. Baker and G. Harman, "The difference between consecutive primes," *Proc. London Math. Soc.* **72** (1996), 261–280.
2. P. Erdős, "On highly composite numbers," *J. London Math. Soc.* **19** (1944), 130–133.
3. P. Erdős and J.L. Nicolas, "Sur la fonction: nombre de diviseurs premiers de *n*," *l'Enseignement Mathématique* **27** (1981), 3–27.
4. N. Feldmann, "Improved estimate for a linear form of the logarithms of algebraic numbers," *Mat. Sb.* **77**(119), (1968), 423–436 (in Russian); *Math. USSR-Sb.* **6** (1968), 393–406.
5. G.H. Hardy and S. Ramanujan, "Asymptotic formulae for the distribution of integers of various types," *Proc. London Math. Soc.* **16** (1917), 112–132. Collected Papers of S. Ramanujan, 245–261.
6. G.H. Hardy and E.M. Wright, *An Introduction to the Theory of Numbers*, 5th edition, Oxford at the Clarendon Press, 1979.
7. J.L. Nicolas, "Ordre maximal d'un élément du groupe des permuatations et highly composite numbers," *Bull. Soc. Math. France* **97** (1969), 129–191.
8. J.L. Nicolas, "Répartition des nombres hautement composés de Ramanujan," *Can. J. Math.* **23** (1971), 116–130.
9. J.L. Nicolas, "Répartition des nombres largement composés," *Acta Arithmetica* **34** (1980), 379–390.
10. J.L. Nicolas, "Nombres hautement composés," *Acta Arithmetica* **49** (1988), 395–412.
11. J.L. Nicolas, "On highly composite numbers," *Ramanujan Revisited* (Urbana-Champaign, Illinois, 1987), Academic Press, Boston, 1988, pp. 215–244.
12. J.L. Nicolas and A. Sárközy, "On two partition problems," *Acta Math. Hung.* **77** (1997), 95–121.
13. S. Ramanujan, "Highly composite numbers," *Proc. London Math. Soc.* **14** (1915), 347–409; Collected Papers, 78–128.
14. S. Ramanujan, *The Lost Notebook and Other Unpublished Papers*, Narosa, New Delhi, 1988.
15. S. Ramanujan, "Highly composite numbers," annotated by J.L. Nicolas and G. Robin, *The Ramanujan Journal* **1** (1997), 119–153.
16. G. Rhin, "Approximants de Padé et mesures effectives d'irrationalité," Séminaire Th. des Nombres D.P.P., 1985–86, Progress in Math. no. 71, Birkhäuser, 155–164.
17. G. Robin, "Méthodes d'optimisation pour un problème de théorie des nombres," *R.A.I.R.O. Informatique théorique* **17** (1983), 239–247.
18. J.B. Rosser and L. Schoenfeld, "Approximate formulas for some functions of prime numbers," *Illinois J. Math.* **6** (1962), 64–94.
19. A. Selberg, "On the normal density of primes in small intervals and the difference between consecutive primes," *Arch. Math. Naturvid.* **47** (1943), 87–105.

THE RAMANUJAN JOURNAL 2, 247–269 (1998)
© 1998 Kluwer Academic Publishers.

Some New Old-Fashioned Modular Identities

PAUL T. BATEMAN   bateman@math.uiuc.edu
Department of Mathematics, University of Illinois at Urbana-Champaign, Urbana, Illinois 61801-2975

MARVIN I. KNOPP
Department of Mathematics, Temple University, Philadelphia, Pennsylvania 19122-2585

Dedicated to the memory of Paul Erdős

Received December 12, 1997; Accepted February 6, 1998

Abstract. This paper uses modular functions on the theta group to derive an exact formula for the sum

$$\sum_{|j| \leq n^{1/2}} \sigma(n - j^2)$$

in terms of the singular series for the number of representations of an integer as a sum of five squares. (Here $\sigma(k)$ denotes the sum of the divisors of k if k is a positive integer and $\sigma(0) = -1/24$.)

Several related identities are derived and discussed.

Two devices are used in the proofs. The first device establishes the equality of two expressions, neither of which is a modular form, by showing that the square of their difference is a modular form. The second device shows that a certain modular function is identically zero by noting that it has more zeros than poles in a fundamental region.

Key words: modular forms, sum-of-divisors function, theta group, sums of squares, singular series

1991 Mathematics Subject Classification: Primary 11F11; Secondary 11E25, 11A25

1. Introduction

In this paper we use modular functions to derive an exact formula for the sum

$$\sum_{|j| \leq n^{1/2}} \sigma(n - j^2),$$

where $\sigma(k)$ denotes the sum of the (positive) divisors of k if k is a positive integer and $\sigma(0) = -1/24$. This formula was conjectured in [2]. We also derive some related identities. (In words used by the late Alexander M. Ostrowski in a conversation with Morris Newman, "Once you enter the modular jungle, you are bound to capture something.")

If instead of σ we consider the function σ^*, where $\sigma^*(n) = \sigma(n)$ when n is a positive integer not divisible by 4 and $\sigma^*(n) = \sigma(n) - 4\sigma(n/4)$ when n is a nonnegative integer divisible by 4 (so that in particular $\sigma^*(0) = 1/8$), then an exact formula for

$$\sum_{|j| \leq n^{1/2}} \sigma^*(n - j^2)$$

was obtained by Hardy [6]. We describe Hardy's results briefly.

For s a positive integer we let $r_s(n)$ denote the number of representations of the nonnegative integer n as a sum of s squares, i.e., the number of solutions of the equation

$$x_1^2 + x_2^2 + \cdots + x_s^2 = n$$

in integers $x_1, x_2, \ldots, x_s$ (positive, zero, or negative). If

$$\vartheta(\tau) = \sum_{n=-\infty}^{\infty} e^{\pi i \tau n^2} \quad (\Im \tau > 0), \tag{1.1}$$

then

$$\vartheta(\tau)^s = \sum_{n=0}^{\infty} r_s(n) e^{\pi i \tau n} = 1 + \sum_{n=1}^{\infty} r_s(n) e^{\pi i \tau n}. \tag{1.2}$$

Inductively, we have

$$r_{s+1}(n) = \sum_{|j| \leq n^{1/2}} r_s(n - j^2). \tag{1.3}$$

For $5 \leq s \leq 8$ Hardy used modular forms on the theta group Γ_ϑ to derive exact formulas for $r_s(n)$ (cf. also [1]). In particular, Hardy proved that

$$r_5(n) = (4\pi^2/3)n^{3/2} \sum_{k=1}^{\infty} A_k(n) \quad (n > 0), \tag{1.4}$$

where $\sum A_k(n)$, the so-called singular series, is given by

$$A_k(n) = \sum_{\substack{h \bmod k \\ (h,k)=1}} (G(h, k)/k)^5 e^{-2\pi i h n/k}, \quad G(h, k) = \sum_{j=1}^{k} e^{2\pi i h j^2/k}. \tag{1.5}$$

If for primes p we put

$$\chi_p(n) = \sum_{j=0}^{\infty} A_{p^j}(n), \tag{1.6}$$

then, since $A_k(n)$ is a multiplicative function of k for fixed n, the result (1.4) may be written

$$r_5(n) = (4\pi^2/3)n^{3/2}\chi_2(n) \sum_{k \text{ odd} > 0} A_k(n) \quad (n > 0). \tag{1.7}$$

Proofs of Hardy's exact formulas can be found in [4, Ch. 13], [5], [7, Ch. 5], [10, Ch. 7], and [11, Section 4.2].

In view of (1.3) and Jacobi's formula

$$r_4(n) = 8\sigma^*(n) \quad (n \geq 0),$$

(1.7) may be restated as

$$\sum_{|j| \leq n^{1/2}} \sigma^*(n - j^2) = (\pi^2/6) n^{3/2} \chi_2(n) \sum_{k \text{ odd} > 0} A_k(n) \quad (n > 0). \tag{1.8}$$

(Jacobi's formula is proved in [1, Section 3], [9, Section 83], and [11, Section 1.2].) In [2] it was conjectured that (for positive integers n) a similar exact formula holds for $\sum \sigma(n - j^2)$, namely

$$\sum_{|j| \leq n^{1/2}} \sigma(n - j^2) = \frac{\pi^2}{6} n^{3/2} \left(\frac{5}{3} - \frac{\chi_2(n)}{3} \right) \sum_{k \text{ odd} > 0} A_k(n) - 2ns(n), \tag{1.9}$$

where $s(n) = 1$ if n is the square of an integer and $s(n) = 0$ otherwise. We shall prove (1.9) in this paper (Corollary to Theorem 1).

Hardy's formula (1.8) is equivalent to the power series identity

$$\vartheta(\tau) \sum_{n=0}^{\infty} 8\sigma^*(n) e^{\pi i \tau n} = 1 + \sum_{n=1}^{\infty} e^{\pi i \tau n} (4\pi^2/3) n^{3/2} \chi_2(n) \sum_{k \text{ odd} > 0} A_k(n). \tag{1.10}$$

Of course, both sides of (1.10) are equal to $\vartheta(\tau)^5$, a modular form of weight $5/2$ with respect to the group Γ_ϑ. The proposed formula (1.9) is similarly equivalent to the power series identity

$$\vartheta(\tau) P(\tau) = 1 - \sum_{n=1}^{\infty} e^{\pi i \tau n} (4\pi^2/3) n^{3/2} (5 - \chi_2(n)) \sum_{k \text{ odd} > 0} A_k(n) + \sum_{n=1}^{\infty} 48 n s(n) e^{\pi i \tau n}, \tag{1.11}$$

where in Ramanujan's notation

$$P(\tau) = E_2(\tau/2) = -24 \sum_{n=0}^{\infty} \sigma(n) e^{\pi i \tau n} = 1 - 24 \sum_{n=1}^{\infty} \sigma(n) e^{\pi i \tau n}. \tag{1.12}$$

The two sides of (1.11) are not modular forms separately, but our proof of (1.11) will involve showing that the square of their difference is a modular form of weight 5 with respect to Γ_ϑ. Since

$$\vartheta'(\tau) = \pi i \sum_{n=-\infty}^{\infty} n^2 e^{\pi i \tau n^2} = \pi i \sum_{n=1}^{\infty} 2ns(n) e^{\pi i \tau n},$$

(1.11) may be rewritten as

$$\vartheta(\tau)P(\tau) = -5G(\tau) + \vartheta(\tau)^5 + 24\vartheta'(\tau)/(\pi i), \qquad (1.13)$$

where

$$G(\tau) = \sum_{n=1}^{\infty} e^{\pi i \tau n} (4\pi^2/3) n^{3/2} \sum_{k \text{ odd} > 0} A_k(n) = \sum_{n=1}^{\infty} e^{\pi i \tau n} r_5(n)/\chi_2(n). \qquad (1.14)$$

Thus, to prove (1.9), (1.11), and (1.13) it suffices to prove the following theorem.

Theorem 1. *For $\Im\tau > 0$ put*

$$S(\tau) = \vartheta(\tau)P(\tau) + 5G(\tau) - \vartheta(\tau)^5 - 24\vartheta'(\tau)/(\pi i), \qquad (1.15)$$

where ϑ is given by (1.1), P is given by (1.12), and G is given by (1.14). Then $S(\tau)$ is identically zero.

Corollary to Theorem 1. *If $s(n) = 1$ when n is a perfect square and $s(n) = 0$ otherwise, then for any positive integer n we have*

$$24 \sum_{|j| \leq n^{1/2}} \sigma(n - j^2) = (4\pi^2/3) n^{3/2} (5 - \chi_2(n)) \sum_{k \text{ odd} > 0} A_k(n) - 48ns(n)$$

$$= \{5/\chi_2(n) - 1\} r_5(n) - 48ns(n)$$

$$= \{5/\chi_2(n) - 1\} \sum_{|j| \leq n^{1/2}} 8\sigma^*(n - j^2) - 48ns(n).$$

We remark that $\chi_2(n)$ can be evaluated in closed form in terms of 4^α, the highest power of 4 dividing n, and the residue class of $n/4^\alpha$ modulo 8. See Lemma 2.6 below.

The following result will appear as a by-product of our proof of Theorem 1.

Theorem 2. *If $\Im\tau > 0$, then*

$$48 \sum_{j=0}^{\infty} (-1)^j \sigma(j) e^{\pi i \tau j} \sum_{m=0}^{\infty} e^{\pi i \tau m(m+1)} + 48 \sum_{m=0}^{\infty} (m^2 + m + 1/4) e^{\pi i \tau m(m+1)}$$

$$= \sum_{\substack{m \equiv 1 \, (\text{mod } 8) \\ m > 0}} r_5(m) e^{\pi i \tau (m-1)/4} - (3/7) \sum_{\substack{m \equiv 5 \, (\text{mod } 8) \\ m > 0}} r_5(m) e^{\pi i \tau (m-1)/4}.$$

Corollary to Theorem 2. *Let $t(k) = 1$ if k is a product of two consecutive integers (i.e., if k is twice a triangular number) and $t(k) = 0$ otherwise. Then for any nonnegative integer n we have*

$$48 \sum_{j=0}^{n} (-1)^j \sigma(j) t(n - j) + 48(n + 1/4) t(n) = \frac{2 + (-1)^n 5}{7} r_5(4n + 1).$$

After giving some preliminaries in Section 2, we shall investigate the behavior of $S(\tau)$ under the substitutions $\tau \to -1/\tau$ and $\tau \to 1 - 1/\tau$ in Sections 3 and 4. In Section 5 we shall use these results to prove Theorems 1 and 2 and their respective corollaries. In Section 6 we shall obtain the following theorem, which is similar to (but distinct from) a result of Henri Cohen in [3]. We thank Ken Ono for calling our attention to Cohen's paper.

Theorem 3. *Suppose D is the discriminant of a real quadratic field, χ_D is the associated residue character given in terms of the Kronecker symbol by*

$$\chi_D(n) = \left(\frac{D}{n}\right),$$

and $L(s, \chi_D)$ is the entire function given by

$$L(s, \chi_D) = \sum_{n=1}^{\infty} \chi_D(n) n^{-s}$$

for $\Re s > 0$. Then

$$L(-1, \chi_D) = \frac{1}{160} r_5(D) - \frac{1}{20} \sum_{|j| < D^{1/2}} \sigma(D - j^2).$$

When combined with Cohen's result, Theorem 3 has the following corollary.

Corollary to Theorem 3. *With D, χ_D, and $L(s, \chi_D)$ as in Theorem 3 we have*

$$r_5(D) = 80 \sum_{\substack{|j| < D^{1/2} \\ D - j^2 \text{ odd}}} \sigma(D - j^2) - 24 \sum_{|j| < D^{1/2}} \sigma(D - j^2), \tag{1.16}$$

$$L(-1, \chi_D) = \frac{1}{2} \sum_{\substack{|j| < D^{1/2} \\ D - j^2 \text{ odd}}} \sigma(D - j^2) - \frac{1}{5} \sum_{|j| < D^{1/2}} \sigma(D - j^2), \tag{1.17}$$

$$L(-1, \chi_D) = \frac{3}{560} r_5(D) - \frac{1}{14} \sum_{\substack{|j| < D^{1/2} \\ D - j^2 \text{ even}}} \sigma(D - j^2). \tag{1.18}$$

Finally, we state without proof a theorem which can be proved by the same method as that used in proving Theorem 3.

Theorem 4. *For $\Im \tau > 0$ we have*

$$G(\tau) = 16 \vartheta(\tau) \sum_{n \text{ odd} > 0} \sigma(n) e^{\pi i \tau n},$$

where G is defined by (1.14) and ϑ is defined by (1.1). In other words

$$(4\pi^2/3)n^{3/2}\sum_{k \text{ odd} > 0} A_k(n) = r_5(n)/\chi_2(n) = 16\sum_{\substack{|j|<n \\ n-j^2 \text{odd}}} \sigma(n-j^2)$$

for every positive integer n.

Recall that the evaluation of $\chi_2(n)$ will be given in Lemma 2.6.

2. Preliminaries

We recall some notation and basic lemmas given, for example, in the exposition of Hardy's method in [1, Sections 1–2]. We also prove two lemmas not found there.

For k a positive integer and h any integer we put $\eta(h, k) = 0$ if $(h, k) > 1$ and

$$\eta(h,k) = \frac{1}{2}k^{-1/2}\sum_{j \bmod 2k} e^{\pi ihj^2/k} = \frac{1}{2}\{1 + (-1)^{hk}\}k^{-1/2}\sum_{j=1}^{k} e^{\pi ihj^2/k} \qquad (2.1)$$

if $(h, k) = 1$. The following lemma summarizes the known evaluations of Gaussian sums in terms of the Legendre-Jacobi symbol.

Lemma 2.1.   *Suppose k is a positive integer and h is any integer. Then $\eta(h, k) = 0$ if $(h, k) > 1$ or if h and k are both odd. If $(h, k) = 1$ and if h and k are of opposite parity, then*

$$\eta(h,k) = \left(\frac{h}{k}\right)e^{-\pi i(k-1)/4} \quad \text{for h even, k odd,}$$

$$\eta(h,k) = \left(\frac{k}{|h|}\right)e^{\pi ih/4} \quad \text{for h odd, k even.}$$

If we put

$$B_k(n) = k^{-5/2}\sum_{h \bmod 2k} \eta(h,k)^5 e^{-\pi ihn/k}, \qquad (2.2)$$

then it is easy to see that

$$A_k(n) = B_k(n) \qquad \text{if k is odd,}$$
$$A_k(n) = 0 \qquad \text{if } 2 \mid k \text{ but } 4 \nmid k,$$
$$A_k(n) = B_{k/2}(n) \qquad \text{if } 4 \mid k.$$

Thus,

$$\sum_{k=1}^{\infty} A_k(n) = \sum_{k=1}^{\infty} B_k(n),$$

so that we have two equivalent forms of the singular series; for the purposes of this paper the latter form is sometimes more convenient. Of course, $B_k(n)$ is multiplicative in k for fixed n.

It is useful to extend the definition of $\eta(h, k)$ to negative values of k, provided $h \neq 0$, by putting

$$\eta(h, k) = \eta(-h, -k)e^{(\text{sign } h)\pi i/2} \quad (h \neq 0, k \neq 0). \tag{2.3}$$

For positive integers h and k we have the familiar reciprocity formula for Gaussian sums $\eta(h, k) = e^{\pi i/4}\eta(-k, h)$. With the convention (2.3) we have the more general reciprocity formula:

Lemma 2.2. *If $h > 0, k \neq 0$ or if $h \neq 0, k > 0$, then*

$$\eta(h, k) = e^{\pi i/4}\eta(-k, h).$$

If $h < 0, k < 0$, then

$$\eta(h, k) = e^{-3\pi i/4}\eta(-k, h).$$

We make an agreement about powers of the form $(hi - ki\tau)^{1/2}$ and $(h - k\tau)^{1/2}$, where h and k are integers and τ is a complex variable in the half-plane $\{\tau : \Im\tau > 0\}$. For $k \neq 0$ the quantity of which we are taking the square root lies in a certain half-plane and thus there is no difficulty in defining the square root. In all cases, even for $k = 0$, we agree to take that branch which has values with positive real part for purely imaginary τ. The case $h = 0, k < 0$ will not occur and consequently our definition is unambiguous. Our choice of branch coincides with the usual principal value of the square root except for the case of $(hi - ki\tau)^{1/2}$ when $k < 0$, for which we take the square root in the upper or lower half-plane according as h is positive or negative. Powers of the form $(hi - ki\tau)^{5/2}$ or $(h - k\tau)^{5/2}$ will be understood to mean the ordinary fifth powers of the square roots defined above. All other powers will have their usual principal values.

It is easy to check that in view of (2.3) and our convention about square roots we have

$$\frac{\eta(h, k)}{(hi - ki\tau)^{1/2}} = \frac{\eta(-h, -k)}{\{(-h)i - (-k)i\tau\}^{1/2}} \quad (h \neq 0, k \neq 0). \tag{2.4}$$

Furthermore, the reciprocity law Lemma 2.2 gives

$$\frac{\eta(h, k)}{(hi - ki\tau)^{1/2}} = \frac{\eta(-k, h)}{(h - k\tau)^{1/2}} \quad (h \neq 0, k \neq 0). \tag{2.5}$$

From the formulas of Lemma 2.1 and the known formula

$$(-1)^{(h+1)k/4}\left(\frac{k}{|h + k|}\right) = \left(\frac{k}{|h|}\right) \quad (k \text{ even} > 0, h \text{ odd}),$$

we readily obtain the following assertion.

Lemma 2.3. *If k is positive integer and h is an odd integer, then*

$$\eta(h+k,k) = \left(\frac{k}{|h|}\right) e^{-\pi i h(k-1)/4},$$

where it is understood that the right-hand side is zero if $(h,k) > 1$.

In view of our convention about square roots, it is easy to obtain the following lemma.

Lemma 2.4. *If h is an odd integer and k is any integer, then the expression*

$$\left(\frac{k}{|h|}\right) \frac{e^{-\pi i h(k-1)/4}}{(hi - ki\tau)^{1/2}}$$

is unchanged if we replace h by $-h$ and k by $-k$.

The following is a special case of a classical formula of Lipschitz (cf. [9, Section 37]).

Lemma 2.5. *If $\Im z > 0$, then*

$$\frac{4\pi^2}{3} \sum_{n=1}^{\infty} n^{3/2} e^{\pi i z n} = \sum_{m=-\infty}^{\infty} \frac{1}{(2mi - iz)^{5/2}}.$$

The 2-adic factor $\chi_2(n)$ in the singular series for five squares (defined by (1.6)) can be evaluated as follows.

Lemma 2.6. *If 4^α is the highest power of 4 dividing the positive integer n, then*

$$\chi_2(n) = \frac{5}{7}\left(1 + \frac{3}{4 \cdot 8^\alpha}\right) \quad \text{if } 4^{-\alpha}n \equiv 2, 3, 6, 7 \,(\mathrm{mod}\, 8),$$

$$\chi_2(n) = \frac{5}{7}\left(1 + \frac{9}{5 \cdot 8^{\alpha+1}}\right) \quad \text{if } 4^{-\alpha}n \equiv 5 \,(\mathrm{mod}\, 8),$$

$$\chi_2(n) = \frac{5}{7}\left(1 - \frac{1}{8^{\alpha+1}}\right) \quad \text{if } 4^{-\alpha}n \equiv 1 \,(\mathrm{mod}\, 8).$$

Proof: This can be proved by using the singular series evaluations in [6] or in [4, Ch. 13]. However, we derive it here from the following formula, proved in [2]:

$$\chi_2(n) = \frac{5}{4} - 3\sum_{k=2}^{\infty} f_{2^k}(n)/2^{2k},$$

where $f_d(n)$ is the number of solutions of the congruence $x^2 \equiv n \,(\mathrm{mod}\, d)$. If $n/4^\alpha \equiv 2, 3 \,(\mathrm{mod}\, 4)$, then $f_{2^k}(n) = 2^{\lfloor k/2 \rfloor}$ for $2 \le k \le 2\alpha + 1$ and $f_{2^k}(n) = 0$ for $k \ge 2\alpha + 2$. If $n/4^\alpha \equiv 5 \,(\mathrm{mod}\, 8)$, then $f_{2^k}(n) = 2^{\lfloor k/2 \rfloor}$ for $2 \le k \le 2\alpha + 2$ and $f_{2^k}(n) = 0$ for $k \ge 2\alpha + 3$.

If $n/4^\alpha \equiv 1 \pmod 8$, then $f_{2^k}(n) = 2^{\lfloor k/2 \rfloor}$ for $2 \leq k \leq 2\alpha + 2$ and $f_{2^k}(n) = 2^{\alpha+1}$ for $k \geq 2\alpha + 3$. A brief calculation in each case then gives the assertion of the lemma. $\qquad\square$

The following lemma shows that for $1 \leq s \leq 7$ the number of solutions of

$$x_1^2 + x_2^2 + \cdots + x_s^2 = 8n + s \qquad (2.6)$$

in integers $x_1, x_2, \ldots, x_s$ has a fixed ratio to the number of solutions of (2.6) in odd integers.

Lemma 2.7.   *If $s = 1, 2, \ldots, 7$, then for any nonnegative integer n we have*

$$r_s(8n + s) = \left[1 + \frac{1}{2} \binom{s}{4} \right] r_s^*(8n + s),$$

where $r_s^*(8n + s)$ is the number of solutions of (2.6) in odd integers $x_1, x_2, \ldots, x_s$ (positive or negative).

Remark.   We have the following table

s	1	2	3	4	5	6	7
$1 + \frac{1}{2}\binom{s}{4}$	1	1	1	3/2	7/2	17/2	37/2

In particular, $r_5(8n + 5) = (7/2)r_5^*(8n + 5)$. The method used in the following proof shows that

$$r_s(8n + s) > \left[1 + \frac{1}{2} \binom{s}{4} \right] r_s^*(8n + s) \quad (s \geq 8).$$

Proof: We merely use the fact that odd squares are congruent to 1 modulo 8 and even squares are congruent to 0 or 4 modulo 8.

If $1 \leq s \leq 3$ and $8n + s$ is expressed as a sum of s squares, then all the squares must be odd. Thus the assertion of the lemma is immediate for $s = 1, 2, 3$.

If $8n + 4$ is expressed as a sum of four squares, then either all the squares are odd or else all the squares are even. Clearly $8n + 4$ is expressible as a sum of four even squares in $r_4(2n + 1)$ ways.

Thus,

$$r_4(8n + 4) = r_4^*(8n + 4) + r_4(2n + 1).$$

Since $r_4(8n + 4) = 24\sigma(2n + 1)$ and $r_4(2n + 1) = 8\sigma(2n + 1)$, it follows that

$$r_4(8n + 4) = 3r_4(2n + 1) = \frac{3}{2}r_4^*(8n + 4), \qquad (2.7)$$

so that the assertion of the lemma is proved for $s = 4$.

If $5 \leq s \leq 7$ and $8n + s$ is a sum of s squares, then either all the squares are odd or else exactly $s - 4$ of the squares are odd. In the second case the four places (out of s) in which the four even squares occur can be selected in $\binom{s}{4}$ ways; if the placement of the four even squares has been selected and if the odd squares are $k_1^2, \ldots, k_{s-4}^2$, then the even squares can be chosen in $r_4(\frac{8n+s-k_1^2-\cdots-k_{s-4}^2}{4})$ ways. Thus,

$$r_s(8n+s) = r_s^*(8n+s) + \binom{s}{4} \sum_{\substack{k_1,\ldots,k_{s-4} \text{ odd} \\ k_1^2+\cdots+k_{s-4}^2 \leq 8n+s}} r_4\left(\frac{8n+s-k_1^2-\cdots-k_{s-4}^2}{4}\right). \quad (2.8)$$

But

$$r_s^*(8n+s) = \sum_{\substack{k_1,\ldots,k_{s-4} \text{ odd} \\ k_1^2+\cdots+k_{s-4}^2 \leq 8n+s}} r_4^*\left(8n+s-k_1^2-\cdots-k_{s-4}^2\right).$$

Since $8n+s-k_1^2-\cdots-k_{s-4}^2 \equiv 4 \pmod 8$, we have from the second equality in (2.7)

$$r_4\left(\frac{8n+s-k_1^2-\cdots-k_{s-4}^2}{4}\right) = \frac{1}{2}r_4^*\left(8n+s-k_1^2-\cdots-k_{s-4}^2\right).$$

Thus, the second term on the right side of (2.8) is equal to $\frac{1}{2}\binom{s}{4}$ times the first term and so

$$r_s(8n+s) = \left[1 + \frac{1}{2}\binom{s}{4}\right] r_s^*(8n+s).$$

Thus, the assertion of the lemma is proved for $s = 5, 6, 7$. This completes the proof of Lemma 2.7.   □

Finally we recall the following basic properties of the function ϑ defined by (1.1):

$$\vartheta(-1/\tau) = (-i\tau)^{1/2}\vartheta(\tau), \quad (2.9)$$

$$\vartheta(1-1/\tau) = (-i\tau)^{1/2} \sum_{n=-\infty}^{\infty} e^{\pi i \tau (n+1/2)^2} = 2(-i\tau)^{1/2} e^{\pi i \tau/4} \sum_{n=0}^{\infty} e^{\pi i \tau n(n+1)}, \quad (2.10)$$

$$\vartheta(\tau) = \prod_{n=1}^{\infty}(1 - e^{2n\pi i \tau})(1 + e^{(2n-1)\pi i \tau})^2, \quad (2.11)$$

$$\vartheta(1-1/\tau) = 2(-i\tau)^{1/2} e^{\pi i \tau/4} \prod_{n=1}^{\infty}(1 - e^{2n\pi i \tau})(1 + e^{2n\pi i \tau})^2. \quad (2.12)$$

These are proved, for example, in [9, Ch. 10].

3. Behavior of $S(-1/\tau)$

Lemma 3.1. *If $\Im\tau > 0$, then*

$$\vartheta\left(-\frac{1}{\tau}\right)P\left(-\frac{1}{\tau}\right) - \frac{24}{\pi i}\vartheta'\left(-\frac{1}{\tau}\right) = (-i\tau)^{5/2}\{-\vartheta(\tau)P(\tau) + 24\vartheta'(\tau)/(\pi i) - 3\vartheta(\tau)^5\},$$

where ϑ is given by (1.1) and P is given by (1.12).

Proof: By formula (64.4) of [9] (with $a = 0$, $b = -1$, $c = 1$, $d = 0$) we have (for $\Im z > 0$)

$$z^{-2}P(-2/z) = P(2z) + 6/(\pi i z).$$

Taking $z = 2\tau$, we get

$$P(-1/\tau) = -4(-i\tau)^2 P(4\tau) + 12\tau/(\pi i).$$

Now Jacobi's formula $r_4(n) = 8\sigma^*(n)$ has the power series interpretation $4P(4\tau) = P(\tau) + 3\vartheta(\tau)^4$. Thus we may rewrite the preceding displayed formula as

$$P(-1/\tau) = (-i\tau)^2(-P(\tau) - 3\vartheta(\tau)^4) + 12\tau/(\pi i).$$

By (2.9) we therefore have

$$\vartheta(-1/\tau)P(-1/\tau) = (-i\tau)^{5/2}(-\vartheta(\tau)P(\tau) - 3\vartheta(\tau)^5) + 12(-i\tau)^{3/2}\vartheta(\tau)/\pi.$$

But by differentiating (2.9) we readily find

$$\frac{24}{\pi i}\vartheta'\left(-\frac{1}{\tau}\right) = \frac{12}{\pi}(-i\tau)^{3/2}\vartheta(\tau) - \frac{24}{\pi i}(-i\tau)^{5/2}\vartheta'(\tau).$$

The assertion of the lemma now follows by subtraction.   □

Lemma 3.2. *If G is defined by (1.14), then for $\Im\tau > 0$ we have*

$$G(\tau) = \frac{1}{(-i\tau)^{5/2}} + \frac{1}{2}\sum_{k \text{ odd}}\sum_{h \neq 0}\frac{\eta(h,k)^5}{(hi - ki\tau)^{5/2}},$$

where $\eta(h,k)$ is defined by (2.1).

Remark. If we were to take the full singular series in the definition of G instead of just the odd terms, then we would get (as in [1, Section 2]):

$$1 + \frac{1}{(-i\tau)^{5/2}} + \frac{1}{2}\sum_{k \neq 0}\sum_{h \neq 0}\frac{\eta(h,k)^5}{(hi - ki\tau)^{5/2}} = \vartheta(\tau)^5.$$

Thus, taking only the odd terms in the singular series in the definition of G has the effect that the Eisenstein series for G consists of half of the terms in the Eisenstein series for $\vartheta(\tau)^5$.

Proof: Since $A_k(n) = B_k(n)$ for k odd, we have by (1.14) and (2.2)

$$G(\tau) = \sum_{n=1}^{\infty} e^{\pi i \tau n} (4\pi^2/3) n^{3/2} \sum_{k \text{ odd} > 0} \sum_{h \bmod 2k} k^{-5/2} \eta(h, k)^5 e^{-\pi i h n/k}.$$

Changing the order of summation gives

$$G(\tau) = \sum_{k \text{ odd} > 0} \sum_{h \bmod 2k} k^{-5/2} \eta(h, k)^5 (4\pi^2/3) \sum_{n=1}^{\infty} n^{3/2} e^{\pi i (\tau - h/k) n}.$$

Using Lemma 2.5 with $z = \tau - h/k$, we get

$$G(\tau) = \sum_{k \text{ odd} > 0} \sum_{h \bmod 2k} \sum_{m=-\infty}^{\infty} \frac{\eta(h, k)^5}{(2mki - ki\tau + hi)^{5/2}}$$

$$= \sum_{k \text{ odd} > 0} \sum_{h=-\infty}^{\infty} \frac{\eta(h, k)^5}{(hi - ki\tau)^{5/2}},$$

where we have used the fact that $\eta(h, k)$ depends only on the residue class of h modulo $2k$. In view of (2.4) and the fact that $\eta(0, 1) = 1$, the assertion of the lemma follows.   □

Lemma 3.3. *If $\Im \tau > 0$, then*

$$G(-1/\tau) = (-i\tau)^{5/2} \{\vartheta(\tau)^5 - G(\tau)\}.$$

Proof: By Lemma 3.2 we have

$$G\left(-\frac{1}{\tau}\right) = \frac{1}{(i/\tau)^{5/2}} + \frac{1}{2} \sum_{k \text{ odd}} \sum_{h \neq 0} \frac{\eta(h, k)^5}{(hi + ki/\tau)^{5/2}}$$

$$= (-i\tau)^{5/2} \left(1 + \frac{1}{2} \sum_{k \text{ odd}} \sum_{h \neq 0} \frac{\eta(h, k)^5}{(k + h\tau)^{5/2}}\right).$$

Replacing h by $-h$ we get

$$G\left(-\frac{1}{\tau}\right) = (-i\tau)^{5/2} \left(1 + \frac{1}{2} \sum_{k \text{ odd}} \sum_{h \neq 0} \frac{\eta(-h, k)^5}{(k - h\tau)^{5/2}}\right)$$

$$= (-i\tau)^{5/2} \left(1 + \frac{1}{2} \sum_{k \text{ odd}} \sum_{h \neq 0} \frac{\eta(k, h)^5}{(ki - hi\tau)^{5/2}}\right),$$

where we have used (2.5). Using (2.4), absolute convergence, and the fact that $\eta(h, k) = 0$ if h and k have like parity, we obtain

$$G\left(-\frac{1}{\tau}\right) = (-i\tau)^{5/2}\left(1 + \sum_{h \text{ even} >0} \sum_k \frac{\eta(k, h)^5}{(ki - hi\tau)^{5/2}}\right).$$

Since $\eta(k, h)$ depends only on the residue class of k modulo $2h$, we obtain

$$G\left(-\frac{1}{\tau}\right) = (-i\tau)^{5/2}\left(1 + \sum_{h \text{ even} >0} \sum_{k \bmod 2h} \eta(k, h)h^{-5/2} \sum_{m=-\infty}^{\infty} \frac{1}{(2mi - i(\tau - k/h))^{5/2}}\right).$$

Using Lemma 2.5 we get

$$G\left(-\frac{1}{\tau}\right) = (-i\tau)^{5/2}\left(1 + \sum_{h \text{ even} >0} h^{-5/2} \sum_{k \bmod 2h} \eta(k, h)^5 \frac{4\pi^2}{3} \sum_{n=1}^{\infty} n^{3/2} e^{\pi i(\tau - k/h)n}\right)$$

$$= (-i\tau)^{5/2}\left(1 + \frac{4\pi^2}{3} \sum_{n=1}^{\infty} e^{\pi i\tau n} n^{3/2} \sum_{h \text{ even} >0} B_h(n)\right)$$

$$= (-i\tau)^{5/2}\left(1 + \frac{4\pi^2}{3} \sum_{n=1}^{\infty} e^{\pi i\tau n} n^{3/2}(\chi_2(n) - 1) \sum_{h \text{ odd} >0} B_h(n)\right)$$

$$= (-i\tau)^{5/2}(\vartheta(\tau)^5 - G(\tau)),$$

so that the lemma is proved. □

Lemma 3.4. *If $\Im\tau > 0$, then*

$$S(-1/\tau) = -(-i\tau)^{5/2}S(\tau),$$

where S is defined by (1.15).

Proof: From the assertion of Lemma 3.3 and the classical identity (2.9) we get

$$5G(-1/\tau) - \vartheta(-1/\tau)^5 = (-i\tau)^{5/2}\{-5G(\tau) + 4\vartheta(\tau)^5\}.$$

Adding this to the identity of Lemma 3.1, we obtain the result of the lemma. □

4. Behavior of $S(1 - 1/\tau)$

Lemma 4.1. *If $\Im\tau > 0$, then*

$$\vartheta\left(1 - \frac{1}{\tau}\right)P\left(1 - \frac{1}{\tau}\right) - \frac{24}{\pi i}\vartheta'\left(1 - \frac{1}{\tau}\right)$$

$$= (-i\tau)^{5/2} e^{\pi i \tau/4} \left[48 \sum_{n=1}^{\infty} (-1)^n \sigma(n) e^{\pi i \tau n} \cdot \sum_{n=0}^{\infty} e^{\pi i \tau n(n+1)} \right.$$

$$\left. + 48 \sum_{n=0}^{\infty} \left(n^2 + n + \frac{1}{4} \right) e^{\pi i \tau n(n+1)} \right]$$

$$= (-i\tau)^{5/2} e^{\pi i \tau/4} (10 - 48 e^{\pi i \tau} + 250 e^{\pi i \tau} - \cdots).$$

Proof: By formula (64.4) of [9] (with $a = 1, b = 0, c = 2, d = 1$) we have (for $\Im z > 0$)

$$P\left(\frac{2z}{2z+1} \right) \frac{1}{(2z+1)^2} = P(2z) + \frac{12}{\pi i (2z+1)}.$$

Taking $z = (\tau - 1)/2$ we get

$$P\left(1 - \frac{1}{\tau} \right) = \tau^2 P(\tau - 1) + (12\tau)/(\pi i).$$

By (2.10)

$$\vartheta\left(1 - \frac{1}{\tau} \right) = 2(-i\tau)^{1/2} \sum_{n=0}^{\infty} e^{\pi i \tau (n+1/2)^2}$$

and so by multiplication we have

$$\vartheta\left(1 - \frac{1}{\tau} \right) P\left(1 - \frac{1}{\tau} \right) = -2(-i\tau)^{5/2} P(\tau - 1) \sum_{n=0}^{\infty} e^{\pi i \tau (n+1/2)^2} + \frac{12\tau}{\pi i} \vartheta\left(1 - \frac{1}{\tau} \right).$$

But by differentiation of (2.10) we find

$$\frac{24}{\pi i} \vartheta'\left(1 - \frac{1}{\tau} \right) = \frac{12\tau}{\pi i} \vartheta\left(1 - \frac{1}{\tau} \right) - 48(-i\tau)^{5/2} \sum_{n=0}^{\infty} \left(n + \frac{1}{2} \right)^2 e^{\pi i \tau (n+1/2)^2}.$$

Subtraction then gives

$$\vartheta\left(1 - \frac{1}{\tau} \right) P\left(1 - \frac{1}{\tau} \right) - \frac{24}{\pi i} \vartheta'\left(1 - \frac{1}{\tau} \right)$$

$$= (-i\tau)^{5/2} \left[-2P(\tau - 1) \sum_{n=0}^{\infty} e^{\pi i \tau (n+1/2)^2} + 48 \sum_{n=0}^{\infty} \left(n + \frac{1}{2} \right)^2 e^{\pi i \tau (n+1/2)^2} \right].$$

But

$$-2P(\tau - 1) \sum_{n=0}^{\infty} e^{\pi i \tau n(n+1)} + 48 \sum_{n=0}^{\infty} \left(n + \frac{1}{2} \right)^2 e^{\pi i \tau n(n+1)}$$

$$= 48 \sum_{n=0}^{\infty} (-1)^n \sigma(n) e^{\pi i \tau n} \cdot \sum_{n=0}^{\infty} e^{\pi i \tau n(n+1)} + 48 \sum_{n=0}^{\infty} \left(n^2 + n + \frac{1}{4} \right) e^{\pi i \tau n(n+1)}$$

$$= (-2 - 48 e^{\pi i \tau} + 144 e^{2\pi i \tau} - \cdots)(1 + e^{2\pi i \tau} + \cdots) + 12 + 108 e^{2\pi i \tau} + \cdots$$

$$= 10 - 48 e^{\pi i \tau} + 250 e^{2\pi i \tau} - \cdots.$$

Thus, the lemma is proved. □

Lemma 4.2. *If* $\Im \tau > 0$, *then*

$$G\left(1 - \frac{1}{\tau} \right) = \frac{1}{7} (-i\tau)^{5/2} \sum_{\substack{m \equiv 5 \,(\mathrm{mod}\, 8) \\ m > 0}} r_5(m) e^{\pi i \tau m/4} - \frac{1}{5} (-i\tau)^{5/2} \sum_{\substack{m \equiv 1 \,(\mathrm{mod}\, 8) \\ m > 0}} r_5(m) e^{\pi i \tau m/4}.$$

Proof: Using Lemmas 3.2 and (2.4) we have (since $\eta(0, 1) = 1$)

$$G(1 + \tau) = \sum_{k \text{ odd} > 0} \sum_{h=-\infty}^{\infty} \frac{\eta(h, k)^5}{(hi - ki - ki\tau)^{5/2}}$$

$$= \sum_{k \text{ odd} > 0} \sum_{h \text{ odd}} \frac{\eta(h + k, k)^5}{(hi - ki\tau)^{5/2}},$$

where we have used the fact that $\eta(h, k) = 0$ if h and k have like parity. By Lemmas 2.3 and 2.4, and absolute convergence we have

$$G(1 + \tau) = \sum_{k \text{ odd} > 0} \sum_{\substack{h \text{ odd} \\ (h,k)=1}} \left(\frac{k}{|h|} \right) \frac{e^{-5\pi i h(k-1)/4}}{(hi - ki\tau)^{5/2}}$$

$$= \sum_{h \text{ odd} > 0} \sum_{\substack{k \text{ odd} \\ (k,h)=1}} \left(\frac{k}{h} \right) \frac{e^{-5\pi i h(k-1)/4}}{(hi - ki\tau)^{5/2}}.$$

In view of our convention about square roots we have (for $h > 0$)

$$(hi + ki/\tau)^{1/2} = e^{\pi i/4} (-hi\tau - ki)^{1/2} / (-i\tau)^{1/2}.$$

Hence,

$$G\left(1 - \frac{1}{\tau} \right) = (-i\tau)^{5/2} \sum_{h \text{ odd} > 0} \sum_{\substack{k \text{ odd} \\ (k,h)=1}} \left(\frac{k}{h} \right) \frac{e^{-5\pi i (hk-h+1)/4}}{(-hi\tau - ki)^{5/2}}$$

$$= (-i\tau)^{5/2} \sum_{h \text{ odd} > 0} e^{5\pi i (h-1)/4} \sum_{\substack{k \bmod 8h \\ (k,8h)=1}} \left(\frac{k}{h} \right) e^{-5\pi i hk/4}$$

$$\times \sum_{m=-\infty}^{\infty} \{ -hi\tau - (k - 8hm)i \}^{-5/2}.$$

Using Lemma 2.5 we have

$$G\left(1 - \frac{1}{\tau}\right) = (-i\tau)^{5/2} \sum_{h \text{ odd} > 0} e^{5\pi i(h-1)/4} \sum_{\substack{k \bmod 8h \\ (k,8h)=1}} \left(\frac{k}{h}\right) e^{-5\pi ihk/4}$$

$$\times (4h)^{-5/2} \frac{4\pi^2}{3} \sum_{m=1}^{\infty} m^{3/2} e^{m\pi i(\tau/4 + k/(4h))}$$

$$= (-i\tau)^{5/2} \sum_{m=1}^{\infty} e^{\pi i\tau m/4} \frac{\pi^2}{6} m^{3/2} \sum_{h \text{ odd} > 0} D_h(m),$$

where

$$D_h(m) = \frac{1}{4} h^{-5/2} e^{5\pi i(h-1)/4} \sum_{\substack{k \bmod 8h \\ (k,8h)=1}} \left(\frac{k}{h}\right) e^{2\pi ik(m-5h^2)/(8h)}$$

$$= \frac{1}{4} h^{-5/2} e^{5\pi i(h-1)/4} \sum_{\substack{j \bmod h \\ (j,h)=1}} \sum_{\substack{g \bmod 8 \\ (g,8)=1}} \left(\frac{hg - 8j}{h}\right) e^{2\pi i(hg-8j)(m-5h^2)/(8h)}$$

$$= \frac{1}{4} h^{-5/2} e^{5\pi i(h-1)/4} \sum_{\substack{j \bmod h \\ (j,h)=1}} \left(\frac{-8j}{h}\right) e^{-2\pi imj/h} \sum_{\substack{g \bmod 8 \\ (g,8)=1}} e^{2\pi ig(m-5h^2)/8}.$$

The inner sum here is zero if $m \not\equiv 1 \pmod 4$ and is equal to $(-1)^{(m-5)/4} 4$ if $m \equiv 1 \pmod 4$. Accordingly, $D_h(m) = 0$ if $m \not\equiv 1 \pmod 4$. Using the fact that $\left(\frac{-1}{h}\right) = e^{-\pi i(h-1)/2}$ we find that for $m \equiv 1 \pmod 4$ we have

$$D_h(m) = (-1)^{(m-5)/4} h^{-5/2} e^{-5\pi i(h-1)/4} \sum_{\substack{j \bmod h \\ (j,h)=1}} \left(\frac{2j}{h}\right) e^{-2\pi imj/h}$$

$$= (-1)^{(m-5)/4} h^{-5/2} \sum_{\substack{j \bmod h \\ (j,h)=1}} \eta(2j, h)^5 e^{-\pi im(2j)/h}$$

$$= (-1)^{(m-5)/4} B_h(m),$$

where we have used Lemma 2.1. Since $A_h(m) = B_h(m)$ when h is odd, we therefore have

$$G\left(1 - \frac{1}{\tau}\right) = (-i\tau)^{5/2} \sum_{\substack{m \equiv 1 \pmod 4 \\ m > 0}} (-1)^{(m-5)/4} e^{\pi i\tau m/4} (\pi^2/6) m^{3/2} \sum_{h \text{ odd} > 0} A_h(m)$$

$$= (-i\tau)^{5/2} \sum_{\substack{m \equiv 1 \pmod 4 \\ m > 0}} (-1)^{(m-5)/4} e^{\pi i\tau m/4} r_5(m)/(8\chi_2(m)),$$

by Hardy's result (1.7). By Lemma 2.6 we have the evaluations $\chi_2(m) = 5/8$ when $m \equiv 1$ (mod 8) and $\chi_2(m) = 7/8$ when $m \equiv 5$ (mod 8), so that the assertion of the lemma follows. □

Lemma 4.3. If $\Im\tau > 0$, then

$$5G\left(1 - \frac{1}{\tau}\right) - \vartheta\left(1 - \frac{1}{\tau}\right)^5$$

$$= \frac{3}{7}(-i\tau)^{5/2} \sum_{\substack{m \equiv 5 \,(\mathrm{mod}\,8) \\ m > 0}} r_5(m)e^{\pi i \tau m/4} - (-i\tau)^{5/2} \sum_{\substack{m \equiv 1 \,(\mathrm{mod}\,8) \\ m > 0}} r_5(m)e^{\pi i \tau m/4}$$

$$= (-i\tau)^{5/2}e^{\pi i \tau/4}\{-10 + 48e^{\pi i \tau} - 250e^{2\pi i \tau} + \cdots\}.$$

Proof: By (2.10)

$$\vartheta(1 - 1/\tau)^5 = (-i\tau)^{5/2}\left\{\sum_{n=-\infty}^{\infty} e^{\pi i \tau(2n+1)^2/4}\right\}^5$$

$$= (-i\tau)^{5/2} \sum_{\substack{m \equiv 5 \,(\mathrm{mod}\,8) \\ m > 0}} r_5^*(m)e^{\pi i \tau m/4}.$$

Hence, by Lemma 2.7

$$\vartheta(1 - 1/\tau)^5 = \frac{2}{7}(-i\tau)^{5/2} \sum_{\substack{m \equiv 5 \,(\mathrm{mod}\,8) \\ m > 0}} r_5(m)e^{\pi i \tau m/4}.$$

Multiplying the result of Lemma 4.2 by 5 and subtracting the preceding equality, we get the first assertion of the Lemma. We get the second assertion by using the values $r_5(1) = 10$, $r_5(5) = 112$, $r_5(9) = 250$. □

Lemma 4.4. If $\Im\tau > 0$, then

$$S\left(1 - \frac{1}{\tau}\right) = (-i\tau)^{5/2}e^{\pi i \tau/4} \sum_{n=0}^{\infty} a_n e^{\pi i \tau n},$$

where $a_0 = a_1 = 0$.

Proof: We add the results of Lemmas 4.1 and 4.3. □

5. Proofs of Theorems 1 and 2 and their Corollaries

Proof of Theorem 1: Let us put $H(\tau) = S(\tau)/\vartheta(\tau)^5$. By (2.9) and Lemma 3.4 $H(-1/\tau) = -H(\tau)$; trivially $H(\tau + 2) = H(\tau)$. Thus $H(\tau)^2$ is invariant under Γ_ϑ. From the definition (1.15) the series for $S(\tau)$ in nonnegative powers of $e^{\pi i \tau}$ has zero constant term, since the power series for both ϑ and P have constant term 1 and the power series for both G and ϑ' have constant term 0. In view of (2.11) it follows that $H(\tau)$ has an expansion in positive powers of $e^{\pi i \tau}$ for $\Im \tau > 0$. On the other hand, by Lemma 4.4 and (2.12) we see that $H(1 - 1/\tau)$ has an expansion in positive powers of $e^{\pi i \tau}$ for $\Im \tau > 0$. Thus $H(\tau)^2$ is invariant under Γ_ϑ and approaches zero as we go to the cusps of the fundamental region for Γ_ϑ. Thus $H(\tau)^2$ is identically zero. $\qquad\square$

Proof of Corollary to Theorem 1:   The assertion of Theorem 1 may be written as

$$-\vartheta(\tau)P(\tau) + \frac{24}{\pi i}\vartheta'(\tau) = 5G(\tau) - \vartheta(\tau)^5$$

or

$$24 \sum_{j=-\infty}^{\infty} e^{\pi i \tau j^2} \sum_{m=0}^{\infty} \sigma(m)e^{\pi i \tau m} + 48 \sum_{n=1}^{\infty} n^2 e^{\pi i \tau n^2}$$

$$= \frac{20\pi^2}{3} \sum_{n=1}^{\infty} e^{\pi i \tau n} n^{3/2} \sum_{k \text{ odd} > 0} A_k(n) - \left(1 + \sum_{n=1}^{\infty} r_5(n)e^{\pi i \tau n}\right).$$

If we use Hardy's result (1.7) we get

$$24 \sum_{j=-\infty}^{\infty} e^{\pi i \tau j^2} \sum_{m=0}^{\infty} \sigma(m)e^{\pi i \tau n} + 48 \sum_{n=1}^{\infty} s(n)n e^{\pi i \tau n}$$

$$= -1 + \sum_{n=1}^{\infty} e^{\pi i \tau n} \frac{4\pi^2}{3} n^{3/2} (5 - \chi_2(n)) \sum_{k \text{ odd} > 0} A_k(n)$$

$$= -1 + \sum_{n=1}^{\infty} \left(\frac{5}{\chi_2(n)} - 1\right) r_5(n)e^{\pi i \tau n}.$$

Equating coefficients of $e^{\pi i \tau n}$ and using the formula

$$r_5(n) = \sum_{|j| \le n^{1/2}} 8\sigma^*(n - j^2),$$

we get the assertion of the Corollary. $\qquad\square$

Proof of Theorem 2 and its Corollary: Once we know that $S(\tau)$ is identically zero, Theorem 2 follows by adding the results of Lemmas 4.1 and 4.3. The Corollary to Theorem 2 follows by comparing coefficients in the identity of Theorem 2. $\qquad\square$

6. Connection with a result of H. Cohen

In 1975 Henri Cohen proved the identity

$$L(-1, \chi_D) = \frac{1}{120} r_5(D) - \frac{1}{6} \sum_{\substack{|j| < \sqrt{D} \\ D - j^2 \text{ odd}}} \sigma(D - j^2), \tag{6.1}$$

where D is the discriminant of a real quadratic field and χ_D is the corresponding quadratic character. (In particular, $D \equiv 0$ or $1 \pmod 4$ and D is not a square. See [3, Corollary 3.2] or [8, Proposition 7, p. 194]. There is a striking similarity between (6.1) and our identity (1.9), which, in light of (1.7), may be given in the form

$$\frac{\pi^2}{18} D^{3/2} \sum_{k \text{ odd} > 0} A_k(D) = \frac{1}{120} r_5(D) + \frac{1}{5} \sum_{|j| < \sqrt{D}} \sigma(D - j^2), \tag{6.2}$$

for these values of D.

The link between (6.2) and (6.1) resides in the underlying modular identities, respectively, (1.15) and the Cohen identity [8, p. 194],

$$H_{5/2}(\tau) = \frac{1}{120} \theta^5(\tau) - \frac{1}{6} \theta(\tau) F(\tau). \tag{6.3}$$

In (6.3),

$$\theta(\tau) = \vartheta(2\tau) = \sum_{n=-\infty}^{\infty} e^{2\pi i n^2 \tau},$$

$$F(\tau) = -\frac{1}{24}[E_2(\tau) - 3E_2(2\tau) + 2E_2(4\tau)]$$

$$= \sum_{n \text{ odd} > 0} \sigma(n) e^{2\pi i n \tau},$$

where $E_2(\tau) = P(2\tau)$ (see (1.12)), and $H_{5/2}(\tau)$ is a linear combination of the two Eisenstein series on $\Gamma_0(4)$ of weight $\frac{5}{2}$, with the multiplier system $\tilde{v}$ belonging to $\theta(\tau)$, a modular form of weight $\frac{1}{2}$ on $\Gamma_0(4)$.

The specific linear combination defining $H_{5/2}(\tau)$ is chosen so that the Dth Fourier coefficient is, in effect, $L(-1, \chi_D)$ for D the discriminant of a real quadratic field. (See [8, Proposition 6, p. 193] for the definition of $H_{5/2}(\tau)$.) The definition shows that $H_{5/2}(\tau)$ is a modular form on $\Gamma_0(4)$, of weight $\frac{5}{2}$ and multiplier system $\tilde{v}$. Also, the transformation law

$$(c\tau + d)^{-2} E_2(M\tau) = E_2(\tau) + \frac{6c}{\pi i (c\tau + d)}, \quad M = \begin{pmatrix} a & b \\ c & d \end{pmatrix} \in \Gamma(1), \tag{6.4}$$

implies that $F(\tau)$ is a modular form on $\Gamma_0(4)$ of weight 2 and multiplier system $\equiv 1$ (see [9, Section 64] for a proof of (6.4); the transformation laws for $P(\tau)$ occurring at the start of the proofs of Lemmas 3.1 and 4.1 are special cases of (6.4).)

We can write the identity (1.15) as

$$G(\tau) = \frac{1}{5}\vartheta^5(\tau) + \frac{6i}{5\pi}\left\{\frac{\pi i}{6}\vartheta(\tau)P(\tau) - 4\vartheta'(\tau)\right\}.$$

Now, as follows directly from Lemma 3.1, $\frac{\pi i}{6}\vartheta(\tau)P(\tau) - 4\vartheta'(\tau)$ is a modular form on $\Gamma(2)$, the principal congruence subgroup of level 2, of weight $\frac{5}{2}$ and multiplier system v^5, where v is the multiplier system belonging to the modular form $\vartheta(\tau)$. (Of course, v and $\tilde{v}$ are closely related, since $\theta(\tau) = \vartheta(2\tau)$.) Thus, (1.15) is an identity between modular forms of weights $\frac{5}{2}$ and multiplier v^5 on $\Gamma(2)$, but it is not hard to show that replacing τ by 2τ transforms (1.15) to an identity of weight $\frac{5}{2}$ and multiplier system $\tilde{v}$ on $\Gamma_0(4)$, that is, to an identity involving the *same group, weight and multiplier system* occurring in (6.3):

$$G(2\tau) = \frac{1}{5}\theta^5(\tau) + \frac{6i}{5\pi}K(\tau), \tag{6.5}$$

with $K(\tau) = \frac{\pi i}{6}\theta(\tau)E_2(\tau) - 2\theta'(\tau)$. In summary, (6.3) expresses $H_{5/2}(\tau)$ as a linear combination of $\theta^5(\tau)$ and $\theta(\tau)F(\tau)$, while (6.5) expresses $G(2\tau)$ as a linear combination of $\theta^5(\tau)$ and $K(\tau)$. We shall obtain a new identity by expressing $H_{5/2}(\tau)$ as a linear combination of $\theta^5(\tau)$ and $K(\tau)$.

To see that such an identity exists, we first obtain information about the expansions of $\theta(\tau)$ at $i\infty$ and 0, two of the three parabolic cusps of $\Gamma_0(4)$. (The third cusp is $-\frac{1}{2}$.) The result we need is

$$\text{(a)} \quad \theta(\tau) = 1 + 2\sum_{n=1}^{\infty} e^{2\pi i n^2 \tau}, \quad \text{at } i\infty;$$

$$\text{(b)} \quad \theta(\tau) = \frac{e^{-\pi i/4}i}{\sqrt{2}}\tau^{-1/2}\left(1 + 2\sum_{n=1}^{\infty} e^{2\pi i n^2(-1/\tau)/4}\right), \quad \text{at } 0. \tag{6.6}$$

Note that (a) is the definition of $\theta(\tau) = \vartheta(2\tau)$, while (b) follows from (a) and the transformation law (2.9) for $\vartheta(\tau)$.

The proof of the identity requires the following lemma.

Lemma 6.1. *The space of cusp forms on $\Gamma_0(4)$, of weight $\frac{5}{2}$ and multiplier system $\tilde{v}$, has dimension 0.*

Proof: We require the explicit formula,

$$\tilde{v}(M) = \left(\frac{c}{d}\right)\epsilon_d^{-1}, \quad M = \begin{pmatrix} a & b \\ 4c & d \end{pmatrix} \in \Gamma_0(4), \tag{6.7}$$

with

$$\epsilon_d = \begin{cases} 1, & d \equiv 1 \pmod 4 \\ i, & d \equiv 3 \pmod 4. \end{cases}$$

(See [8], p. 178]; (6.7) follows easily from [7, Theorem 3, p. 51].) We recall that $\Gamma_0(4)$ has three parabolic cusps: $i\infty$, 0 and $-\frac{1}{2}$. The cusp at $i\infty$ has width 1, the cyclic subgroup of $\Gamma_0(4)$ fixing $i\infty$ generated by $S = \left(\begin{smallmatrix} 1 & 1 \\ 0 & 1 \end{smallmatrix}\right)$. By (6.7), $\tilde{v}(S) = 1$. The cusp at 0 has width 4, with the corresponding cyclic subgroup of $\Gamma_0(4)$ generated by $P_1 = \left(\begin{smallmatrix} 1 & 0 \\ -4 & 1 \end{smallmatrix}\right)$. By (6.7), $\tilde{v}(P_1) = 1$. The cusp at $-\frac{1}{2}$ has width 1, with corresponding cyclic subgroup generated by $P_2 = \left(\begin{smallmatrix} 3 & 1 \\ -4 & -1 \end{smallmatrix}\right)$. By (6.7), $\tilde{v}(P_2) = e^{\pi i/2}$.

Suppose that $f(\tau)$ is a cusp form on $\Gamma_0(4)$, of weight $\frac{5}{2}$, with multiplier system $\tilde{v}$. By the remarks about $\tilde{v}$ in the previous paragraph, $f(\tau)$ has a zero of order $n_1 \in Z^+$ (in the local variable) at $i\infty$, a zero of order $n_2 \in Z^+$ at 0, and a zero of order $n_3 + \frac{1}{4}$ ($n_3 \in Z$, $n_3 \geq 0$) at $-\frac{1}{2}$.

On the other hand, (6.6) shows that $\theta(\tau)$ is nonzero at $i\infty$ and 0. Furthermore, by modifying the calculation of [7, p. 48, last paragraph] we derive the expansion of $\theta(\tau)$ at the cusp $-\frac{1}{2}$:

$$\theta(\tau) = \left(\tau + \frac{1}{2}\right)^{-1/2} \sum_{n=0}^{\infty} a_n e^{2\pi i(n+1/4)\tau/(2\tau+1)}, \quad a_0 \neq 0.$$

The function $f(\tau)/\theta^5(\tau)$ is now a modular function (invariant) on $\Gamma_0(4)$, holomorphic in $\Im\tau > 0$ (since $\theta(\tau) \neq 0$ for $\Im\tau > 0$), with zeros of positive integral order at $i\infty$ and 0, and a pole of integral order at most 1 at $-\frac{1}{2}$. It follows that $f(\tau)/\theta^5(\tau)$ has more zeros than poles in a fundamental region for $\Gamma_0(4)$. This implies $f(\tau) \equiv 0$, and the proof of Lemma 6.1 is complete.   □

Proof of Theorem 3:   We wish to prove an identity of the form

$$H_{5/2}(\tau) = \alpha\theta^5(\tau) + \beta K(\tau), \tag{6.8}$$

with complex numbers α, β. But (6.8) will follow if we choose α and β so that the constant terms of the two sides are equal in their expansions at the cusps $i\infty$ and 0. For, such a choice forces $H_{5/2}(\tau) - \alpha\theta^5(\tau) - \beta K(\tau)$ to be a cusp form (since $\tilde{v}(P_2) = e^{\pi i/2} \neq 1$) and hence zero by Lemma 6.1.

We shall obtain the constant terms for $H_{5/2}(\tau)$ from (6.3), so we need the corresponding constant terms of

$$F(\tau) = \sum_{\substack{n=1 \\ n \text{ odd}}}^{\infty} \sigma(n)e^{2\pi in\tau}. \tag{6.9}$$

This expression shows that the constant term of $F(\tau)$ at $i\infty$ is 0. To obtain the constant term at the cusp 0, we apply the special case $M = T = \left(\begin{smallmatrix} 0 & -1 \\ 1 & 0 \end{smallmatrix}\right)$ of (6.4) to the definition

$$F(\tau) = -\frac{1}{24}\{E_2(\tau) - 3E_2(2\tau) + 2E_2(4\tau)\}.$$

This yields the expansion

$$F(\tau) = \tau^{-2}\left[-\frac{1}{64} + \sum_{n=1}^{\infty} b_n e^{2\pi i n(-1/\tau)/4} \right]. \tag{6.10}$$

Now, (6.3), (6.6), (6.9) and (6.10) imply that $H_{5/2}(\tau)$ has constant term $\frac{1}{120}$ at $i\infty$ and constant term $\frac{ie^{-\pi i/4}}{\sqrt{2}} \frac{1}{1920}$ at 0. Furthermore, it is not too hard to show from the transformation law

$$\tau^{-2} E_2(-1/\tau) = E_2(\tau) + \frac{6}{\pi i \tau}$$

(which is a special case of (6.4)) and (6.6) that the basis function

$$K(\tau) = \frac{\pi i}{6}\theta(\tau) E_2(\tau) - 2\theta'(\tau)$$

has constant term $\pi i/6$ at $i\infty$ and constant term $-\frac{\pi}{6\sqrt{2}} e^{-\pi i/4}$ at 0.

Thus, in order to prove (6.8) it suffices to choose α and β subject to the two linear relations

$$\frac{1}{120} = \alpha + \frac{\pi i}{6}\beta, \quad \frac{ie^{-\pi i/4}}{1920\sqrt{2}} = -\frac{e^{-\pi i/4}i}{4\sqrt{2}}\alpha - \frac{\pi e^{-\pi i/4}}{6\sqrt{2}}\beta.$$

The solution is $\alpha = \frac{1}{160}$, $\beta = -i/(80\pi)$, and so (6.8) becomes

$$H_{5/2}(\tau) = \frac{1}{160}\theta^5(\tau) - \frac{i}{80\pi}K(\tau). \tag{6.11}$$

Comparing expansions at $i\infty$ on the two sides of (6.11) yields the identity

$$h(n) = \frac{1}{160}r_5(n) - \frac{1}{20}\left\{ \sum_{|j|\le\sqrt{n}} \sigma(n-j^2) + 2ns(n) \right\},$$

where $h(n)$ is the coefficient of $e^{2\pi i n\tau}$ in $H_{5/2}(\tau)$, and $s(n) = 1$ if n is a square and $s(n) = 0$ otherwise. (See the Corollary to Theorem 1.) Combining this result with Cohen's calculation of $h(n)$ ([3, pp. 272–273] or [8, Proposition 6, p. 193]), we have

$$L(-1, \chi_D) = \frac{1}{160}r_5(D) - \frac{1}{20}\sum_{|j|<\sqrt{D}} \sigma(D-j^2), \tag{6.12}$$

when D is the discriminant of a real quadratic field and χ_D is the corresponding quadratic character. Thus Theorem 3 is proved.   □

The assertions of the Corollary to Theorem 3 are obtained by taking suitable linear combination of (6.1) and (6.12). Multiplying both (6.1) and (6.12) by 480 and subtracting, we get (1.16). Multiplying (6.1) by 3 and (6.12) by 4 and subtracting, we get (1.17). Multiplying (6.1) by 3/7 and (6.12) by 10/7 and subtracting, we get (1.18).

References

1. P.T. Bateman, "On the representations of a number as the sum of three squares," *Trans. Amer. Math. Soc.* **71** (1951), 70–101.
2. P.T. Bateman, "The asymptotic formula for the number of representations of an integer as a sum of five squares," *Analytic Number Theory: Proceedings of a Conference in Honor of Heini Halberstam*, Birkhaüser, vol. 1, pp. 129–139, 1996.
3. Henri Cohen, "Sums involving the values at negative integers of L-functions of quadratic characters," *Math. Ann.* **217** (1975), 271–285.
4. Leonard E. Dickson, *Studies in the Theory of Numbers*, University of Chicago Press, 1930.
5. T. Estermann, "On the representations of a number as a sum of squares," *Acta Arith.* **2** (1936), 47–79.
6. G.H. Hardy, "On the representation of a number as the sum of any number of squares, and in particular of five," *Trans. Amer. Math. Soc.* **21** (1920), 255–284; **29** (1927), 845–847.
7. Marvin I. Knopp, *Modular Functions in Analytic Number Theory*, Markham, 1970 or Chelsea, 1993.
8. Neal Koblitz, *Introduction to Elliptic Curves and Modular Forms*, Springer-Verlag, 1984, 1993.
9. Hans Rademacher, *Topics in Analytic Number Theory*, Springer-Verlag, 1973.
10. Robert A. Rankin, *Modular Forms and Functions*, Cambridge University Press, 1977.
11. Arnold Walfisz, *Gitterpunkte in mehrdimensionalen Kugeln*, Monografie Matematyczne, Tom 33, 1957.

References

THE RAMANUJAN JOURNAL 2, 271–281 (1998)

Linear Forms in Finite Sets of Integers

SHU-PING HAN
Department of Mathematics, CUNY Graduate School, New York, NY 10036

CHRISTOPH KIRFEL   chrisk@lsv.hib.no
Department of Mathematics, Bergen College, N-5030 Landås, Bergen, Norway

MELVYN B. NATHANSON*   nathansn@alpha.lehman.cuny.edu
Department of Mathematics, Lehman College (CUNY), Bronx, NY 10468

Dedicated to Paul Erdős

Received January 5, 1998; Accepted February 6, 1998

Abstract. Let $A_1, \ldots, A_r$ be finite, nonempty sets of integers, and let $h_1, \ldots, h_r$ be positive integers. The linear form $h_1 A_1 + \cdots + h_r A_r$ is the set of all integers of the form $b_1 + \cdots + b_r$, where b_i is an integer that can be represented as the sum of h_i elements of the set A_i. In this paper, the structure of the linear form $h_1 A_1 + \cdots + h_r A_r$ is completely determined for all sufficiently large integers h_i.

Key words: sums of sets of integers, sumsets, growth in semigraphs, additive number theory

1991 Mathematics Subject Classification: Primary 11B05; Secondary 11B13, 11B75, 11P99

1. Sums of sets of integers

Let A be a nonempty set of integers. For every positive integer h, the sumset hA is the set of all integers that can be represented as the sum of exactly h not necessarily distinct elements of A. For example,

$$2\{1, 2, 4\} = \{2, 3, 4, 5, 6, 8\}$$

and

$$3\{0, 2, 5\} = \{0, 2, 4, 5, 6, 7, 9, 10, 12, 15\}.$$

We define $hA = \{0\}$ for $h = 0$. For any set A and integers a_0 and δ, we define

$$a_0 + A = \{a_0 + a : a \in A\},$$
$$a_0 - A = \{a_0 - a : a \in A\},$$

*The work of M.B.N. was supported in part by grants from the PSC-CUNY Research Award Program and the National Security Agency Mathematical Sciences Program.

and

$$\delta * A = \{\delta a : a \in A\}.$$

Let $[x, y]$ denote the interval of integers n such that $x \leq n \leq y$.

A finite set A of integers is called *normalized* if it consists of 0 and a nonempty set of relatively prime positive integers. If A is a finite set of integers with $|A| \geq 2$, we can normalize A as follows. Let a_0 be the least element of A, and let δ be the greatest common divisor of the positive integers of the form $a - a_0$ for $a \in A$. The normalized form of A is the set

$$A^{(N)} = \left\{ \frac{a - a_0}{\delta} : a \in A \right\}.$$

Then

$$A = a_0 + \delta * A^{(N)}$$

and

$$hA = ha_0 + \delta * hA^{(N)}. \tag{1}$$

Note that A is normalized if and only if $A = A^{(N)}$.

Nathanson [5, 6] completely determined the structure of the sumset hA for all nonempty, finite sets A and all sufficiently large integers h. By (1), it suffices to describe the structure of sumsets of normalized sets.

Theorem 1 (Nathanson). *Let A be a normalized finite set of integers, and let a^* be the greatest element of A. There exist integers c and d and sets $C \subseteq [0, c-2]$ and $D \subseteq [0, d-2]$ such that, for h sufficiently large,*

$$hA = C \cup [c, ha^* - d] \cup (ha^* - D).$$

In this paper we generalize this result to linear forms in finite sets of integers. Let $r \geq 1$. If $A_1, \ldots, A_r$ are nonempty sets of integers and $h_1, \ldots, h_r$ are positive integers, then the sumset

$$h_1 A_1 + \cdots + h_r A_r \tag{2}$$

is the set of all integers that can be represented in the form $b_1 + \cdots + b_r$, where $b_i \in h_i A_i$ for $i = 1, \ldots, r$. The sumset (2) is called a *linear form* in the sets $A_1, \ldots, A_r$. We shall describe explicitly the structure of linear forms in finite sets of integers for all sufficiently large values of $h_1, \ldots, h_r$.

2. The structure of linear forms

The system of sets $A_1, \ldots, A_r$ is *normalized*, if each A_i is a finite set of nonnegative integers, if $0 \in A_i$ for $i = 1, \ldots, r$, and if $\bigcup_{i=1}^{r} A_i \backslash \{0\}$ is a nonempty set of relatively prime positive integers. For example, the sets $A_1 = \{0, 6\}$, $A_2 = \{0, 10\}$, and $A_3 = \{0, 15\}$ are a normalized system, since $(A_1 \cup A_2 \cup A_3) \setminus \{0\} = \{6, 10, 15\}$, and $(6, 10, 15) = 1$.

Let $A_1, \ldots, A_r$ be nonempty, finite sets of integers such that $|A_i| \geq 2$ for all i. We shall normalize this system of sets as follows. Let $a_{i,0}$ be the smallest element in A_i. Let δ be the greatest common divisor of the integers in the set

$$\bigcup_{i=1}^{r} \{a_{i,j} - a_{i,0} : a_{i,j} \in A_i\}.$$

Let

$$A_i^{(N)} = \left\{ \frac{a_{i,j} - a_{i,0}}{\delta} : a_{i,j} \in A_i \right\}.$$

The system of sets $A_1^{(N)}, \ldots, A_r^{(N)}$ is normalized, and

$$A_i = a_{i,0} + \delta * A_i^{(N)}$$

for all $i = 1, \ldots, r$. For any positive integers $h_1, \ldots, h_r$, we have

$$\sum_{i=1}^{r} h_i A_i = \left(\sum_{i=1}^{r} h_i a_{i,0} \right) + \delta * \sum_{i=1}^{r} h_i A_i^{(N)}. \tag{3}$$

By (3), it suffices to describe the structure of sums of normalized systems of finite sets of integers.

Let A be a set of nonnegative integers that contains 0, let $\gcd(A)$ denote the greatest common divisor of the elements of A, and let

$$a^* = \max(A).$$

We define the *reflected set*

$$\hat{A} = a^* - A = \{a^* - a : a \in A\}.$$

Then $\hat{A}$ is also a set of nonnegative integers that contains 0,

$$\max(\hat{A}) = \max(A) = a^*,$$
$$\gcd(A) = \gcd(\hat{A}),$$

and

$$\hat{\hat{A}} = A.$$

For any positive integer h, we have $0 \in hA$, and

$$\max(hA) = ha^*,$$

and so

$$hÂ = \left\{ \sum_{j=1}^{h}(a^* - a_j) : a_j \in A \right\}$$

$$= ha^* - \left\{ \sum_{j=1}^{h}a_j : a_j \in A \right\}$$

$$= ha^* - hA$$

$$= \widehat{hA}.$$

Lemma 1. *Let $A_1, \ldots, A_r$ be a normalized system of finite sets of integers, and let $a_i^* = \max(A_i)$ for $i = 1, \ldots, r$. The reflected sets $Â_1, \ldots, Â_r$ also form a normalized system. For any integer x,*

$$x \in \sum_{i=1}^{r} h_i Â_i$$

if and only if

$$\sum_{i=1}^{r} h_i a_i^* - x \in \sum_{i=1}^{r} h_i A_i.$$

Moreover,

$$\left[d, \sum_{i=1}^{r} h_i a_i^* - d' \right] \subseteq \sum_{i=1}^{r} h_i Â_i,$$

if and only if

$$\left[d', \sum_{i=1}^{r} h_i a_i^* - d \right] \subseteq \sum_{i=1}^{r} h_i A_i.$$

Proof: For $i = 1, \ldots, r$, let

$$d_i = \gcd(A_i).$$

If the system $A_1, \ldots, A_r$ is normalized, then

$$1 = \gcd\left(\bigcup_{i=1}^{r} A_i \setminus \{0\} \right) = (d_1, \ldots, d_r).$$

Since

$$\gcd(Â_i) = \gcd(A_i) = d_i,$$

it follows that

$$\gcd\left(\bigcup_{i=1}^{r}\hat{A}_i\setminus\{0\}\right)=(d_1,\ldots,d_r)=1,$$

and so the system $\hat{A}_1,\ldots,\hat{A}_r$ is also normalized.

If

$$x\in\sum_{i=1}^{r}h_i\hat{A}_i=\sum_{i=1}^{r}\widehat{h_iA_i}=\sum_{i=1}^{r}(h_ia_i^*-h_iA_i),$$

then there exist integers $b_i\in h_iA_i$ such that

$$x=\sum_{i=1}^{r}(h_ia_i^*-b_i),$$

and so

$$\sum_{i=1}^{r}h_ia_i^*-x=\sum_{i=1}^{r}b_i\in\sum_{i=1}^{r}h_iA_i.$$

The proof in the opposite direction is similar.

We observe that

$$x\in\left[d',\sum_{i=1}^{r}h_ia_i^*-d\right],$$

if and only if

$$\sum_{i=1}^{r}h_ia_i^*-x\in\left[d,\sum_{i=1}^{r}h_ia_i^*-d'\right],$$

and this suffices to prove the last part of the Lemma. $\qquad\square$

Theorem 2. *Let $A_1,\ldots,A_r$ be a normalized system of finite sets of integers. Let $a_i^*=\max(A_i)$ for $i=1,\ldots,r$. There exist integers c and d and finite sets*

$$C\subseteq[0,c-2]$$

and

$$D\subseteq[0,d-2]$$

and there exist integers $h_1^*,\ldots,h_r^*$ such that, if $h_i\geq h_i^*$ for all $i=1,\ldots,r$, then

$$h_1A_1+\cdots+h_rA_r=C\cup\left[c,\sum_{i=1}^{r}h_ia_i^*-d\right]\cup\left(\sum_{i=1}^{r}h_ia_i^*-D\right).$$

Proof: For $i = 1, \ldots, r$, let

$$A_i = \{a_{i,0}, a_{i,1}, \ldots a_{i,k_i-1}\},$$

where

$$k_i = |A_i|$$

and

$$0 = a_{i,0} < a_{i,1} < \cdots < a_{i,k_i-1}.$$

Renumbering the sets A_i, we can assume that

$$a^* = \max\{a_i^* : i = 1, \ldots, r\} = a_r^* = a_{r,k_r-1}.$$

For any integers c and m^* with $m^* \geq a^*$, we have

$$[c, c + m^* - 1] + A_j = [c, c + m^* - 1 + a_j^*] \tag{4}$$

for $j = 1, \ldots, r$.

Since $\bigcup_{i=1}^{r} A_i \setminus \{0\}$ is a nonempty set of relatively prime positive integers, it follows that for every integer n there exist integers $x_{i,j}'$ such that

$$n = \sum_{i=1}^{r} \sum_{j=1}^{k_i-1} x_{i,j}' a_{i,j}.$$

For each pair $(i, j) \neq (r, k_r - 1)$, we can choose an integer $x_{i,j}$ such that

$$x_{i,j}' \equiv x_{i,j} \pmod{a^*}$$

and

$$0 \leq x_{i,j} \leq a^* - 1.$$

There exist integers $t_{i,j}$ such that

$$x_{i,j}' = x_{i,j} + t_{i,j}a^*.$$

Then

$$n = \sum_{i=1}^{r} \sum_{j=1}^{k_i-1} x_{i,j}' a_{i,j}$$

$$= \sum_{i=1}^{r} \sum_{\substack{j=1 \\ (i,j)\neq(r,k_r-1)}}^{k_i-1} (x_{i,j} + t_{i,j}a^*)a_{i,j} + x_{r,k_r-1}' a_{r,k_r-1}$$

$$= \sum_{i=1}^{r} \sum_{j=1}^{k_i-1} x_{i,j} a_{i,j},$$

where

$$x_{r,k_r-1} = x'_{r,k_r-1} + \sum_{i=1}^{r} \sum_{\substack{j=1 \\ (i,j)\neq(r,k_r-1)}}^{k_i-1} t_{i,j}a_{i,j}.$$

If

$$n \geq (a^* - 1) \sum_{i=1}^{r} \sum_{\substack{j=1 \\ (i,j)\neq(r,k_r-1)}}^{k_i-1} a_{i,j},$$

then

$$x_{r,k_r-1} \geq 0.$$

Therefore, every sufficiently large integer is a nonnegative integer linear combination of the elements of $\bigcup_{i=1}^{r} A_i$. Let c be the smallest integer such that every integer $n \geq c$ can be represented in the form

$$n = \sum_{i=1}^{r} \sum_{j=1}^{k_i-1} x_{i,j}(n)a_{i,j},$$

where the coefficients $x_{i,j}(n)$ are nonnegative integers. Then

$$c - 1 \notin \sum_{i=1}^{r} h_i A_i$$

for all nonnegative integers $h_1, \ldots, h_r$. For each $i = 1, \ldots, r$, we define

$$h_i^{(1)} = \max \left\{ \sum_{j=1}^{k_i-1} x_{i,j}(n) : n = c, c+1, \ldots, c+a^* - 1 \right\}.$$

Then

$$[c, c + a^* - 1] \subseteq \sum_{i=1}^{r} h_i^{(1)} A_i.$$

It follows that

$$c + a^* - 1 \leq \max \left(\sum_{i=1}^{r} h_i^{(1)} A_i \right) = \sum_{i=1}^{r} h_i^{(1)} a_i^*,$$

and so

$$c' = \sum_{i=1}^{r} h_i^{(1)} a_i^* - (c + a^* - 1) \geq 0.$$

We shall prove that if $h_i \geq h_i^{(1)}$ for all $i = 1, \ldots, r$, then the sumset $\sum_{i=1}^{r} h_i A_i$ contains the interval of integers

$$\left[c, \sum_{i=1}^{r} (h_i - h_i^{(1)}) a_i^* + c + a^* - 1 \right] = \left[c, \sum_{i=1}^{r} h_i a_i^* - c' \right].$$

The proof is by induction on

$$\ell = \sum_{i=1}^{r} (h_i - h_i^{(1)}).$$

If $\ell = 0$, then $h_i = h_i^{(1)}$ for all $i = 1, \ldots, r$, and the assertion is true.

Let $\ell \geq 1$, and assume that the statement holds for $\ell - 1$. Then $h_j \geq h_j^{(1)} + 1$ for some j. By the induction assumption, we have

$$\sum_{\substack{i=1 \\ i \neq j}}^{r} h_i A_i + (h_j - 1) A_j \supseteq \left[c, \sum_{\substack{i=1 \\ i \neq j}}^{r} h_i a_i^* + (h_j - 1) a_j^* - c' \right]$$

$$= \left[c, \sum_{i=1}^{r} (h_i - h_i^{(1)}) a_i^* - a_j^* + c + a^* - 1 \right].$$

Applying (4) with

$$m^* = a^* + \sum_{i=1}^{r} (h_i - h_i^{(1)}) a_i^* - a_j^* \geq a^*,$$

we obtain

$$\sum_{i=1}^{r} h_i A_i = \left(\sum_{\substack{i=1 \\ i \neq j}}^{r} h_i A_i + (h_j - 1) A_j \right) + A_j$$

$$\supseteq \left[c, \sum_{i=1}^{r} (h_i - h_i^{(1)}) a_i^* - a_j^* + c + a^* - 1 \right] + A_j$$

$$= \left[c, \sum_{i=1}^{r} (h_i - h_i^{(1)}) a_i^* + c + a^* - 1 \right]$$

$$= \left[c, \sum_{i=1}^{r} h_i a_i^* - c' \right].$$

This completes the induction.

If the system of sets $A_1, \ldots, A_r$ is normalized, then the system of reflected sets $\hat{A}_1, \ldots, \hat{A}_r$ is also normalized. Applying the previous argument to the reflected system, we obtain integers $d, d', h_1^{(2)}, \ldots, h_r^{(2)}$ such that d is the largest integer with the property that $d - 1$ cannot be written as a nonnegative integral linear combination of the elements of $\bigcup_{i=1}^{r} \hat{A}_i$, and

$$\left[d, \sum_{i=1}^{r} h_i a_i^* - d' \right] \subseteq \sum_{i=1}^{r} h_i \hat{A}_i$$

if $h_i \geq h_i^{(2)}$ for $i = 1, \ldots, r$. By Lemma 1,

$$\left[d', \sum_{i=1}^{r} h_i a_i^* - d \right] \subseteq \sum_{i=1}^{r} h_i A_i$$

and

$$\sum_{i=1}^{r} h_i a_i^* - d + 1 \notin \sum_{i=1}^{r} h_i A_i$$

for all nonnegative integers $h_1, \ldots, h_r$.

Choose $h_i^{(3)} \geq \max\{h_i^{(1)}, h_i^{(2)}\}$ such that

$$c' + d' \leq \sum_{i=1}^{r} h_i^{(3)} a_i^*.$$

If $h_i \geq h_i^{(3)}$, then

$$\left[c, \sum_{i=1}^{r} h_i a_i^* - d \right] \subseteq \sum_{i=1}^{r} h_i A_i.$$

Since

$$c - 1 \notin \sum_{i=1}^{r} h_i A_i$$

and

$$\sum_{i=1}^{r} h_i a_i^* - d + 1 \notin \sum_{i=1}^{r} h_i A_i$$

for all nonnegative integers $h_1, \ldots, h_r$, it follows that if

$$h_i \geq h_i^* = \max\left\{h_i^{(3)}, c, d\right\},$$

then there exist sets $C \subseteq [0, c-2]$ and $D \subseteq [0, d-2]$ such that

$$\sum_{i=1}^{r} h_i A_i = C \cup \left[c, \sum_{i=1}^{r} h_i a_i^* - d \right] \cup \left(\sum_{i=1}^{r} h_i a_i^* - D \right).$$

This completes the proof. $\square$

Theorem 3. *Let $A_1, \ldots, A_r$ be a normalized system of finite sets of integers, and let $a_i^* = \max(A_i)$ for $i = 1, \ldots, r$. Let B be a finite set of nonnegative integers with $0 \in B$ and $b^* = \max(B)$. There exist integers c and d and finite sets*

$$C \subseteq [0, c-2]$$

and

$$D \subseteq [0, d-2]$$

such that

$$B + h_1 A_1 + \cdots + h_r A_r = C \cup \left[c, b^* + \sum_{i=1}^{r} h_i a_i^* - d \right] \cup \left(b^* + \sum_{i=1}^{r} h_i a_i^* - D \right)$$

for all sufficiently large intergers h_i.

Proof: This is a simple consequence of Theorem 2. $\square$

3. The cardinality of linear forms

Theorem 3 immediately implies the following estimate for the size of a sumset of integers.

Theorem 4.   *Let $A_1, \ldots, A_r$ be a normalized system of finite sets of integers, and let B be a nonempty, finite set of nonnegative integers. There exist positive integers $a_1^*, \ldots, a_r^*$ and nonnegative integers b^* and Δ such that*

$$|B + h_1 A_1 + \cdots + h_r A_r| = \sum_{i=1}^{r} a_i^* h_i + b^* + 1 - \Delta$$

for all sufficiently large integers h_i.

Theorem 4 shows that the cardinality of the sumset $B + h_1 A_1 + \cdots + h_r A_r$ is a linear polynomial in the variables $h_1, \ldots, h_r$. This is a special case of the following very general result. Let S be an arbitrary abelian semigroup, written additively, and let $B, A_1, \ldots, A_r$ be finite, nonempty subsets of S. We can define the sumset $B + h_1 A_1 + \cdots + h_r A_r$ in S exactly as we defined sumsets in the semigroup of integers. Extending results of Khovanskii [1, 2] for the case $r = 1$, Nathanson [7] proved that there exists a polynomial $p(x_1, \ldots, x_r)$ such that

$$|B + h_1 A_1 + \cdots + h_r A_r| = p(h_1, \ldots, h_r)$$

for all sufficiently large integers h_i. For an arbitrary semigroup S, it is not known how to compute this polynomial, nor even to determine its degree.

References

1. A.G. Khovanskii, "Newton polyhedron, Hilbert polynomial, and sums of finite sets," *Funktsional. Anal. i Prilozhen.* **26** (1992), 276–281.
2. A.G. Khovanskii, "Sums of finite sets, orbits of commutative semigroups, and Hilbert functions," *Funktsional. Anal. i Prilozhen.* **29** (1995), 102–112.
3. V.F. Lev, "Structure theorem for multiple addition and the Frobenius problem," *J. Number Theory* **58** (1996), 79–88.
4. V.F. Lev, "Addendum to 'Structure theorem for multiple addition'," *J. Number Theory* **65** (1997), 96–100.
5. M.B. Nathanson, "Sums of finite sets of integers," *Amer. Math. Monthly* **79** (1972), 1010–1012.
6. M.B. Nathanson, *Additive Number Theory: Inverse Problems and the Geometry of Sumsets*, volume 165 of *Graduate Texts in Mathematics*. Springer-Verlag, New York, 1996.
7. M.B. Nathanson, "Growth of sumsets in abelian semigroups," Preprint, 1997.
8. Ö.J. Rödseth, "On h-bases for n," *Math. Scand.* **48** (1981), 165–183.

THE RAMANUJAN JOURNAL 2, 283–298 (1998)

A Binary Additive Problem of Erdős and the Order of 2 mod p^2

ANDREW GRANVILLE*   andrew@math.uga.edu
Department of Mathematics, University of Georgia, Athens, Georgia 30602, USA

K. SOUNDARARAJAN   skannan@math.princeton.edu
Department of Mathematics, Princeton University, Princeton, New Jersey 08544, USA

We'd like to thank Paul Erdős for the questions

Received February 24, 1998; Accepted March 4, 1998

Abstract. We show that the problem of representing every odd positive integer as the sum of a squarefree number and a power of 2, is strongly related to the problem of showing that p^2 divides $2^{p-1} - 1$ for "few" primes p.

Key words: Fermat quotients, order mode p^2, squarefree numbers, powers of 2

1991 Mathematics Subject Classification: Primary: 11A15, Secondary: 11A07, 11B05, 11B34, 11N25, 11N36, 11N69, 11P99

Introduction

It is frustrating that there is no plausible known approach to the question of determining whether there are infinitely primes p for which p^2 does not divide $2^{p-1} - 1$. We know very little of consequence; only the computational result [1] that p^2 divides $2^{p-1} - 1$ for just the primes 1093 and 3511 of all $p \leq 4 \cdot 10^{12}$. Naive heuristics suggest that the number of primes up to x, for which p^2 divides $2^{p-1} - 1$, should be $\sim \log \log x$; and so we believe that there are infinitely many primes for which p^2 divides $2^{p-1} - 1$, and infinitely many primes p for which p^2 does not divide $2^{p-1} - 1$.

In 1910, Wieferich [15] showed that if there are integers x, y, z satisfying $x^p + y^p = z^p$ and $(p, xyz) = 1$ then p^2 divides $2^{p-1} - 1$; such primes p are thus known as *"Wieferich primes"*. Of course, Fermat's Last Theorem is now proved [16] so this result has become a (delightful) historical curiosity.

Recently, Paul Erdős has made the following, seemingly unrelated, conjecture (see Section A19 in [9]):

Conjecture 1 (Erdős). *Every odd positive integer is the sum of a squarefree number and a power of 2.*

*The author is a Presidential Faculty Fellow. He is also supported, in part, by the National Science Foundation. The second author is supported by an Alfred P. Sloan dissertation fellowship.

Remark. There is no significant loss of generality in Erdős' restriction to odd integers n. For if $n = m + 2^j$ then m is odd, so $2n = 2m + 2^{j+1}$, and vice versa; and if $4n = m + 2^j$ then 4 divides m so it cannot be squarefree.

In this note, we show that these questions are indeed related.

Theorem 1. *Suppose that every odd positive integer can be written as the sum of a squarefree number and a power of* 2. *Then there are infinitely many primes p for which p^2 does not divide $2^{p-1} - 1$. In fact there then exists a constant $c > 0$ such that there are arbitrarily large values of x for which*

$$\#\{ \text{primes } p \le x : 2^{p-1} \not\equiv 1 \pmod{p^2} \} \ge c\#\{ \text{primes } p \le x \}.$$

In the other direction we prove, at the suggestion of Neil Calkin:

Theorem 2. *Assume that there are $\le 2 \log x / (\log \log x)^2$ primes $p \le x$ for which p^2 does divide $2^{p-1} - 1$, whenever $x \ge 3$. Then all but $O(x/\log x)$ of the odd integers $n \le x$ can be written as the sum of a squarefree number and a power of* 2.

Remark. Assuming $\sum_{p^2 | 2^{p-1}-1} 1/\text{ord}_p(2) \le 5/8$ we can make the same deduction by the same proof.

It would be nicer to have an "if and only if" statement of some kind, rather than our two results above, which would probably require some strengthening of both of these results. We hope the reader will embrace this challenge.

Erdős' conjecture has been verified for all odd integers up to 10^7 by Andrew Odlyzko.

In Proposition 3 we give a result giving conditions under which we can guarantee that almost all integers, in certain arithmetic progressions, are the sum of a squarefree number and an element of a given sequence $\mathcal{A}$. This implies Theorem 2 and various other results. For example, there are no known primes for which $2^{p-1} \equiv 3^{p-1} \equiv 1 \pmod{p^2}$. If this is true, then we deduce:

Corollary 1. *Suppose that there does not exist a prime p for which p^2 divides both $2^{p-1} - 1$ and $3^{p-1} - 1$. Then almost all integers coprime to 6 are the sum of a squarefree number and an integer which is the product of a power of 2 and a power of 3.*

One might try to justify Erdős' conjecture by the following heuristic argument. The probability that a random odd integer is squarefree is $\prod_{p \ge 3}(1 - 1/p^2) = 8/\pi^2$. Thus the probability that none of $n - 2, n - 4, n - 8, \ldots, n - 2^r$ is squarefree (under the assumption that these events are independent) is $(1 - 8/\pi^2)^r \asymp n^{-c}$ where $c = -\log(1 - 8/\pi^2)/\log 2$ (since $r = \log n / \log 2 + O(1)$). Since $8/\pi^2 < 4/5$, we have $c > \log 5 / \log 2 > 2$. Hence, we 'deduce', by the Borel-Cantelli Lemma, that at most finitely many n fail to be the sum of a squarefree number and a power of 2. In fact, one can deduce from similar reasoning that if $r(n)$ denotes the number of positive integers i for which $n - 2^i$ is a positive squarefree integer, then $r(n) \sim (8/\pi^2) \log n / \log 2$ for almost all integers n (we write that $r(n)$ has "normal order" $(8/\pi^2) \log n / \log 2$).

However, this reasoning is highly dubious, since the proof of Theorem 1 (in fact, of Proposition 1 below) rests, appropriately interpreted, on the fact that the events $n - 2, n - 4, \ldots, n - 2^r$ being squarefree, are not independent. By studying the first two moments of $r(n)$, we show below that $r(n)$ does not have a normal order. In fact our analysis extends to $r_A(n)$, the number of ways of writing $n = m + a_i$ with m a positive squarefree number and $a_i \in \mathcal{A}$, where $\mathcal{A} = \{a_1 < a_2 < \cdots\}$ is a sparse sequence of positive integers. Define $A(x)$ to be the number of $a_i \leq x$.

Theorem 3. *Suppose we are given a sequence $\mathcal{A}$ of distinct positive integers for which $A(2x) \sim A(x)$, and an arithmetic progression $a \pmod{q^2}$ with $(a - a_i, q^2)$ squarefree for all $a_i \in \mathcal{A}$. Then $r_A(n)$ has mean $\sim c_q A(x)$ when averaging over the integers $n \leq x$, for which $n \equiv a \pmod{q^2}$, where $c_q := \prod_{p \nmid q}(1 - 1/p^2)$.*

Moreover, these $r_A(n)$ have normal order $c_q A(n)$ if and only if $\mathcal{A}$ is equidistributed amongst the arithmetic progressions $\pmod{d^2}$, for every integer d which is coprime to q (that is, there are $\sim A(x)/d^2$ integers $a_i \leq x$ with $a_i \equiv l \pmod{d^2}$ for each l).

Remark. Note that the condition $(a - a_i, q^2)$ is squarefree for all $a_i \in \mathcal{A}$ ensures that it is feasible that $n - a_i$ is squarefree for each a_i.

Take $\mathcal{A} = \{2, 4, 8, \ldots\}$ with $q = 2$ and $a = 1$ or 3. Since the powers of 2 are not equi-distributed in residue classes $\pmod{d^2}$ for any odd d, we deduce by Theorem 3 that $r(n) = r_A(n)$ cannot have a normal order.

We now give an example of a set $\mathcal{A}$ which is sparser than the powers of 2, but for which $r_A(n)$ has a normal order.

Corollary 2. *Almost all integers are the sum of a squarefree number and an integer of the form $1^1 + 2^2 + \cdots + k^k$. In fact, if $n \equiv 2, 3 \pmod 4$ the number of such representations has normal order $(8/\pi^2) \log n / \log \log n$; and if $n \equiv 0, 1 \pmod 4$, the number of such representations has normal order $(4/\pi^2) \log n / \log \log n$.*

The genesis of Erdős' conjecture is from de Polignac's (incorrect) claim [10] (retracted in the second reference [10]) that every odd integer is the sum of a prime and a power of 2. The first counterexample is 127, though Euler had noted the counterexample 959 in a letter to Goldbach. In 1934, Romanoff [12] showed that a positive proportion of odd integers can be represented in this way, and in 1950 van der Corput [14] and Erdős [5] showed that a positive proportion of odd integers cannot be represented in this way. Romanoff's proof uses the Cauchy-Schwarz inequality, estimating the mean of the number of representations, and bounding the mean square; this last upper bound follows from Brun's sieve followed by showing that $\sum_d 1/d \operatorname{ord}_d(2) < \infty$. Erdős invented the elegant notion of a "covering system of congruences", which we describe in detail in the next section, to find an infinite arithmetic progression of odd values of n that cannot be written as a prime plus a power of 2. It is still an open question, of Erdős, as to whether there is a precise proportion of the odd integers that are so representable (asymptotically), and then even an informed prediction of what that proportion is (of course the Romanoff and Erdős results can be used to get non-trivial upper and lower bounds on that proportion).

Next one might perhaps replace "prime" by "squarefree number" in the above problem (as Erdős did); an alternative is to replace "a power of 2" by "two powers of 2". Unfortunately, Crocker [2] observed that for any odd integer $n = 2^{2^m} - 1$ with $m \geq 3$, the numbers $n - 2^a - 2^b$ with $1 \leq a < b < 2^m$ are never prime. To see this let 2^k be the highest power of 2 dividing $b - a$. Then $2^{2^k} + 1$ divides $2^{b-a} + 1$, which divides $2^b + 2^a$. Moreover $k \leq m - 1$ so that $2^{2^k} + 1$ divides n, and so $2^{2^k} + 1$ divides $n - 2^a - 2^b$. If these were equal then $2^{2^m} = 2^b + 2^a + 2^{2^k} + 2$, and this can be seen to be impossible by considering this equation mod 16 to restrict a and b. Thus it is not the case that every odd integer is the sum of a prime and two powers of 2, though Erdős predicted that perhaps almost all odd integers n can be so described. We conjecture that all odd integers > 1 are the sum of a prime and at most three powers of 2. Along these lines, Gallagher [8] showed that the proportion of odd integers that can be written as a prime plus the sum of k powers of two, tends to 1 as $k \to \infty$.

We take a lead from this line of investigation to discuss whether one can prove that almost all odd n are the sum of a squarefree integer plus at most k powers of 2, for some k. We prove:

Theorem 4. *Assume that $\sum_{p^2 | 2^{p-1}-1} 1/\mathrm{ord}_p(2) < \infty$. Then there exists an integer k such that almost every odd integer can be written as the sum of a squarefree number plus no more than k distinct powers of 2.*

It is completely straightforward to prove the analogy to Gallagher's result: Every integer up to 2^k is the sum of at most k powers of two. The number of integers amongst $n - 1$, $n - 2, \ldots, n - 2^k$ divisible by the square of a prime $p < 2^k$ is $\leq \sum_{p < 2^k}(2^k/p^2 + 1) < 0.49 \times 2^k + O(2^k/k) < 2^{k-1}$ for k sufficiently large. Of the integers $n \leq x$, the number for which more than 2^{k-1} of the integers $n - 1, n - 2, \ldots, n - 2^k$ are divisible by the square of a prime $p > 2^k$ is

$$\leq 2^{1-k} \sum_{n \leq x} \sum_{\sqrt{x} > p > 2^k} \sum_{\substack{1 \leq i \leq 2^k \\ p^2 | n-i}} 1 \ll \sum_{\sqrt{x} > p > 2^k} \frac{x}{p^2} \ll \frac{x}{k2^k}.$$

Thus, all but $O(x/k2^k)$ of the odd integers $\leq x$ can be written as the sum of a squarefree number and k powers of two.

There are many other intriguing questions of this type asked by Erdős (see Section A19 of [9]): Erdős conjectured that 105 is the largest integer for which $n - 2^k$ is prime whenever $2 \leq 2^k < n$ (analogously it was shown in [3] that 210 is the largest integer $2n$ for which $2n - p$ is prime for every prime p, $n \leq p < 2n$). He showed that there exist n for which $n - 2^k$ is prime for $\gg \log \log n$ such values of k, and asked whether this could be improved. He also conjectured that for infinitely many n, all of the integers $n - 2^k, 2 \leq 2^k < n$ are squarefree. Erdős conjectured that there are arbitrarily large gaps between consecutive odd numbers that can be represented as the sum of a prime and a power of 2.

Erdős asked whether there are $\gg x^\epsilon$ odd integers $n \leq x$ that are not equal to a prime plus two powers of two. By modifying Crocker's construction slightly, this is easily shown for arbitrarily large x if infinitely many Fermat numbers $F_k = 2^{2^k} + 1$ are composite, and for all x if $\{k_{i+1} - k_i\}$ is bounded where F_{k_i} is the sequence of composite Fermat numbers.

Notation. Henceforth A will always denote a sequence $\{a_1 < a_2 < \cdots\}$ of positive integers. We will let $A(x)$ be the number of $a_i \leq x$, and $A(x; d, b)$ be the number of $a_i \leq x$ for which $a_i \equiv b \pmod{d}$.

Covering systems and all that

The more detailed proof is that of Theorem 1, which stems from some modifications of constructions due to Paul Erdős. We will discuss here these constructions, beginning with the idea behind Erdős' disproof of de Polignac's "conjecture" [5].

A *covering system* for the integers is a finite set of arithmetic progressions, such that every integer belongs to at least one of these arithmetic progressions. For example 0 (mod 2); 1 (mod 2) or 1 (mod 2); 1 (mod 3); 0 (mod 6); 2 (mod 6).

Now suppose that we can find a covering system with arithmetic progressions like a_i (mod $\text{ord}_{p_i}(2)$), for $i = 1, 2, \ldots, k$, where the p_i are distinct, odd primes. Let n_0 be the smallest odd integer satisfying $n_0 \equiv 2^{a_i} \pmod{p_i}$ for each i (which is a well-defined integer (mod $2 \prod_{1 \leq i \leq k} p_i$) by the Chinese Remainder Theorem). For any $n \equiv n_0 \pmod{2 \prod_{1 \leq i \leq k} p_i}$ and for any integer j, we have that $j \equiv a_i \pmod{\text{ord}_{p_i}(2)}$ for some i (since we have a covering system above), and so $n - 2^j \equiv n_0 - 2^{a_i} \equiv 0 \pmod{p_i}$. Therefore $n - 2^j$ is composite provided it is $> p_i$; and if this is so for each j with $2^j < n$, then n cannot be written as the sum of a prime and a power of 2.

There are several ways to get around the problem that we might have $n - 2^j = p_i$ for some i and j. Usually one imposes an extra congruence on n. Alternately, note that if $n < 2^{m+1}$ can be so represented then $j \leq m$ and $1 \leq i \leq k$, so that there are $\leq mk$ such exceptional n. However the number of integers $n \equiv n_0 \pmod{2 \prod_{1 \leq i \leq k} p_i}$ in this range is $\geq 2^m / \prod_{1 \leq i \leq k} p_i - 1$, so we certainly have non-exceptional such n if m is chosen sufficiently large.

The question reduces to producing such a covering system. This is easily done by taking 0 (mod $\text{ord}_3(2) = 2$), 0 (mod $\text{ord}_7(2) = 3$), 1 (mod $\text{ord}_5(2) = 4$), 3 (mod $\text{ord}_{17}(2) = 8$), 7 (mod $\text{ord}_{13}(2) = 12$), 23 (mod $\text{ord}_{241}(2) = 24$); so we end up with the arithmetic progression $n \equiv 7629217 \pmod{11184810}$.

Define $\omega(p)$ to be the order of 2 (mod p^2). If we were able to construct a covering system out of arithmetic progressions with moduli $\omega(p)$, then we could give a similar disproof of Conjecture 1.

Theorem 5. *Suppose that there exists a covering system $\{a_i \pmod{\omega(p_i)}\}_{1 \leq i \leq m}$ where the p_i are distinct, odd primes. Then a positive proportion of the odd integers $n \leq x$ cannot be written as the sum of a squarefree number and a power of 2.*

Proof: Let n be any odd integer $\equiv 2^{a_i} \pmod{p_i^2}$ for $1 \leq i \leq m$ (the density of such odd integers is $1/(\prod_{1 \leq i \leq m} p_i)^2 > 0$ amongst the odd integers). For any positive integer j, select i, $1 \leq i \leq m$ so that $j \equiv a_i \pmod{\omega(p_i)}$ (which is possible by hypothesis), and thus $n - 2^j$ is divisible by p_i^2. Therefore $n - 2^j$ is never squarefree and so n cannot be written as the sum of a squarefree number and a power of 2.   □

However, we do not believe that such a covering system can exist:

Conjecture 2. *There is no finite set of distinct, odd primes* $\{p_1, p_2, \ldots, p_m\}$ *and integers* $a_1, a_2, \ldots, a_m$ *such that every integer belongs to at least one of the congruence classes* $a_i \pmod{\omega(p_i)}$.

Erdős remarked to us that, given that Theorems 1 and 2 suggest that it will probably be difficult to prove Conjecture 1, we might try for the weaker result that "almost all" odd n can be represented as the sum of a squarefree number and a power of 2. In Theorem 5 we saw that to prove this we will at least need to show that Conjecture 2 is true, which looks difficult. One encouraging remark is that, by a slight strengthening of Conjecture 2, we are able to deduce that "almost all" odd n can be represented as the sum of a squarefree number and a power of 2:

Conjecture 3. *There exists a constant* $\delta > 0$ *such that, for any finite set of primes* $\{p_1, p_2, \ldots, p_m\}$ *and any choice of integers* $a_1, a_2, \ldots, a_m$, *the proportion of positive integers which belong to at least one of the congruence classes* $a_i \pmod{\omega(p_i)}$, *is* $< 1 - \delta$.

Theorem 6. *Assume that Conjecture 3 is true. Then "almost all" odd integers* $n \le x$ *can be written as the sum of a squarefree number and a power of 2.*

Proof: Fix integer K and consider odd n in the range $2^K < n \le 2^{K+1}$. Select $y = \log K$, so that the number of integers $n - 2^i, 1 \le i \le K$ not divisible by the square of any prime $\le y$ is $> \delta K - 2^{\pi(y)} > \delta K/2$, using Conjecture 3 and the combinatorial sieve. Moreover, the total number of pairs (n, i) in these ranges, for which $n - 2^i$ is divisible by the square of a prime $> y$ is $\ll K \sum_{p > y} 2^{K+1}/p^2 \ll K 2^K/y \log y$; and thus there are $< \delta K/2$ such values of i for all but $O(2^K/\log K \log \log K)$ of the values of n in our range. For these n, we have some i with $n - 2^i$ squarefree, and the number of failures is thus $\ll x/\log \log x$. □

Note that the condition $\sum_p 1/\omega(p) < 1$ implies Conjecture 3 (with $\delta = 1 - \sum_p 1/\omega(p)$), which implies Conjecture 2.

Lemma 1. *If Conjecture 3 is true then* $\sum_p 1/\omega(p) < \infty$.

Proof: Select a_p's by induction as follows: Let S_p be the set of positive integers $\le b_p := \operatorname{lcm}[\omega(q) : 2 < q < p]$ which do not belong to any of the arithmetic progressions $a_q \pmod{\omega(q)}$ for $q < p$, and note that $\delta_p := |S_p|/b_p$ is exactly the proportion of all positive integers not belonging to any of the congruence classes $a_q \pmod{\omega(q)}$ with $q < p$. Given the choices of a_q for each $q < p$, we select $a_p \pmod{\omega(p)}$ so that this arithmetic progression contains as many integers in S_p as possible. Evidently there must be one such arithmetic progression containing $\ge |S_p|/\omega(p)$ such integers, and thus $\delta_\ell \le (1 - 1/\omega(p))\delta_p$, where ℓ is the smallest prime $> p$. Starting with $a_3 = 0$, and then iterating the above procedure, we find that $\delta_p \le \prod_{q < p}(1 - 1/\omega(q))$. Now, Conjecture 3 implies that $\delta_p > \delta > 0$ for all p, and so $\sum_q 1/\omega(q)$ must converge. □

Deduction of Theorems 1 and 2 from technical propositions

Theorems 1 and 2 follow from stronger, but more technical, propositions.

Proposition 1. *Let $\omega(p)$ be the order of 2 (mod p^2). Fix $\varepsilon > 0$. If there exist arbitrarily large values of y for which*

$$\prod_{p \leq y}\left(1 - \frac{1}{\omega(p)}\right) \leq \left(\frac{\log 2}{4} - \varepsilon\right)\frac{1}{\log y}$$

then there are infinitely many odd integers n which cannot be written as the sum of a squarefree number and a power of 2.

In the other direction we show

Proposition 2. *Suppose that*

$$\sum_{p}\frac{1}{\omega(p)} < 1.$$

Then all but $O(x/\log x)$ of the odd integers $n \leq x$ can be written as the sum of a squarefree number and a power of 2.

One can see that there is a lot of difference in these hypotheses in their requirements for the average size of $1/\omega(p)$. Improving this paper would necessitate closing that gap.

Deduction of Theorem 1 from Proposition 1. Suppose that the conclusion of Theorem 1 is false so that

$$\#\{ \text{primes } p \leq x : 2^{p-1} \equiv 1 \quad (\text{mod } p^2)\} = o(\#\{ \text{primes } p \leq x\}).$$

Thus, almost all primes p satisfy $2^{p-1} \equiv 1 \pmod{p^2}$, implying that $\omega(p) \leq p - 1$. Moreover if $p \equiv \pm 1 \pmod 8$ then 2 is a quadratic residue (mod p) so that $\omega(p) \leq \frac{p-1}{2}$. Now, as is well-known,

$$\prod_{p \leq y}\left(1 - \frac{1}{p-1}\right) \ll \frac{1}{\log y}$$

and

$$\prod_{\substack{p \leq y \\ p \equiv \pm 1 \, (\text{mod } 8)}}\left(1 - \frac{1}{p-2}\right) \ll \frac{1}{(\log y)^{1/2}}.$$

Therefore,

$$\prod_{p \le y}\left(1 - \frac{1}{\omega(p)}\right) \le \prod_{\substack{p \le y \\ 2^{p-1} \equiv 1 \,(\mathrm{mod}\, p^2)}}\left(1 - \frac{1}{p-1}\right) \cdot \prod_{\substack{p \le y,\, p \equiv \pm 1 \,(\mathrm{mod}\, 8) \\ 2^{p-1} \equiv 1 \,(\mathrm{mod}\, p^2)}}\left(1 - \frac{1}{p-2}\right)$$

$$\ll \frac{1}{(\log y)^{3/2+o(1)}}$$

The condition in Proposition 1 is thus satisfied and so there are infinitely many integers n which cannot be expressed as the sum of a squarefree number and a power of 2. Theorem 1 follows.   □

Remark. This argument can be used to show that one can take any constant c, in the range $1/4 > c > 0$, in Theorem 1. One can improve this by considering the appropriate products over those primes p for which 2 is a cubic residue mod p, or a quartic residue, or quintic residue, etc. The density of such primes is determined by the Cebotarev Density Theorem. By such methods we were able to show that $(\sum_{p \le y} 1/\mathrm{ord}_p(2))/\log \log y \to \infty$ as $y \to \infty$. This leads us to ask:

$$\text{What is the true order of magnitude of } \prod_{p \le y}\left(1 - \frac{1}{\mathrm{ord}_p(2)}\right)?$$

As far as the averaged order of 2 (mod p^2) is concerned, we certainly believe that $\sum_p \frac{1}{\omega(p)} < \infty$, and that even the hypothesis of Proposition 2 is true.

Deduction of Theorem 2 from Proposition 2.   We write

$$\sum_{p \text{ prime}} \frac{1}{p\,\mathrm{ord}_p(2)} = \sum_{n \ge 2} \frac{1}{n} \sum_{\substack{p \text{ prime} \\ \mathrm{ord}_p(2)=n}} \frac{1}{p}$$

Now, if $\mathrm{ord}_p(2) = n$ then $p \equiv 1 \,(\mathrm{mod}\, n)$. Thus $p > n$ and so the total number of such primes is $\le \log(2^n - 1)/\log(n+1) < n \log 2 / \log n$.

Therefore, for $m = [n \log 2/ \log n]$

$$\sum_{\substack{p \text{ prime} \\ \mathrm{ord}_p(2)=n}} \frac{1}{p} \le \sum_{k=1}^{m} \frac{1}{kn+1} < \frac{1}{n}\sum_{k=1}^{m} \frac{1}{k}$$

$$\le \frac{1}{n}(\log m + 1) \le \frac{1}{n}\log\left(\frac{ne\log 2}{\log n}\right) < \frac{\log n}{n}$$

for $n \ge 7 > 2^e$. Thus for $N \ge 6$,

$$\sum_{n \ge N+1} \frac{1}{n} \sum_{\substack{p \text{ prime} \\ \mathrm{ord}_p(2)=n}} \frac{1}{p} < \sum_{n \ge N+1} \frac{\log n}{n^2} \le \int_N^\infty \frac{\log t}{t^2}\, dt = \frac{1 + \log N}{N}.$$

Using Maple, we have determined that

$$\sum_{n \leq 100} \frac{1}{n} \sum_{\substack{p \text{ prime} \\ \text{ord}_p(2) = n}} \frac{1}{p} = 0.31586267847633\ldots$$

Also, $(1 + \log 100)/100 = 0.05605170185988\ldots$ so that

$$\sum_{p \text{ prime}} \frac{1}{p \, \text{ord}_p(2)} < 0.372$$

Note that, if p^2 divides $2^{p-1} - 1$ then $\omega(p) = \text{ord}_p(2)$; otherwise $\omega(p) = p \, \text{ord}_p(2)$. Therefore,

$$\sum_{p \text{ prime}} \frac{1}{\omega(p)} = \sum_{p \text{ prime}} \frac{1}{p \, \text{ord}_p(2)} + \sum_{p^2 | 2^{p-1} - 1} \frac{p-1}{p \, \text{ord}_p(2)}$$

$$< 0.372 + \sum_{p^2 | 2^{p-1} - 1} \frac{1}{\text{ord}_p(2)}.$$

Now, we know that $\text{ord}_p(2) > \log p / \log 2$ and that only $p = 1093$ and $p = 3511$ satisfy $p^2 \mid 2^{p-1} - 1$ when $p < 4 \cdot 10^{12}$. Also, $\#\{p \leq x : p^2 \mid 2^{p-1} - 1\} \leq 2 \log x / (\log \log x)^2$ by the hypothesis of Theorem 2, so that

$$\sum_{p^2 | 2^{p-1} - 1} \frac{1}{\text{ord}_p(2)} \leq \frac{1}{\text{ord}_{1093}(2)} + \frac{1}{\text{ord}_{3511}(2)} + \sum_{\substack{p > 4 \cdot 10^{12} \\ p^2 | 2^{p-1} - 1}} \frac{\log 2}{\log p}$$

$$\leq \frac{1}{364} + \frac{1}{1755} + \frac{\log 2}{\log(4 \cdot 10^{12})} \frac{2 \log(4 \cdot 10^{12})}{(\log \log(4 \cdot 10^{12}))^2}$$

$$+ \int_{4 \cdot 10^{12}}^{\infty} \frac{\log 2}{\log t} \, d\left(\frac{2 \log t}{(\log \log t)^2}\right)$$

$$\leq 0.00332 + \frac{2 \log 2}{(\log \log(4 \cdot 10^{12}))^2} + \int_{4 \cdot 10^{12}}^{\infty} \frac{2 \log 2}{t \log t (\log \log t)^2} \, dt$$

$$\leq 0.1255361175 + \frac{2 \log 2}{\log \log(4 \cdot 10^{12})} = 0.5371568161\ldots$$

Thus

$$\sum_{p} \frac{1}{\omega(p)} < 0.9091\ldots$$

and the hypothesis of Proposition 2 is satisfied.

The proof of Proposition 1: two constructions of Erdős

Let $m = \prod_{p \leq 2y} p$ and select N so that $2y = (\frac{1}{2} - \varepsilon) \log N$. We shall construct an arithmetic progression $\ell \pmod{m^2}$, such that if $n \leq N$ and $n \equiv \ell \pmod{m^2}$ then $(n - 2^i, m^2)$ is divisible by a square of a prime for $1 \leq i \leq r$, where $2^r \leq N < 2^{r+1}$. Since $m^2 \ll N^{1-\varepsilon}$ (by the Prime Number Theorem) there exist such integers n, and the result follows.

We shall actually select arithmetic progressions $a_p \pmod{\omega(p)}$ for each prime $p \leq 2y$ so that every integer in $[1, r]$ belongs to at least one of these arithmetic progressions (rather like in the "Erdős-Rankin method" [4, 11]). Then we select $\ell \equiv 2^{a_p} \pmod{p^2}$ for each $p \leq 2y$, constructing $\ell \pmod{m^2}$ by the Chinese Remainder Theorem (rather like in Erdős' use of covering congruences [5] in the de Polignac problem). If we can do all this then we have proved our result; for if $1 \leq i \leq r$ then $i \equiv a_p \pmod{\omega(p)}$ for some prime $p \leq 2y$. But then $2^i \equiv 2^{a_p} \pmod{p^2}$ and so $n - 2^i \equiv \ell - 2^{a_p} \equiv 0 \pmod{p^2}$.

We shall partition the odd primes $\leq 2y$ into the sequence of odd primes $p_1 = 3$, $p_2 = 5, \ldots, p_k \leq y$ and the set Q of odd primes in the range $(y, 2y]$. We select our a_p's in the style of the Erdős-Rankin method:

Let S_1 be the set of positive integers $\leq r$. For $j = 1, \ldots, k$, select $a_{p_j} \pmod{\omega(p_j)}$ so that $\#\{n \in S_j : n \equiv a_{p_j} \pmod{\omega(p_j)}\}$ is maximized and then let $S_{j+1} = S_j \setminus \{n \in S_j : n \equiv a_{p_j} \pmod{\omega(p_j)}\}$. Evidently, $|S_{j+1}| \leq |S_j|(1 - \frac{1}{\omega(p_j)})$ so that

$$|S_{k+1}| \leq r \cdot \prod_{j=1}^{k} \left(1 - \frac{1}{\omega(p_j)}\right) \leq \frac{\log N}{\log 2} \prod_{p \leq y} \left(1 - \frac{1}{\omega(p)}\right)$$

$$< \left(\frac{1}{2} - 2\varepsilon\right) \frac{\log N}{2 \log y} < (1 - \varepsilon) \frac{y}{\log y} < |Q|$$

by hypothesis. We complete our construction by selecting, for each integer $a \in S_{k+1}$, a different prime $p_a \in Q$ and taking $a_{p_a} \equiv a \pmod{\omega(p_a)}$.

Remark. By a slight modification of the above proof one can prove an analogous result about representing n in any arithmetic progression of odd numbers (that is $n \equiv 2a + 1 \pmod{2q}$ for any positive integers a, q).

The proof of Proposition 2: easy sieving

The proof of Proposition 2 is based on a simple sieving procedure. We develop this in a very general form, as it will be useful in proving Theorem 4 and other generalizations.

Proposition 3. *Suppose that for a given sequence of positive integers $\mathcal{A}$, there is an absolute constant $c > 0$ such that for any sufficiently large x, for every prime p there is a non-empty set of arithmetic progressions $M_p(x)$, each with modulus p^2, and an absolute constant $\delta_p > 0$, such that*

$$A(x; p^2, m) \leq \delta_p A(x) + c \quad \text{for all } m \in M_p(x).$$

Let $N(x)$ be the set of all integers n in the interval $x < n \le 2x$ for which $n \in M_p(x)$ for every prime p; and assume that $|N(x)| \gg x$. If $\sum_p \delta_p < 1$, then almost all integers $n \in N(x)$ can be written in $\gg A(x)$ different ways as the sum of a squarefree number and some a_i in the sequence.

Proof: Consider $n \in N(x)$, and let $y := A(x)$. We shall try to write $n = m + a_i$ where $a_i \le x$ and m is squarefree. Then the number of integers $a_i \le x$ for which p^2 does not divide $n - a_i$ for any prime $p < y$, is $> y - \sum_{p<y}(\delta_p y + c)$, by the conditions on the sequence $\{a_i\}$ above. This amount is $> (1 - \sum_{p<y} \delta_p - O(1/\log y))y \gg y$, by hypothesis.

On the other hand, the number of n in the range $x < n \le 2x$, for which $n - a_i$ is divisible by p^2 for some prime $p \ge y$, is, (noting that we must have $p^2 \le 2x$),

$$\le \sum_{y \le p \le 2x^{1/2}} \sum_{n \le 2x, \, p^2 | n - a_i} 1 \ll \sum_{y \le p \le 2x^{1/2}} x/p^2 \ll x/y \log y.$$

Thus, there are $O(x/y \log y)$ integers n in the range $x < n \le 2x$, for which there are $\gg y$ values of $a_i \le x$ with $n - a_i$ divisible by the square of a prime $> y$. The result follows. $\qquad\square$

Deduction of Proposition 2. We choose our sequence A to be the powers of 2; and take $N(x)$ to be all the odd numbers less than x. Thus, $\delta_2 = 0$ and $\delta_p = 1/\omega(p)$ for all odd primes p. Proposition 2 now follows from Proposition 3, and from its proof noting that $A(x) \asymp \log x$.

Corollary 1 follows from Proposition 3 by taking $\delta_p = 1/\max\{\omega_p(2), \omega_p(3)\}$, which will be $\le \log 2/p \log p$ under hypothesis, and by taking $N(x)$ to be all integers coprime to 6.

A variant, easily proved by modifying the argument in Proposition 3, is

Corollary 3. Suppose that $\sum_p 1/\omega(p) < \infty$. Then almost all odd integers are the sum of a power of two and the product of a squarefree number and a bounded powerful number.

Deduction of Theorem 4

Lemma 2. Given A and squarefree integer v, assume that for any prime p which does not divide v, there are infinitely many integers in A which are not divisible by p. Then, for any given arithmetic progression $b \pmod{d}$ with $(d, v) = 1$ there is a finite subset of elements of A whose sum is $\equiv b \pmod{d}$ and $\equiv 0 \pmod{v}$.

Proof: Suppose p^α is the exact power of p which divides d. By the pigeonhole principle there exists a congruence class $\beta \pmod{vd}$, with $p \nmid \beta$, such that there is an infinite subsequence of integers in A which are all $\equiv \beta \pmod{vd}$. Let k_p be a positive integer $\equiv b/\beta \pmod{p^\alpha}$ and $\equiv 0 \pmod{vd/p^\alpha}$. The sum of the first k_p integers of our subsequence is $\equiv b \pmod{p^\alpha}$ and $\equiv 0 \pmod{vd/p^\alpha}$. We do this for each prime p dividing d, in turn, omitting from the sequence A those elements already used. The sum of all of these subsequences is thus $\equiv b \pmod{d}$ and $\equiv 0 \pmod{v}$, as required. $\qquad\square$

Proposition 4. *Suppose that we are given a sequence of positive integers A and a square-free integer v. If prime p does not divide v then assume there are infinitely many integers in A which are not divisible by p. For all primes p assume that $A(x; p^2, m) \leq \delta_p A(x)$ for all m, if x is sufficiently large, for some absolute constant $\delta_p > 0$. If $\sum_p \delta_p < \infty$ then, for some integer $k \geq 1$ and all large x, almost all integers n, which are coprime to v, can be written as the sum of a squarefree number plus at most k distinct elements from A.*

Proof: Let x be sufficiently large and prime $q > v$ such that $\sum_{p>q} \delta_p < 1/2$. Let dv be the product of the primes $\leq q$. If $p > q$ then let M_p be all residue classes (mod p^2); if $p \leq q$ and $p \nmid v$ then let $M_p(x)$ be that residue class m (mod p^2) for which $A(x; p^2, m)$ is minimal. For primes p dividing v we select M_p to be all congruence classes b (mod p^2) where p does not divide b.

By Proposition 3 (with δ_p as above for $p > q$, and equal to $1/p^2$ for $p \leq q$) we find that there is an arithmetic progression B (mod d), such that almost every integer $n \equiv B$ (mod d) with $(n, v) = 1$ and $x < n \leq 2x$ can be written in $\gg A(x)$ different ways as the sum of a squarefree number and some a_i in the sequence.

Now, select any congruence class b mod d, and consider integers n in this congruence class which are coprime to v and in the range $x < n \leq 2x$. By Lemma 2 there is a finite subset of elements of A whose sum, s, is $\equiv b - B$ (mod d) and $\equiv 0$ (mod v). Thus, $n - s \equiv B$ (mod d) and is coprime to v. Since s is absolutely bounded (as a function of the set A), we may use the result in the paragraph above to deduce that almost all such $n - s$ may be written as the sum of a squarefree number and some a_i in the sequence. The result follows. $\square$

Theorem 4 is an immediate corollary with $v = 2$ and $\delta_p = 1/\omega(p)$.

Unconditional results

As is well-known, there are $\sim 4x/\pi^2$ odd integers $n \leq x$ such that $n - 2$ is squarefree. Thus, a positive proportion of odd integers can be written as the sum of a squarefree number and a power of 2. With a little work we can show that there are $c_r x/2$ odd integers $n \leq x$ such that $n - 2^i$ is squarefree, for some i in the range $1 \leq i \leq r$, where $c_r > 0$ is a computable constant. For example, defining $d_k = \prod_{p \geq 3}(1 - k/p^2)$:

$$c_1 = d_1 = 8/\pi^2 = 0.810569\ldots$$
$$c_2 = 2d_1 - d_2 = 0.975870\ldots$$
$$c_3 = 3d_1 - 3d_2 + d_3 = 0.997851\ldots$$
$$c_4 = 4d_1 - 6d_2 + 4d_3 - d_4 = 0.999860\ldots$$
$$c_5 = 5d_1 - 10d_2 + 10d_3 - 5d_4 + d_5 = 0.999993\ldots$$
$$c_6 = 6d_1 - 15d_2 + 20d_3 - 15d_4 + 6d_5 - d_6 = 0.999999\ldots.$$

In fact, c_r will always be the sum of multiples of such Euler products, and those multiples are easily determined for a given r. However, the multiples will not persist in being binomial coefficients, as above, since the order of 2 (mod p^2) will play a significant role for $r \geq 7$

(since then 3^2 can feasibly divide both $n - 2$ and $n - 2^7$). Thus we are unable to prove that $c_r \to 1$ as $r \to \infty$, for much the same reasons as those behind the proof of Theorem 5, although the numerical evidence above is striking. Of course, if we could prove that $c_r \to 1$ as $r \to \infty$, this would allow us to deduce unconditionally that "almost all" odd integers may be written as a sum of a squarefree number and a power of 2.

Normal order

Proof of Theorem 3: First note that the condition $A(2x) \sim A(x)$ implies that $A(x) \sim A(x/2) \sim A(x/4) \sim \cdots \sim A(x/2^r)$ for any fixed r. Thus, $\sum_{a_i \leq x} a_i \leq A(x/2^r)x/2^r + (A(x) - A(x/2^r))x \ll xA(x)/2^r$, and therefore $\sum_{a_i \leq x} a_i = o(xA(x))$.

We have

$$\sum_{\substack{n \leq x \\ n \equiv a \pmod{q^2}}} r_A(n) = \sum_{a_i \leq x} \sum_{\substack{a_i < n \leq x \\ n \equiv a \pmod{q^2}}} \mu^2(n - a_i) = \sum_{a_i \leq x} \sum_{\substack{m \leq x - a_i \\ m \equiv a - a_i \pmod{q^2}}} \mu^2(m).$$

Since p^2 does not divide $a_i - a$ for all i, and all primes p dividing q, and by using the combinatorial sieve, the sum over m above is $\sim c(x - a_i)/q^2 + O(1)$, where $c = c_q$. Inserting this above, using the fact that $\sum_{a_i \leq x} a_i = o(xA(x))$, gives that the average order of $r(n)$ with $n \leq x, n \equiv a \pmod{q^2}$, is $\sim cA(x)$.

Note that, $r_A(n) \leq A(x)$, so that $|r_A(n) - cA(x)| \leq A(x)$. Therefore, if $r_A(n)$ is to have "normal order" then it must be $\sim cA(x)$. Moreover, this is so if and only if the mean square of $|r_A(n) - cA(x)|^2$ is $o(A(x)^2)$.

To compute the mean of the second moment of $r_A(n)$ we proceed as above:

$$\sum_{\substack{n \leq x \\ n \equiv a \pmod{q^2}}} r_A(n)^2 = \sum_{a_i, a_j \leq x} \sum_{\substack{\max(a_i, a_j) < n \leq x \\ n \equiv a \pmod{q^2}}} \mu^2(n - a_i)\mu^2(n - a_j).$$

Again using the hypothesis and the combinatorial sieve, the inner sum over n is

$$\sim \left(\frac{x - \max(a_i, a_j)}{q^2} + O(1)\right) \prod_{p \nmid q} \left(1 - \frac{2}{p^2}\right) \prod_{\substack{p^2 | a_i - a_j \\ p \nmid q}} \frac{p^2 - 1}{p^2 - 2}$$

$$= \left(\frac{x - \max(a_i, a_j)}{q^2} + O(1)\right) \prod_{p \nmid q} \left(1 - \frac{2}{p^2}\right) \sum_{\substack{a_i \equiv a_j \pmod{d^2} \\ (d, q) = 1}} f(d),$$

where $f(d) = \mu^2(d)/\prod_{p | d}(p^2 - 2)$. Inserting this above, and using the fact that $\sum_{a_i \leq x} a_i = o(xA(x))$, gives that the mean of the second moment of $r(n)$ is

$$\sim \prod_{p \nmid q} \left(1 - \frac{2}{p^2}\right) \sum_{\substack{d=1 \\ (d,q)=1}}^{\infty} f(d) \sum_{\substack{a_i, a_j \leq x \\ a_i \equiv a_j \pmod{d^2}}} 1 + o(A(x)^2).$$

Now

$$\sum_{\substack{a_i, a_j \le x \\ a_i \equiv a_j \pmod{d^2}}} 1 = \left(\frac{A(x)}{d}\right)^2 + \sum_{l=1}^{d^2} \left(A(x; d^2, l) - \frac{A(x)}{d^2}\right)^2,$$

so that the mean of the second moment is

$$\sim (cA(x))^2 + \prod_{p \nmid q} \left(1 - \frac{2}{p^2}\right) \sum_{\substack{d=1 \\ (d,q)=1}}^{\infty} f(d) \sum_{l=1}^{d^2} \left(A(x; d^2, l) - \frac{A(x)}{d^2}\right)^2.$$

Now, the contribution to the sum, for each d, is $\le f(d)A(x)^2$; and so the contribution of the terms with $d > D$ is $\ll \sum_{d>D} A(x)^2/d^2 \ll A(x)^2/D$. Therefore, taking D to be a large integer, we get

$$\frac{1}{x/q^2} \sum_{\substack{n \le x \\ n \equiv a \pmod{q^2}}} |r_{\mathcal{A}}(n) - cA(x)|^2 \asymp \sum_{\substack{d \le D \\ (d,q)=1}} \frac{\mu^2(d)}{d^2} \sum_{l=1}^{d^2} \left(A(x; d^2, l) - \frac{A(x)}{d^2}\right)^2$$

$$+ O\left(\frac{A(x)^2}{D}\right).$$

Now, if $A(x; d^2, l) \sim A(x)/d^2$ for each l and $(d, q) = 1$, then the right hand side is $o(A(x)^2)$, letting $D \to \infty$. On the other hand if $A(x; d^2, l) \not\sim A(x)/d^2$ then this one term contributes $\gg A(x)^2$ to the right hand side. The theorem is thus proved. $\quad\square$

Deducing Corollary 2. Let $\mathcal{A}$ to be the integers $a_k := 1^1 + 2^2 + \cdots + k^k$ and let $q = 2$. The results of [13] imply that the hypotheses of Theorem 3 are satisfied for $a = 2$ or $a = 3$, and that if d is odd then there are $\sim A(x)/d^2$ integers $a_i \le x$ with $a_i \equiv l \pmod{d^2}$, for every l. Corollary 2 follows in these cases noting that $A(x) \sim \log x/\log\log x$.

Every $a_{4k-1} \equiv a_{4k} \equiv 0 \pmod 4$ and $a_{4k+1} \equiv a_{4k+2} \equiv 1 \pmod 4$. Thus, when considering $a = 0$ or 1, we restrict attention to $\mathcal{A}_{\text{odd}}$ or $\mathcal{A}_{\text{even}}$, respectively, the odd or even, elements of $\mathcal{A}$. Once again, the results of [13] imply that the hypotheses of Theorem 3 are satisfied, and that if d is odd then there are $\sim A(x)/d^2$ integers $a_i \le x$ with $a_i \equiv l \pmod{d^2}$, for every l. Corollary 2 follows in these cases now noting that $A(x) \sim \log x/2\log\log x$.

Squarefree numbers plus powers of odd primes

Lagarias asks about analogous results with powers of 3 or larger primes. One needs to be a little careful here since the analogy to Conjecture 2, and thus the obvious analogy to Conjecture 1, are often false: For example, if $n \equiv 1 \pmod 4$ then $n - 5^i$ is divisible by $4 = 2^2$ for every positive integer; in other words, $\omega_5(2) = 1$ so we get the "covering system" of congruences $0 \pmod{\omega_5(2)}$. Another, less trivial example, would be that if $n \equiv 17 \pmod{36}$ then $n - 71^i$ is divisible by 4 whenever i is odd, and $n - 71^i$ is divisible by 9 whenever

i is even. Thus we need to avoid the integers n in those arithmetic progressions that arise from counterexamples to the analogy of Conjecture 2:

For given integer $q > 1$, define $\omega(p) = \omega_q(p)$ to be the order of q (mod p^2) for each prime p that does not divide q. Let S_q be the set of finite lists of arithmetic progressions $\{a_i \pmod{\omega_q(p_i)}, i = 1, 2, \ldots, m\}$ which form a "covering system", but for which no sublist forms a "covering system".

Define a set T_q of arithmetic progressions, as follows: First we include each a (mod q) where $(a, q) > 1$. Next, for each list $\{a_i \pmod{\omega_q(p_i)}, i = 1, 2, \ldots, m\}$ in S_q, let $B = \prod_{1 \le i \le m} p_i^2$ and take $A \equiv q^{a_i} \pmod{p_i^2}$ for each i; we then include A (mod B) in T_q. Note that if $n \equiv A$ (mod B) then $n - q^j$ is never squarefree, because $j \equiv a_i \pmod{\omega_q(p_i)}$ for some i (since we have a covering system of congruences), and thus p_i^2 divides $n - q^j$.

Note that Conjecture 2 implies that $T_2 = \{0 \pmod 2\}$.

Conjecture 4. *Fix squarefree integer $q > 1$. Then S_q is finite so that T_q is also finite. Further, if n is a sufficiently large integer, which does not belong to any of the arithmetic progressions in T_q, then n can be written as the sum of a squarefree positive integer and a power of q.*

Note that, if T_q is finite as conjectured, then the set of integers n, which can possibly be written as the sum of a squarefree integer and a power of q, can be partitioned into a finite set of arithmetic progressions. Restrict n to one of these "good" arithmetic progresions and argue as in Theorem 1 and Proposition 1. The main differences are that we now restrict the product in Proposition 1 to be only over primes coprime to the modulus of our arithmetic progression, and we replace the constant there by some sufficiently small constant, depending on q. In the deduction of the appropriate analogue of Theorem 1 we take our second product to be over those primes p for which q is a quadratic residue (mod p). In this way, we obtain:

Theorem 7. *Fix squarefree integer $q > 1$, and suppose that Conjecture 4 is true. Then there are arbitrarily large values of x for which*

$$\#\{\text{prime } p \le x : q^{p-1} \not\equiv 1 \pmod{p^2}\} \ge c\#\{\text{primes } p \le x\},$$

where c is a positive constant.

In the other direction we might again ask how often integers n (in a good arithmetic progression) can be written as the sum of a squarefree number and a power of q. One can prove (using Proposition 3) conditional results analogous to Theorems 2 and 4. For these, we require, at the very least, the following conjecture:

Conjecture 5. *Fix squarefree integer $q > 1$. Let $\omega_q(p)$ be the order of q (mod p^2) for each prime p that does not divide q. Then $\sum_{p \nmid q} 1/\omega_q(p)$ is bounded.*

Acknowledgments

We would like to thank Neil Calkin, Jeff Lagarias, Tauno Metsankyla, Andrew Odlyzko and Carl Pomerance for helpful remarks incorporated into this paper, and the referee for the rapid and useful report.

References

1. R. Crandall, K. Dilcher, and C. Pomerance, "A search for Wieferich and Wilson primes," *Math. Comp.* **66** (1997), 433–449.
2. R. Crocker, "On a sum of a prime and two powers of two," *Pacific J. Math.* **36** (1971), 103–107.
3. J.-M. Deshouillers, A. Granville, W. Narkiewicz, and C. Pomerance, "An upper bound in Goldbach's problem," *Math. Comp.* **617** (1993), 209–213.
4. P. Erdős, "On the difference of consecutive primes," *Quart. J. Pure and Appl. Math.*, Oxford **6** (1935), 124–128.
5. P. Erdős, "On integers of the form $2^k + p$ and some related problems," *Summa. Brasil. Math.* **2** (1950), 113–123.
6. P. Erdős, "On some problems of Bellman and a theorem of Romanoff," *J. Chinese Math. Soc.* **1** (1951), 409–421.
7. P. Erdős, "On the sum $\sum_{d|2^n-1} d^{-1}$," *Israel J. Math.* **9** (1971), 43–48.
8. P.X. Gallagher, "Primes and powers of 2," *Invent. Math.* **29** (1975), 125–142.
9. R.K. Guy, *Unsolved Problems in Number Theory*, 2nd edition, Springer-Verlag, New York, 1994.
10. A. de Polignac, "Recherches nouvelles sur les nombres premiers," *C.R. Acad. Sci. Paris Math.* **29** (1849), 397–401, 738–739.
11. R.A. Rankin, "The difference between consecutive prime numbers, V," *Proc. Edinburgh Math. Soc.* **13**(2) (1962/63), 331–332.
12. N. Romanoff, "Über einige Sätze der additiven Zahlentheorie," *Math. Ann.* **57** (1934), 668–678.
13. K. Soundararajan, "Primes in a sparse sequence," *J. Number Theory* **43** (1993), 220–227.
14. J.G. van der Corput, "On de Polignac's conjecture," *Simon Stevin* **27** (1950), 99–105.
15. A. Wieferich, "Zum letzten Fermat'schen Satz," *J. Reine Angew. Math.* **136** (1909), 293–302.
16. A. Wiles, "Modular curves and Fermat's last theorem," *Annals of Mathematics* **141** (1995), 443–551.